Curtis Wilson
Selected Writings

Dean's Lectures and Other Writings for St. John's College

Edited by
Chaninah Maschler and Nicholas Maistrellis

General Editor, William Pastille

St. John's College Press
Annpolis, Maryland

Published by
St. John's College Press,
60 College Avenue,
Annapolis, MD
21401, USA

ISBN 978-0-692-83269-1 pbk.

Library of Congress Cataloging-in-Publication Data

Names: Wilson, Curtis. | Maschler, Chaninah, 1931-2014, editor. |
Maistrellis, Nicholas, editor. | Pastille, William Alfred, 1954- editor.
Title: Curtis Wilson : selected writings : dean's lectures and other writings
for St. John's College / edited by Chaninah Maschler and Nicholas
Maistrellis; general editor, William Pastille.
Description: Annpolis, Maryland: The St. John's College Press, [2017] |
Includes bibliographical references.
Identifiers: LCCN 2017015289 | ISBN 9780692832691 (pbk. : alk. paper)
Subjects: LCSH: Astronomy--History. | Science--History. | St. John's College
(Annapolis, Md.)--Officials and employees.
Classification: LCC QB15 .W55 2017 | DDC 520--dc23
LC record available at https://lccn.loc.gov/2017015289

Table of Contents

Biographical Note

Curtis Alan Wilson was born in Los Angeles, California in 1921. After graduation from the University of California, Los Angeles, he received both his master's and doctoral degrees from Columbia University, New York. At Columbia he was a University Fellow. He was appointed to the St. John's faculty in 1948, even before having obtained his doctorate. In 1950-51 he received a Fullbright Fellowship to complete work toward his dissertation at the University of Padua. His doctoral dissertation became the source of his first book, *William Heytesbury: Medieval Logic and the Rise of Mathematical Physics.*

Ten years after joining the St. John's faculty, in 1958, he was chosen Dean of the College and served in this capacity for four years, until 1962. In 1962-63 he was a visiting research fellow at Birbeck College of the University of London, and in 1964 he was one of a group of tutors who became the core of the faculty at the new campus of St. John's in Santa Fe, New Mexico. Beginning in 1966 he spent seven years as a professor at the University of California in San Diego. He returned to St. John's in 1973 to begin a second term as Dean of the College. During his two deanships, St. John's made, under his guidance, two major curricular revisions, the institution of preceptorials and the reorganization of the mathematics and science programs. During the autumn semesters of 1982 and 1986 he was a member of the Institute for Advanced Study at Princeton. Following his retirement in 1988 he was a visiting professor in 1991 at the University of Toronto.

Mr. Wilson's research in the history of science centered on the writings of Kepler, Newton, Horrocks, Delambre, Euler, and other scientists important in the early foundations of modern physics and astronomy. In recognition of his contributions he became the first recipient in 1998 of the LeRoy E. Doggett Prize for writings in the history of astronomy awarded by the American Astronomical society. Between 1979 and 2001 he served on the editorial board of the Archive for the History of the Exact Sciences. He was also editor of the second volume of *The General History of Astronomy: Planetary Astronomy from the Renaissance to the Rise of Astrophysics.* In 1995 he published his translation of Francois de Gandt's *Force and Geometry in Newton's Principia.* In 2010, more than seven years of study culminated in the publication of his final book, *The*

Hill-Brown Theory of the Moon's Motion: Its Coming-to-be and Short-lived Ascendancy (1877-1984). Mr. Wilson was a member of the International Academy of the History of Science, and of Commission 41 of the International Astronomical Union.

He was a man of wide interests. He played both the piano and the harpsichord. He knew many languages and had deep and abiding affinity for literature and philosophy. He died on Aug. 24, 2012 at the age of 91. He is survived by his wife, the former Rebecca Marston, two sons, John Wilson of Blacksburg, Virginia, and Christopher Wilson of Pueblo West, Colorado, four grandchildren, and a number of nephews and nieces.

Editors' Preface

In addition to being a tutor at St. John's College, Curtis Wilson was an eminent historian of science. He published many detailed studies in the history of astronomy and physics. Our collection of his writings purposely omits all this purely scholarly work, limiting itself to spoken or written discourses produced for the St. John's College community, of which he was a member as tutor, tutor emeritus, and dean for fifty-four years.

The volume begins with the eight lectures he delivered as Dean of the College in 1958-1961, and 1973-1976. In one way or another, each of these lectures speaks about the place of the natural sciences within the liberal arts. They are followed by three other lectures and an article co-authored with Chaninah Maschler that appeared in *The St. John's Review* 54.1 (Fall 2012). These, in turn, are followed by two book reviews also written for *The St. John's Review.* All of this work exhibits Mr. Wilson's sustained thinking about the sciences and invites the layman to join him.

Those who wish to see one of Mr. Wilson's more accessible scholarly works might read his contribution to *Beyond Hypothesis: Newton's Experimental Philosophy,* a conference held at St. John's College, March 19-21, 1999, and published in a separate volume of *The St. John's Review,* 45.2 (Spring 1999). It bears the title, "Redoing Newton's Experiment for Establishing the Proportionality of Mass and Weight."

Mr. Wilson also spoke when the College gathered to honor those who had died, and to address the graduating seniors. We have included some of his eulogies of colleagues, commencement addresses to graduating seniors, and a toast to the seniors at the annual senior dinner.

Finally, we conclude this volume with eulogies for Curtis given by his colleagues.

The editors want to thank Catherine Dixon, Library Director, Cara Sabolcik, Public Services Librarian, and Andrew Hastings for their invaluable help in the production of this volume. They helped us find all of Curtis Wilson's writings by producing a catalog of those of his publications available in our Greenfield Library. Cara Sabolcik and Cathy Dixon also helped us track down the photographs of Curtis Wilson held by the library. Finally, Cathy and Cara helped us with perceptive, competent, and freely offered advice about a number of matters important to this volume, including showing two amateurs how to find things on the internet. We would also like to thank Desiree Arnaiz, Allison Tretina, Frederick Nesfield, Kathryn Trojanowski, and Sawyer Neale, editorial assistants for *The St. John's Review,* who helped with text entry, layout, and proofreading.

And finally, we would like to express our gratitude to Daniel Sullivan for permission to use his original artwork for "Reflections on the Idea of Science."

Photographs of Mr. Wilson are used by permission of the Greenfield Library Archives, St. John's College, Annapolis, Maryland.

As far as we know, all other visual examples and figures are in the public domain. If we are mistaken in this regard, we would appreciate being informed about the proper credits so that corrections can be made in future editions.

Dean's Lectures

The Archimedean Point and the Liberal Arts
(1958)

The subject of the lecture is, in accordance with tradition, the liberal arts, liberal artistry. And I wish to point out, to begin with, that these words "art" and "liberty" are difficult words; they do not designate anything you can point at with the index finger; they belong, not to the order of motion and perception, but to the order of action and idea. They are, as I shall try to explain later on, dialectical words. And the question arises, how are such words to be defined? Where should one begin? What *standpoint* should be taken in setting out to define these words? These are not merely theoretical questions; wars are fought between those who understand words like "liberty," "justice," "right," "obligation," in different ways.

Archimedes, the mathematician, is said to have said: Give me a fixed point on which to stand, and I shall move the world. He was referring, of course, to the power of the lever; to the law according to which the ratio of the two forces is the same as the inverse ratio of the lever arms. All that Archimedes requires, then, is a fixed pivot or fulcrum, a lever of extraordinary length, and a place to stand, and he will be able to move the earth.

This claim would not be stated in quite the same terms by modern scientists, beginning with Newton; I shall not go into the modifications required, but only state that they are required precisely because, in a sense that is both real and figurative, man has now discovered the

Ever since Stringfellow Barr and Scott Buchanan instituted the New Program at St. John's College in Annapolis, classes began on Thursday nights, with the first meeting of Freshman, Sophomore, Junior, and Senior Seminars. The day thereafter, Friday, all members of the community gather to hear a lecture followed by a question period. When Jacob Klein, who had joined the faculty in 1938, took on the Deanship in 1948, it became customary for the Friday Night Lecture to be delivered by the Dean. Inspection of the college catalogue, which used to record the names of the Friday Night lecturers, shows that this custom was adhered to thereafter. The Friday Night Lecture for 1958 was, accordingly, delivered by the 37-year-old new dean, Curtis Wilson. It was called "The Archimedian Point and the Liberal Arts." This, like most of Wilson's Dean's Lectures, was published in a College publication. However, the dates of publication of the lectures do not correspond to the chronology of their conception and delivery. Moreover, the final Dean's Lecture, "On Knowing How and Knowing What," was never previously published.

Archimedean point, the point outside the earth, the knowledge of which permits us to unhinge the earth. And this point, being a place to stand, is also a standpoint from which man attempts to view himself. Kafka somewhere says that, while man has discovered the Archimedean point, he uses it against himself; that it seems that he was permitted to find it only under this condition.

Modern science, beginning with the Copernican or heliocentric theory, is a return to Archimedes, and was so regarded by its founders, particularly Galileo. Copernicus discovers in the sun the fixed point from the stand point of which the earth moves. He looks upon the earth as though he were actually an inhabitant of the sun. He lifts himself, by an act of the mathematical imagination, by means of ratios and geometrical diagrams, to a point from which the earth and its earthbound inhabitants can be viewed from the outside.

The Archimedean point is shifted yet once again, or rather made infinitely mobile, when Giordano Bruno announces the infinity of the universe. What is characteristic of the thought of Bruno is the fact that the term "infinity" changes its meaning. In classical thought the word "infinity" is understood negatively. The infinite is the indeterminate, the boundless; it has no limit or form, and is inaccessible to human reason which lives in the realm of form. But according to Bruno the word "infinity" no longer means a mere negation of form. It means rather the immeasurable and inexhaustible abundance of the extended universe, and the unrestricted power of the human intellect. Man no longer lives in the world as a prisoner enclosed within the narrow walls of finite ordered cosmos. He can traverse the air and break through the imaginary boundaries of the celestial spheres. The human intellect becomes aware of its own infinity through measuring its powers by the infinite universe.

Einstein has insisted that we may assume with equal validity that the earth turns round the sun or that the sun turns round the earth; that both assumptions are in agreement with observed phenomena, and that the difference is only a difference of the chosen point of reference. Thus the Archimedean point is moved a step farther away from the earth to an imaginary point in the universe where neither earth nor sun is a center. We are no longer to be bound even to the sun, but move freely in the universe, choosing our point of reference wherever it may be convenient for a specific purpose.

This shift of standpoint, from the earth to a point outside the earth, received a certain kind of corroboration in Galileo's telescopic discoveries, the discovery of the moons of Jupiter and of the phases of Venus. These discoveries did not *prove* the truth of the Copernican theory; theories are never proved, only confirmed. And in fact, if we accept the theory of Einstein, we can no longer ask about the truth of the Copernican theory, for the Archimedean point becomes infinitely shiftable. But for those of Galileo's contemporaries who already accepted the Copernican theory, his telescopic discoveries were a confirmation of the power of the human intellect, which, by means of man-made instruments and mathematical theories, can free itself from the earth, and break down the age-old barrier between the sublunar and the celestial spheres.

One cannot fail to note, in the works of Kepler and Galileo, a certain exhilaration, a sense of the power of the human mind. According to Galileo, "the understanding is to be taken in two ways, that is, *intensively, or extensively;* and *extensively,* that is, as to the multitude of intelligibles, which are infinite, the understanding of man is as nothing, though he should understand a thousand propositions; for a thousand in respect of infinity is but as a cypher: but taking the understanding *intensively,* I say that human wisdom understandeth some propositions so perfectly, and is as absolutely certain thereof, as Nature herself; and such are the pure Mathematical sciences, namely, Geometry and Arithmetick: in which Divine Wisdom knows infinite more propositions, because it knows them all; but I believe that the knowledge of those few comprehended by human understanding equalleth the divine, as to the objective certainty, for that it arriveth to comprehend the necessity thereof, than which there can be no greater certainty."

By mathematics Galileo understood implicitly the science of physics, since the book of nature, as he says, "is written in mathematical characters." For both Kepler and Galileo, man becomes a god, travelling through space, able to calculate for his own displacement, and so to arrive at knowledge which, intensively considered, is perfect.

I shall not attempt to retrace the vicissitudes of this scientific faith through the last three centuries. It would be a complex story, I would even say a dialectical story, a romantic biography, as it were, of a recently deceased friend. The aim was to express qualities through figure, to substitute a geometrical configuration for each primordial quality, to explain all things by figure and movement considered as situated in an infinite matrix of time and space. The doctrine of atomism was part and parcel of this scientific faith, the notion of an inert matter or stuff cut up into tiny shapes. But it was soon found necessary to attribute occult qualities to the matter of the atoms, mysterious dynamic qualities like gravitation, and the atoms were gradually transformed beyond recognition. The geometrized space of Galileo has become with Einstein a symbolic space-time matrix. The development of theoretical structures has been constantly in a direction away from the simple geometrical object, which the mind's eye can see with the certainty that it *is* there. Einstein has to deny that at a definite, present instant all matter is simultaneously real. Whatever theoretical physics is talking about today, it is not something which is imaginable with the eye of geometrical imagination. I am told that you can learn the fundamentals of quantum mechanics in about six months; then it takes another six months to understand that you understand it, though you cannot imagine what the theory is supposed to be about. The tension between the empirically given and the imaginable on the one hand, and the content of theoretical physics on the other, has increased to the breaking point. The mirror of nature that scientific faith endeavored to build has been shattered, and the scientist finds himself looking straight out into the unknown.

Already in the seventeenth century the new conception of the world—the world as viewed from the Archimedean point—had given rise to a reaction of doubt and fear. "The eternal silence of these infinite spaces frightens me," says Pascal. Pascal's distinction between the *esprit de géometrie* and the *esprit de finesse* is directed against the geometrical and astronomical view of the world. The geometrical spirit excels in all those subjects that are capable of a perfect analysis into simple elements. It starts with axioms and from them derives propositions by universal logical rules. Its excellence lies in the clarity of its principles and the logical necessity of its deductions. But, Pascal would say, there are things which because of their subtlety and variety defy the geometrical spirit, which can be comprehended, if at all, only by the *esprit de finesse,* the acute and subtle spirit. And if there is anything which thus defies the geometrical spirit, it is the nature and mind of man. Pascal holds that contradiction is the very essence of human existence. Man has no "nature," no homogeneous being; he is a mixture of being and non-being. His precarious place is midway between these poles.

The discovery of the Archimedean point produced a crisis in man's knowledge of himself. Self-knowledge has almost always been recognized as the highest aim of philosophic inquiry. It is impossible to penetrate into the secret of nature unless one also penetrates into the secret of man. The discovery of the Archimedean point demanded that man view himself from a totally alien standpoint, that he understand himself ultimately in terms of geometrical figure and the impact of atoms. The seventeenth-century philosophers were fully aware of this crisis, and attempted to meet it in different ways. But the most obvious and crucial step, the step which was already implicit in the Copernican shift of standpoint, was taken by Descartes: the removal of the Archimedean point into the mind of man, so that he could carry it with him wherever he went, and thus free himself entirely from the human condition of being an inhabitant of the earth.

Descartes says: I think, therefore I exist; or: I doubt, therefore I exist. Beginning with the idea of universal doubt, he concludes that there must be something which doubts, which thinks; and this something is what he is. He identifies himself as a mind, a thinking thing. And this thinking thing is the fixed and immovable point from which all else must be derived, the existence of God and of other minds, and of things which have extension or occupy space. Descartes himself recognizes the connection of his thought with the Copernican revolution, for he states that if the earth does not move, then all of his doctrines are false.

Unfortunately the Cartesian removal of the Archimedean point into the mind of man fails to assuage the Cartesian doubt. Descartes's argument, *cogito ergo sum,* I think, therefore I exist, is faulty. For, as Nietzsche pointed out, it ought to read: *cogito ergo cogitationes sunt,* I think therefore there are thoughts. It does not establish the existence of a something which thinks; it can only end with what it begins, namely thinking. And the questioning and doubting remain; universal doubt, not in the sense of really doubting everything all at once, which is impossible, but in the sense of the indefinite possibility of doubting things one by one as they occur in thought. And so one becomes a question to himself, asking who he is—which is one sort of question—and what he is, which is another sort of question.

Both the successes and failures of the scientific revolution have resulted in an anarchy of thought with regard to the nature of man. The Archimedean standpoint leads to theories of man in terms of impulses, forces which are analogous to mechanical forces, the sexual instinct for Freud, the economic instinct for Marx. But the different theories contradict one another. No age previous to ours was ever so favorably placed with regard to the empirical sources of knowledge of human nature, and yet never was there so little conceptual agreement. And it becomes the task of modern man, if he is to avoid the piecemeal response of dissipation, and the one-track response of fanaticism, to inquire once again into the being that he is, and that he can become.

I make a new start, not from the Archimedean point, but right in the middle of things.

And let me begin this time with the obvious, with the observation that man is a linguistic animal. He speaks; also, he uses writing as a substitute for speech. The word "linguistic" is derived by a metaphorical extension from the word for a bodily organ, the tongue. The tongue is used in articulating the voice. The Homeric epithet for men was *hoi meropes anthrōpoi; meropes* is from *merizō,* the verb meaning *to divide;* and the phrase means *those who divide or articulate their voice.*

This does not, of course, tell us what a language is, or in particular what human language is. A chimpanzee can articulate most of the sounds used in human speech; his tongue and lips can be used to artic-ulate sound in the same way as the human tongue and lips; but he is not a linguistic animal in the same sense as man is. The chimpanzee uses gesture and voice to express rage, terror, despair, grief, pleading, desire, playfulness, pleasure; he expresses emotions. Man also utters cries expressive of distress, pleasure, and so on, but these *interjections,* as they are called, are quite frequently vocal signs of a higher order, the use of which as interjections comes about by a degradation from their proper use; in fact, they are quite frequently vocal signs borrowed from the language of theology.

Man articulates his voice with the conscious intention of signify-ing, or *sign*-ifying, something to somebody. The notion of a sign is, or-dinarily, wider than that of language. With respect to the relation between sign and thing signified, we can distinguish three kinds of signs. First, indexical signs, or indices. Here the sign is *causally* con-nected with that which it signifies; thus smoke is a sign of fire, because it is produced by fire; the direction of a weather vane is a sign of the direction of the wind, because the direction of the weather vane is de-termined by the direction of the wind; and the position of a speedome-ter needle indicates the speed of the automobile, because it is causally connected with the rotation of the wheels. Secondly, there are *iconic signs,* or *icons,* which are significant of something to somebody because they are similar to that thing in some respects. Examples of such signs are photographs, replicas, geometrical diagrams, images of every kind. Finally, there are conventional signs, often called symbols; and under this heading fall most of the words of human language. Symbols are all those signs which are signs *only* because they are interpreted as such by some organism or mind; there is no other connection between sign and thing signified, as there is in the case of indices and icons.

Sometimes the word "language" is taken in a broad sense, as any set or system of objects or events which are significant for some being, or which are such that certain combinations of them are meaningful or significant for some being. In this case, we should have to include as special cases the language of looks and glances, the language of the bees, and the language of the stars.

The incredible navigation feats of migratory birds, such as the white-throated warblers which migrate bet ween northern Europe and Africa, have been shown recently to depend on celestial navigation, a reading of the stars as indexical signs of latitude and longitude. The experiments were performed in a planetarium, and it was shown that during the migratory period the birds decide, on the basis of the look of the sky and an inner time sense, exactly in what direction to point in order to be aiming toward their destination. If they are so far put off course as to have, say at midnight, the midnight appearance of the Siberian sky over their heads, they know in what direction to point in order to regain their course.

The language of the bees, on the other hand, consists in significant actions which are mostly iconic. As the researches of von Frisch have shown, a honey bee that has returned after successful foraging for food goes through a strange and complicated dance, and this dance is so designed, by the direction of the step and tempo, as to show to the other members of the hive both the direction and distance of the find.

None of these systems of signs is strictly comparable to human language, which differs in essential respects. But all of them consist of signs, and a sign is a very special sort of thing, which would not come into focus if we stood at the Archimedean point.

Wherever there is a sign, there is a relation which is at least triadic in complexity, that is, a relation which relates at least three things. The sign stands *to* somebody *for* something. The something may be called the *object* of the sign; but it should not be supposed that the object is

9

always, or even ordinarily, what we call a physical object or thing, something that is spatially bounded, capable of existing for a stretch of time, and movable. The object of the sign is just whatever the sign signifies, which might be redness or horizontality or justice. The somebody, human or not, for whom the sign is a sign, *interprets* the sign as signifying the object; or we may say that the sign produces in this somebody an *interpretant* or *thought*. Thus the three things related in the sign relation are (1) the sign, which will be a physical object or event in any particular case; (2) the object, or thing signified; and (3) the interpretant.

A triadic relation, such as we have in the sign-relation, cannot be reduced to any sum of dyadic relations, that is, relations relating two things. Dyadic relations can be diagrammed by means of a letter with two tails, thus: - R - . It is understood that something has to be written in at the ends of the tails, to indicate the two things related. Hitting is a case of such a dyadic relation, as when we say "a hits b." Triadic relations, on the other hand, have to be diagrammed by means of a letter with three tails, thus: - R - . An example of such a relation is the giving involved when John gives the book to Mary; the giving is a relation between three things, John, the book, and Mary. Similarly a sign signifies something to somebody.

Now it is easy to show that the combination of dyadic relations only leads to further dyadic relations; for instance, by combining the relation "uncle of" (- U -) with the relation "cousin of" (- C -), we only obtain the relation "uncle of cousin of" (- U - C-) or "cousin of uncle of" (- C - U -); and the diagram shows that the combined relation has only two tails. Therefore triadic relations cannot be built up out of dyadic relations. Hence the sign relation is not reducible to anything involving only dyadic relations. As a consequence, no theory about the world which seeks to account for everything in terms of dyadic relations, such as we have in the impact of atoms or gravitational attraction, is adequate to account for sign relations. Lucretius, for instance, is wrong.

Now given the irreducibility of signs to things which are not signs, we have yet to advance another step before we reach the level of human language. We have, in the first place, to understand the distinction between a sign which is a *signal* or *operator,* and a sign which is a *designator* or *name;* between a sign which serves as stimulus to a motor response, and a sign which serves as an instrument of reflective thought. Man is a *naming* animal.

Human language is characterized by a freedom of naming. Man can devise a vocal name for anything that he can identify or distinguish as. being, in some way, *one*. This freedom of naming depends, for one thing, on the manifoldness of the sounds which the human voice can produce and which the human ear can distinguish; and it depends for another on leisure and reflectiveness. Among peoples whose mode of life grants them little leisure, the naming of things may be very restricted; thus Malinowski found that among the Trobrianders of the South Seas there are no special names for the various trees or bushes which provide no edible fruit, but all of these are alike called by a name we may translate "bush." When leisure intervenes, however, the reflective botanist or zoologist, or in general the reflective *namer,* makes his appearance, and nothing is safe from being named, not even the Nameless, which after all has that name.

The identification that goes with naming is like the drawing of a circle, which separates all that is outside the circle from all that is inside. Among the words which name, I include adjectives and verbs as well as nouns, for all such words have the general function of identification, of signifying something that is, in some way, one. Corresponding to every adjective or verb there exists, or can be invented, a corresponding noun; in English, for instance, we frequently turn adjectives into nouns by adding N-E-S-S, and verbs into nouns by adding T-I-O-N. All such words are called, traditionally, *categorematic* terms. And the key to the categorematic terms, the key noun or noun of nouns, is "monad" or "unit," which Euclid defines as that in accordance with which each of the things that are is said to be one; and which was also defined by the ancients as the form of forms, *eidōn eidos.*

The categorematic terms are of different sorts, depending on the kinds of thing they designate. Some of the things they designate can be simply located in space and time, and others cannot.

There are, for instance, terms for simple qualities like "red," "bitter," "shrill." The awareness of such a quality, considered by itself, is unanalyzable and incommunicable; it is just what it is and nothing else. I can never know that my neighbor's awareness of the redness of the curtain is the same as mine; if he uses the words "red" and "blue" on the same occasions as I do, this only means his *classification* of colors corresponds to mine. The identification of qualities by name presupposes acts of comparison and classification.

There are names for physical objects, "horse," "chariot," "Hektor." A physical object has unity insofar as it is bounded in space, persistent for a stretch of time, and capable of moving or being moved. It is identified as an invariant within a spatial and temporal framework. The character of the spatial framework is determined by the character of possible motions. Motions are reversible, so that one can return to his starting-point; and motions are associative, so that one can change direction, add motion to motion. In other words, motions form what mathematicians call a *group of operations*. The character of possible motions implies that space is homogeneous, that it constitutes a uniform background against which physical objects can manifest their unity and invariance, that is, their boundedness, persistence in time, and mobility.

There are names for materials, such as gold and water, and names for such strange beings as rivers and streams; whatever a stream or river is, it is something you can step into twice, though the water is never the same.

There are names for happenings, events, motions; running, grasping, twisting, leaping, coronation, assassination.

All the kinds of name I have mentioned so far designate things that can be pointed at. But the meaning of such words cannot be defined simply by pointing; pointing by itself is totally ambiguous; if the pointing is to be understood, something else must be understood at the same time. For instance, we have to understand that it is a physical object that is meant, or a color, or a shape, or a material, or a motion. The tree is not only a tree, but is green-leaved, tall, branched, and so on. In whatever direction one points, there is manyness, plurality of aspect.

All such naming, then, presupposes and implies an act of comparison and classification, the isolation of something from a matrix or background of possible meanings. Man is an animal who compares, finds ratios; he is a *rational* animal.

There are names for things which cannot be localized in time and space, names like "law," "liberty," "art," "nature," "justice," "knowledge," "wisdom." Such words belong, not to the order of perception and motion, but to the order of action and idea. We ascend here to a new level which, once again, is not discernible from the Archimedean point. These words cannot be defined through classification, through specification of genus and differentia. They are polar or dialectical words, which take up their meanings in relation to the meanings of other words of the same kind. The word "freedom" presents different facets to the word "tyranny" and to the word "slavery"; and anyone of these words requires the services of the others.

Most if not all of the dialectical words are borrowed from the realm of the corporeal, visible, and tangible; the original reference is forgotten, and only the metaphorical extension survives. Both the Greek *dikē*, justice, and the Chinese word *i,* morality, originally meant *a way of life,* that is to say, a particular way of life. But there are many ways of life, and the adjudication between rival opinions requires a universal meaning. The universal is then grasped *in* the particular. The definition of the dialectical words depends on representative images or anecdotes, like the Hobbesian state of nature, or the state constructed in Plato's dialogue, *Republic.*

In all cases, naming involves the location of a kind of commonness, law, regularity, invariance—something on the basis of which one might classify or predict. And in all cases the commonness, law, regularity, invariance, makes its appearance in a matrix of relations. Whenever anyone has managed to grasp such an invariance or regularity or commonness, he has thereby in some measure released himself from the tyranny of diversity. As Aristotle says, the soul is so constituted as to be capable of this process. And he adds that it is like a rout in battle, stopped by first one man making a stand and then another, until the original formation has been restored.

The human freedom of linguistic formation is not limited to naming. Human language is combinational; it permits the combination of sign with sign to form a complex sign called the sentence, the proposition, the affirmation or denial, or—to use the Greek word—the *logos.* In order really to say something, one must say something about something. The fundamental type of expression with complete or independent meaning is the sentence; a meaning is completely specified only if it is imbedded in an affirmation or denial, something that could be an answer to a question.

Words that, in a broad sense, name or identify, can be answers to questions. They have a certain *possible* completeness of meaning, which becomes actual when they are uttered in a context of other words or in a non-verbal situation which serves to specify the way in which they are being used. The single word "fire," for instance, may have different meanings depending on the situation in which it is uttered; whether, say, by a neighbor whose house has caught on fire, or by an artillary officer, or by Pascal in his study, in an attempt to express a theological truth.

Or to take a case where the context is verbal: the meaning of the word "man" in the sentence "Some man is a liar" is not entirely the same as its meaning in the sentence "Man is mortal," and is different again from its meaning in the sentence "Man is a species." In "Man is mortal" the word stands for all things which it is capable of signifying, all men who ever were, or are, or will be, this man and that man and so on. In "Man is a species" the word stands for a certain *nature* which it signifies; and it is not possible to descend to individuals, to assert that this man or that man is a species. In "Some man is a liar," the word "man" stands not for all things it is capable of signifying, but only for an indeterminate individual, this man or that man. We may say in general that while any categorematic term is capable of signifying, the precise way in which it signifies is determined by its use in an assertion, a sentence.

Every sentence contains, besides categorematic terms, other signs which are called syncategorematic signs, words like "if," "with," "by," "the," "is," "every," "because," "not," and signs which consist of inflectional endings or word order. These signs are not names; they determine the range of meaning of other terms, or the mode of con nection of terms in sentences; they express instrumentality, the modalities of the possible or probable, tense, negation, conditionality, and so on. In translating from one language to another, these signs present the greatest difficulty, for they are most likely not to translate into a completely analogous form in the second language. The conditional "if," for instance, can be expressed in German by a mere inversion of the order of subject and verb; Greek and Latin can express the instrumental "by" or "with" by means of case endings of nouns; and Latin somehow—though not very happily—manages to get along without a definite article. Nevertheless, we can expect that any adequate language will supply the connective and determining functions in some way.

The crucial syncategorematic sign is the sign of assertion itself. In the Indo-European languages this is supplied by the finite verb form;

the verb has, in addition to its function of naming or identifying, the function of indicating that something is to be affirmed of something. In Chinese there is no verb "to be," and instead there is a little particle "yeh," which may be translated "indeed." Thus one says "Tail long indeed," meaning *The tail is long,* and "Boat wooden-thing indeed," meaning *Boats are made of wood.* The particle "yeh" may be taken as an epitome of the business of the sentence, to assert or declare.

There is, then, a freedom of linguistic formation in human language, freedom in the formation of names and sentences. And this freedom extends to the subject-matter of language itself; we can talk *about* language, use language to describe language. This peculiar atop-the-atopness is characteristic of human capacity. Thus we can make machines which make tools, which are used in turn to make macliines. And according to Kant, man is the only animal who can read a sign *as* sign. This implies that man is the only animal who can make signs *of* signs; the only animal that has a hierarchical or self-reflexive language. And it implies also that he can become aware, as by a sidelong glance, of his own linguistic activity, and raise it to the level of conscious artfulness, liberal artistry.

Because of the triadic relation between the sign, the object, and the thought or interpretant, we can distinguish three branches of linguistic artfulness. Grammar will deal with linguistic formation, with the conditions which any sequence of signs, and in particular any sentence, must satisfy if it is to be meaningful. Logic will deal with the conditions which any sequence of signs must satisfy if it is to be true of any object, and in particular with linguistic transformations which preserve truth, with the derivation of one sentence from another in such a way that if the first sentence is true of *any object* or objects, then so is the second. Rhetoric will deal with linguistic transformations that are persuasive, with the conditions under which one thought or interpretant leads to another in the mind of the interpreter. The focal topic of grammar is the sentence; of logic, the argument; of rhetoric, the trope or figure of speech.

Grammar has to do with the conditions of meaningfulness, or conversely, with the avoidance of nonsense. Meaninglessness or nonsense is to be distinguished from absurdity. A word heap like *king but or similar and* is meaningless, and so is Gertrude Stein's *A rose is a rose is a rose,* unless a comma be inserted after the second occurrence of "rose";

an expression like *round square* or *All squares have five corners,* is absurd or countersensical, though meaningful. The avoidance of nonsense is the business of grammar; the avoidance of absurdity is the business of logic.

Grammar has to do with the recognition and distinction of forms and modifications of meaning which any adequate language must be capable of expressing, the existential sentence, the hypothetical antecedent, the generic sense of a common noun, negation, the plural, the modalities of the possible and probable, past, present, and future, and so on. If a language is to mirror truly, in its verbal materials, the various kinds of possible meanings, then it must have control over grammatical forms which permit the giving of a sensuously distinguishable "expression" to all distinguishable forms of meanings. Different languages may differ with respect to their adequacy. It is the task of the grammatical art to *see through* the grammatical forms of particular languages to essential distinctions of meaning, and to the ways in which meanings may be combined so as to result in the completed meaning of the sentence.

Logic is concerned with relations between sentences, with transformations of sentences yielding new sentences, in such a way that if the original sentences be true of any objects of thought, then so are the derived sentences. Wherever logic is being employed, the logical function will be expressible in terms of a sequence of sentences, of which one or more will be regarded as antecedent, and one or more as consequent.

Among sentences, some are denials or contradictions of others; in fact every sentence has a denial, and the denial of a denial is the same as the original sentence denied. Everyone who cares to speak or assert anything, has to take it as a rule that a given sentence cannot be both truly affirmed and truly denied; on pain of contradiction, we say, he cannot both affirm and deny something of something at the same time and in the same respect. This principle, called the law of non-contradiction, cannot be proved. Anyone who dares or cares to deny it cannot be talked with without absurdity, for his very denial would imply a denial of his denial. He is, as Aristotle says, no better than a vegetable.

There are sentences which are consistent or compatible with one another, so that one can be denied or affirmed without our having, on pain of contradiction, to affirm or deny the other.

And there are sentences which are related as antecedent and consequent, where the affirmation of the one requires us, on pain of contradiction, to affirm the other. In this case, the antecedent is said to *imply* the consequent.

Implication always depends on syncategorematic words, words which do not name, but which connect or modify the meanings of names, words like "and," "or," "if then," "all," "every," "some," and so on. For instance, if p and q are two sentences, and if I assert the sentence "If p then q," and also assert the sentence "p," then I am forbidden on pain of contradiction to deny the sentence "q." Or if A, B, and C arc objects of thought, and if I assert that all A is B, and that no B is C, then I am forbidden on pain of contradiction to deny that no A is C.

In all applications of logic there are signs—either categorematic terms or sentences—which occur vacuously; all that is required of them is that their meaning should remain self-identical. The implication depends solely on the connective and determining words, the syncategorematic signs.

The logical art enables us to pass from sentence to sentence, to draw out the consequences of what has previously been asserted, to construct the tremendous deductive sciences of mathematics and theoretical physics. An omniscient being would have no need for such an art, but man is a discursive animal, who can only pass from truth to truth in some consecutive order, in time.

Rhetoric has to do with the ways in which one thought leads to another. As rhetorician, one is concerned with linguistic transformations which occur in daydreams and reveries, in jokes and poems and myths, in the formation of opinion, in the coming about of discoveries and insights. While the task of logic is to look through signs, so to speak, toward the self-identical character of objects of thought, the task of rhetoric is to look through signs toward the polar character of thoughts.

Every identification of meaning involves the drawing of a circle which includes and excludes. Every sentence involves affirmation or negation. The fundamental polarity in thought is that between same and other.

There is an ancient Pythagorean table of opposites, contrarieties, polarities: odd-even, unity-plurality, right left, male-female, rest-motion, light-dark, good-bad, and so on. These polarities rest not only on the law of contradiction, but on the polarized character of man's life, the erotic character of his linear voyage through time and space. The other polarities become invested with Eros, the desire for pleasure, for honor and power, for community, and for knowledge.

Wherever there are poles, there are tropics. The word "pole" comes from the Greek word *polos,* meaning pivot. Wherever there are pivots, one expects to find something that turns; and the Greek word *tropos,* from which we derive the word "tropic;' means a turning. Thus the tropics of the earth turn round the poles. Wherever there are polar oppositions of terms, one may expect to find what are called *tropes,* that is, turns or figures of speech, similes, metaphors, metonymies, ironies.

In the fifteenthth book of the *Iliad,* there is a point at which Hector is seeking to break the ranks of the Achaians, but is unable, we are told, for they endured like a tower, "just as a rock in the sea endures despite wind and waves." The rock in the sea is a simile, of course, for the endurance and courage of the Achaians. The polarity here is between man and rock. I read into the rock the human endurance, and then I turn round and read into the human endurance the steadfastness of the rock. I look at each from the standpoint of the other; I use each to obtain a perspective of the other. The movement is from man to rock and back to man. I obtain an echo of man from the rock.

As I pointed out earlier, the words for moral notions and for the activities of the mind are derived by metaphor from words for visible or tangible things and motions. Poetry involves a regaining of the original relation in reverse, a metaphorical extension back from the intangible into a tangible equivalent. It involves the discovery of what T. S. Eliot calls an *objective correlative* of the interior life; that is, the finding of a set of objects, a situa tion, a chain of events, which will be the formula of a particular feeling or thought, so that when the external facts are given, the feeling or thought is immediately evoked.

Modern science can also be viewed, on the theoretical side, as a gigantic trope or series of tropes, a series of models or images whose meanings are drawn out by logical inferences. Thus one may conceive electric current after the analogy of a river, or electric oscillations after the analogy of mechanical oscillations, and other aspects of electricity suggest other metaphors which in turn acquire corresponding mathematical formulation. Modern mathematics and mathematical physics overlies a mass of disjunct imagery which it does not appear possible to unify; instead, imagery is used dialectically to transcend imagery, in successive stages of formalization.

Finally, let me not fail to mention the trope of irony, the dialectical trope *par excellence.* Irony is an elusive trope; its essence lies in simulation or dissimulation, in the use of the tension between what appears and what is. It can be savagely or gently mocking, but it also contains the seeds of humility. When Newton (the hymn-writer) sees a criminal being led to the gallows and says "There but for the grace of God go I," he is not congratulating himself on not being a criminal; he reads himself in the other and the other in himself, and the irony lies in this peculiar combination of "yes" and "no," as these two are connected by means of the God-term. When Socrates says "I know that I do not know," he combines affirmation and denial in such a way as to produce a peculiar transcendence. Irony is here the net of the educator.

The possibility of irony rests on the tension between what appears and what is. Man exists at the horizon between appearance and idea; his being is an intermediate, a metaxy, as Plato would say (*to metaxu*). And the task of education, starting in the middle of things, is to use the appearances, the images, the names and the sentences, to produce a development toward hierarchy and wholeness which uses all the terms.

I have but a few more words to say. Man is a being who is constantly in search of himself; this is the human condition. Socratically speaking, he is a questioning animal, a being who, when asked a rational question, can give a rational answer. So questioning and responding, both to himself and others, man becomes a *responsible* being, a moral being. In the image of the *Republic,* the movement of dialectical or dialogical thought, as guided by Socratic irony, is upward, from darkness into light, from partiality to wholeness, from appearance to intellectual vision. The Socratic irony produces a transformation of terms, a hierarchy, a perspective of perspectives, in which the contradictions of political life, and of the soul which is an inner political life, are resolved by becoming hierarchially related to the idea of knowledge. The Socratic irony punctures pretense, and points beyond, to the unity of knowledge and to the great dialectical interchange which has yet to be carried out.

Groups, Rings, and Lattices
(1959)

The words "group," "ring," and "lattice," as used in modern mathematics, are names for certain kinds of formal structure. Let me say at once that this will not be a lecture about mathematics; I am going to attempt to discuss the nature of intellectual work, and to use the mathematical structures, especially the group structure, as models or paradigms of structures which, so I shall claim, are rather generally present, either implicitly or explicitly, in the exercise of intelligence.

To describe what goes on in the exercise of intelligence is no easy task. In a rough analogy, one might compare thinking to riding in an airplane, and now and then going up front to do a bit of steering; the riding and even the steering are possible without understanding what keeps the airplane aloft and moves it forward.

This difficulty of description is rooted in the fact that thinking involves a *temporal process*. The fact has several aspects.

In the first place, whatever is accomplished in intellectual work, whatever is grasped or understood, is grasped or understood through successive steps, by running through connections. At a certain moment, I believe myself justified in saying: "Now I understand the situation which I previously did not understand." What has gone on in the interim? Well, I take it that whenever I understand anything, whenever I grasp anything, what is understood or grasped is a complex of elements, with their properties and relations; if it were only a solitary thing, without any internal complexity or any relation to anything else, we would speak not of understanding but perhaps of trance. In other words, understanding is always understanding of something which is somehow many. In the interim, then, I have been presumably tracing out the relations between elements of the situation, presumably one by one, and then I say: "I understand now." But at this moment in which I say that I understand, it does not seem possible that all these relations are present to me at once, in their full significance; and it becomes a problem as to *how* they are present. And in any case, it is clear that the acquisition of any understanding involves *necessarily* a kind of evanescence; different aspects of the situation to be understood have to fall successively into the background, into the past; and when I try to understand my understanding, to grasp reflectively what has gone on in the process of understanding, it seems that I must either reactivate the original process, step by step, or else I am liable to fall into superficiality or false generalization.

This problem of evanescence goes beyond any single process leading to a single act of understanding. *All* intellectual work is based on previous acquisitions which have become as though embedded and submerged in one's thinking. Previous acquisitions, in order to become transmissible from one person to another, or even to remain accessible to one person, have in general to be framed in words, written or spoken. And written or spoken words exercise a seductive power; increasing familiarity with certain words and patterns of words makes possible a certain kind of passive and superficial understanding; which carries us forward to another stage without our grasping the full meaning, without our having gotten to the roots of what has been presented. Even thinking which has seemed satisfying and adequate always involves an interlacing of what is grasped centrally and with a degree of clarity and distinctness, and what is accepted passively as pre-given, often without the awareness that there is an embedded structure which needs to be brought to light. Learning never starts from a zero-situation, complaining members of the teaching profession to the contrary notwithstanding.

This state of affairs with regard to past acquisitions is inevitable, is an essential aspect of the human quest for knowledge. I am not the first member of the human race, nor can I actually put myself in the *position* of the first member of the human race. What happened in the past of the race cannot be resurrected and re-lived just as it was in fact; first because the past presents us only with a few documents and monuments, fragmentary end-products of processes whose factual character remains inaccessible; secondly because whatever I understand of the past is understood in my own present, and in the light of my own interests and preoccupations. I cannot even re-live my own past as it was in fact, for recollection differs essentially from the original experience; I cannot abolish the fact that I *now* know the outcome. Moreover, I do not even remember my own birth—I suspect that no human being ever does—nor do I remember how I began to emerge from the buzzing and booming confusion of the sensations which first bombarded me. Whatever its cause, whether it is because we forget what is painful, as Freud says, or because we forget what is useless, as Bergson says, this childhood amnesia seems to be universal. We are all in the situation of Adam, who according to William James— James may not be quite the proper authority to refer to here—was created with a navel, and must have been rather puzzled by it, if no one told him what it meant.

The situation with regard to past acquisitions, this engulfment in time, is mirrored by a corresponding situation with regard to the future. Whatever I accomplish in intellectual work remains open to modification or qualification by a series of future investigators, including me. If I claim to understand or grasp anything with any kind of completeness—and the nature of this completeness is just our problem—there yet remains the open possibility of grasping further relations, determinations, connections, so that what has thus far been grasped appears as a special case of something else.

The difficulty of describing the mind at work lies just in the fact that the being of the mind is its work, it is what it does, and this doing involves temporal succession, a coming-to-be and passing-away of moments, with a bewildering complexity of structure which is constantly being modified, or fading into the past.

So much for the difficulty. What I propose to do is to start with an example of thinking, one in which there is an advance from passive acceptance of the pre-given to an active grasp of a situation and its parts and relations. And then I shall try to frame a generalization on the basis of the example, to arrive at a model or paradigm that can be applied and tested in other cases.

Now for an example.* Suppose I am asked to find the sum of $1 + 2 + 3 + 4 + 5 + 6$. (The correct answer is 21). What do you do when you add? Ordinarily when one is asked what the sum of $4 + 5$ is, I suspect the answer "9" comes immediately—we even say "without thought;" one has been drilled in the repetition of the addition tables since childhood, the associations are built into one's memory, and the answer comes automatically when a situation requires it. But what if the series to be summed were much longer? Suppose one were asked to find the sum of the first 201 whole numbers? Adding them up successively would be tedious, and one might have to check the additions several times to be sure of having the correct result, which is 20,301.

Is there a shorter way? The reader may know that there is a formula for the sum of any such series; it is $n(n + 1)/2$, where n is the last term of the series. Use of the formula involves merely a recognition of the

*This example is given in Max Wertheimer, *Productive Thinking* (New York: Harper Collins, 1959).

cases to which it is applicable, a substitution for n, and a multiplication and a division, based on memorized tables.

But someone will undoubtedly ask: how do we know that this formula is correct? The answer is, of course, that we can prove it. There are several proofs, but let me give one which can be stated very briefly. I write the series down twice, once in the usual way and the second time, just underneath, in reverse order:

$$
\begin{array}{ccccccc}
1 & 2 & 3 & & (n\text{-}2) + (n\text{-}1) & + n \\
n & + (n\text{-}1) + (n\text{-}2) + & \ldots + & 3 & 2 & 1
\end{array}
$$

$$(n+1) + (n+1) + (n+1) + \ldots + (n+1) + (n+1) + (n+1)$$

Then if I add the two terms in each vertical column, I find the sum in each case to be (n+1). There are n such sums, that is, just as many as there are terms in the series. So the sum of the series taken twice is n(n+1), and the sum of the series taken just once is half that, or n(n+1)/2.

All right. I have gone through the proof, nodding in assent at each step, and my conclusion is that the formula is true. I may still be left with a vague sense of dissatisfaction, as though a neat trick had been performed which I did not fully understand; insight may still be lacking. I can still ask why the formula is correct; more specifically, what is the connection between the form of the series and the sum of its terms? Why is just this particular formula the formula for this particular kind of series, a series of this form?

There is a clue to the answer in the proof I have just given, but let me return to a particular case, the series 1+2+3+4+5+6. First I note that the series has a direction of increase going from left to right:

$$1+2+3+4+5+6$$

Next, an obvious remark, but one that will prove decisive. The increase from left to right involves a corresponding decrease from right to left:

$$\overset{\longleftarrow}{1+2+3+4+5+6}$$

If I go from left to right, from the first number to the second, there is an increase of one; if I go from right to left, from the last number at

$$1+2+3+4+5+6$$

the right to the next preceding, there is a decrease of one. Hence the sum of the first and last numbers must be the same as the sum of the

$$1+2+3+4+5+6$$

next inner pair. And this must be true throughout:

What we grasp now can be symbolized by two arrows, meeting in the center:

There remains only the question: how many pairs are there? Obviously the number of pairs is one-half of all the numbers, hence of the last number; so we get (6/2). 7 as the sum, or in general $(n/2)(n+1)$. Here $(n+1)$ represents the value of each pair, $(n/2)$ the number of pairs.

If one knew the formula only blindly, then expressions of the forms $(n^2+n)/2$ and $(n+1)(n/2)$ would be completely equivalent. But in view of the derivation just completed, the meaningful form is $(n/2)(n+1)$: we have a sum for each pair, namely $(n+1)$, and then we multiply by the number of pairs. The two factors of the product have different functions.

The formula applies equally when the series ends with an odd number, for example:

$$1+2+3+4+5+6+7$$

What is to be done with the number in the middle which cannot be paired? Well, it turns out to be half a pair, that is, $(n+1)/2$, so that all in all we have 3+1/2 or n/2 pairs, and the formula does not change. Or better: just take the center term and multiply it by the number of terms in the series: $7×4$ or $n×(n+1)/2$. In the case of a series ending with an even number, the corresponding thing would be to take the average of the middle two terms. In each case a central value is taken, and multiplied by n.

Now I believe we can say at this point that we have more than a formula, a way of getting the correct answer. I grasp, I have insight into, the relation between the formula and the form of the series, and I can go on to work out the sums for series of different types from the type just considered. For example, I see almost at once that the sum of the numbers: 96+97+98+102+103+104 is 600. For the terms are grouped symmetrically about 100, and there are six terms; so six times the central value, 100, gives 600. The formula $(n/2)(n+1)$ now appears as a special case only. The important thing is the basic relationship: some series show a clear relation between their principle of construction and their sum. The relationship uncovered here, the notion of a balance in the whole, compensation among the parts, symmetry, has numerous applications; it is fundamental, for instance, in the integral calculus.

Let me try to restate and generalize what is involved in the example. In the first place, I am presented with a manifold, a many-ness of elements. The elements in this case are numbers. In the second place, the elements have a certain kind of order among themselves. In the example this is what is called linear or serial order. It is grasped initially as a relation between each term and its immediate successor and predecessor. In the third place, in the course of approaching this order from opposite sides, I come to grasp the symmetry of the situation, the possibility of pairing off the terms symmetrically about a central position in the series.

I would propose that any piece of intellectual work, leading to understanding or insight of any kind, necessarily involves a consideration of a *multiplicity,* a *many-ness,* a *manifold* of elements and things, with their properties and relations. By an "element" or "thing" here I mean just any possible object of thought, anything that is somehow one; by a relation, any kind of connectedness between things, any characteristic of a thing that can be specified *only* through the intermediary of another thing.

Let me add just one more term: "operation." The consideration of a manifold of things and their relations involves the performance of operations of some kind, carried out on images or symbols, or perhaps even on physical objects. The distinction between a relation and an operation is just that a relation holds between two or more elements of a manifold simply in virtue of *what* and *how* the elements are, while an operation is something that *we* can perform, an action that we may *will* to carry out. Equality, for instance, is a relation; addition is an operation.

25

I would maintain, then, that different kinds of objects or elements and relations or orderings of things can be isolated and grasped only with the development of a corresponding set of operations. For instance, consider the linear order in our example: $1 < 2 < 3 < 4 < 5 < 6$. The order here is like that which I may construct by arranging different sticks, A, B, C, D, and so on, in the order of their lengths: $A < B < C < D < \ldots$. The ordering relation is asymmetrical—B is shorter than C but C is not shorter than B—and transitive—B is not only shorter than C but also than D or E or any stick farther along in the series. The construction of such a series presupposes, first, the operation of comparing any two of the objects and noting the difference, the assymmetry. Then, as one learns to proceed systematically, rather than just comparing parts of the objects at random, a further operation is presupposed: one will try to find the smallest of the elements first, then the smallest of those left over, and so on. Here one is coordinating two inverse relations: C is larger than A or B, and is shorter than D or any longer object. In the case of the series of numbers, the ordering presupposes the more specific operation of getting from one number to the next by adding 1; in fact, the understanding of what the numbers themselves are presupposes my awareness of the ability always to add one more, and there is a sense in which one can say that one successively *constructs* or *reconstructs* the numbers by this process. The discovery of symmetry in the example depended on the possibility of applying this operation reversibly, the correlation of increase with decrease.

Let me say a few words about kinds of order other than linear order. Linear order is an instance of a more general kind of order called "lattice" order, which is an instance of a yet more general kind of order called "partial" order. In general any set of a finite number of elements, in which the elements are related by a single asymmetrical relation, can be presented by a diagram; the elements a, b, c, and so on, can be represented by small circles, and an ascending line from a to b will mean that a is less than b, or is included in b or is, so to speak, at the lower pole of the asymmetrical relation between a and b. In the case of linear order, we have simply a series of circles placed one above the other and connected by vertical lines.

By drawing different diagrams one can illustrate very different kinds of order. For instance, suppose we have a

26

classification, like the classifications used in biology, or like the one presented in Plato's dialogue *The Sophist*. We start with a class of things—call it "A"— characterized by some property. Then we subdivide A—for simplicity's sake I shall assume that we subdivide it only into two parts, B and B', and so on. A system of class inclusions of this kind presupposes a number of reversible operations. For instance, there is the formation of the union of D and D', the result being C': $D + D' = C'$; $C + C' = B$; and so on. Reversely, $C' - D' = D$; $B - C' = C$; and so on. Also, there is the operation of forming the *intersection* of any two classes, that is, forming the class of all those elements which are in *both* of the given classes; thus I write $D \times D' = 0$, because there are no elements common to both D and D'; and $C \times B = C$, because the only members which are in both B and C are those in C. This system of class inclusions forms a semi-lattice. Roughly speaking, in a complete lattice one would have to have the diagram end not only in a single circle above, but also in a single circle below. Different sorts of ordered systems will presuppose a number of different sorts of operations, and the understanding of such systems will involve implicitly the performance of such operations.

Ordered systems such as I have described, in which there is a set of elements and a single, asymmetrical relation, constitute one of the simplest kinds of mathematical structure. A more complex structure may involve several relations. Also, there is another possibility. An operation, such as addition, may be drawn from its hiding place behind a relation, be given a symbol, and be incorporated *explicitly* as part of a mathematical structure. Thus another simple kind of mathematical structure will consist of a set of elements together with a single operation; the most important example of this kind is group structure. We can complicate matters now by constructing systems in which there are sets of elements, and *both* relations and explicit operations. There is still another possibility. Operations may be incorporated in a structure not only *as* operations, but also as elements. That is, the character of operations may be grasped reflectively, the operations may themselves be made into objects of thought, and sets of operations may be found to have an objective structure which can be described. The structure may be that of a group, and in this case we speak of a *group of transformations*.

Let me summarize. I am proposing that intellectual work consists in a consideration of a manifold of elements and relations. At any given stage of intellectual work, some of the elements and relations are taken as pre-given, others are not given but are progressively isolated by means of operations. What is taken as pregiven at any stage itself involves sets of related operations, which may be embedded in one's thought, but which can be unearthed by a kind of retrogressive inquiry.

The interest in this connection of the mathematical theory of groups of transformations lies first in the fact that the operations, which belong initially to the subjective side, are here *objectivated;* one is no longer performing them, one is viewing them. In the second place, the mathematical theory of groups of transformations brings to the fore the question of *invariance.* Given a group of transformations, one asks what remains invariant or unaltered under this group of transformations. Or given a presumptive invariant, the problem is to discover the group of transformations under which this something remains invariant. The importance of the notion of an invariant under a group of operations lies simply in the fact that the most general or universal aim of intellectual work is the discovery of invariants, of that which is not time-bound in a fluctuant world. Every piece of intellectual work, I should say—and I believe the remark is nothing extraordinary—aims at the discovery of an invariant structure of some kind. Mathematics, in the most general sense, is the study of formal structure. Group structure is just one such structure. But it suggests itself as a kind of paradigm for intellectual work generally, because it involves an explicit consideration of operation and invariant, the two poles, subjective and objective, of intellectual work. To keep these two poles in an articulate and conscious relation with one another, I would suggest, constitutes the liberal dimension of intellectual work. consideration of operation and invariant, the two poles, subjective and objective, of intellectual work. To keep these two poles in an articulate and conscious relation with one another, I would suggest, constitutes the *liberal* dimension of intellectual work.

In presenting the mathematical notion of a group I shall follow the standard textbook expositions. The notion of a mathematical group first received explicit formulation in a letter written by Evariste Galois in 1832, on the night before he was killed, at the age of twenty, in a duel which had nothing to do with his mathematical interests. Since that time the theory of groups has been found to have wide ramifications; it is applied, for instance, in relativity theory and in quantum mechanics. To begin with, I shall describe a simple example, having to do with the symmetry of a rectangle, then go on to a formal definition.

Imagine a cardboard rectangle, placed against the chalkboard so that two of its edges are horizontal and two are vertical. I label the corners of the cardboard A, B, C, D. If I rotate the cardboard clockwise through 180° about its center, that is, the intersection of its diagonals, the cardboard will then cover exactly the same rectangular spot on the chalkboard as before, but now the corner A will have taken the former place of C, C will have taken the former place of A, and also the positions of B and D will have been interchanged. In the mathematician's way of speaking, the rectangle has been *carried into itself* by a clockwise rotation of 180° Let us call this rotation R.

What about a clockwise rotation of 360°? This has the same final effect as not rotating the cardboard at all, and so does a rotation through any integral number of whole revolutions. Before considering these, let me ask also about counter-clockwise rotations. 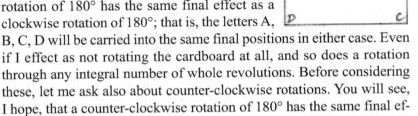 You will see, I hope, that a counter-clockwise rotation of 180° has the same final effect as a clockwise rotation of 180°; that is, the letters A, B, C, D will be carried into the same final positions in either case. Even if I effect as not rotating the cardboard at all, and so does a rotation through any integral number of whole revolutions. Before considering these, let me ask also about counter-clockwise rotations. You will see, I hope, that a counter-clockwise rotation of 180° has the same final effect as a clockwise rotation of 180°; that is, the letters A, B, C, D will be carried into the same final positions in either case. Even if I rotate the rectangle through 360° plus 180° in either direction, or through 720° plus 180° in either direction, the final result is always the same. Because of this state of affairs, it will be well if we change our definition of R; henceforth let us mean by R *any* rotation which interchanges the position of A with the position of C, and the position of B with the position of D. R refers to anyone of a whole class of rotations, or in mathematical language, designates a transformation. A transformation is specified completely by its initial and final positions; the path from the one to the other is unimportant for specifying the transformation.

The transformation so far described exhibits the *rotational symmetry* of the rectangle. There are two additional transformations which carry the rectangle into itself, and which exhibit the *reflective symmetry* of the rectangle. Thus we may reflect the rectangle about a horizontal axis running through its center. This transformation carries A into D and D into A, and it carries B into C and C into B. Let us call this transformation H. Similarly, we may reflect the rectangle about the vertical line through its center; let us call this transformation V.

What happens if I perform two of these transformations in succession? For instance, suppose I first perform R, and then V. Starting from the initial position $\boxed{\begin{smallmatrix}AB\\DC\end{smallmatrix}}$ I first obtain $\boxed{\begin{smallmatrix}CD\\BA\end{smallmatrix}}$ and then $\boxed{\begin{smallmatrix}DC\\AB\end{smallmatrix}}$. But this is the same final result I would obtain if I simply performed H. To symbolize the situation, I write RV = H, where the letters R and V in succession, written like an algebraic product, mean that I first perform R and then V. The same sort of result occurs in other cases: if I perform two of the transformations in succession, the result is always the same as the third transformation. In symbols, RV = H = VR, VH = R = HV, HR = V = RH.

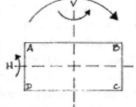

What if I perform R twice in succession? The result is a total rotation of 360°, and I have simply returned to the starting-point. I get the same result if I perform H twice, flipping the rectangle twice over the horizontal axis, or if I perform V twice. In order to be able to write equations in these cases also, we do something a bit strange; we speak of any rotation or motion which carries the cardboard from its initial position back into its initial position as the *identity* transformation—although it is not really a transformation at all—and we symbolize it by "I." The transformation I has the same sort of function as zero in the addition of numbers, and the reason for introducing the one is about the same as the reason for introducing the other. I can now write:

$$R^2 = H^2 = V^2 = I.$$

(The superscript "2" just means that the transformation is performed twice in succession.)

Adding *I* now to our list of transformations, we have four all told; there are no other transformations, distinguishable from these four, which carry the rectangle into itself; and I say that these four transformations, along with the ways in which they combine, tell us all there is to say about the symmetry of the rectangle. To say a figure is symmetrical is to say that there is a set of transformations which carry the figure into itself. In the case of a square there are eight transformations in the set; in the case of a cube there are 48; and so on.

So much for the example; I have now to show that it exemplifies

30

group structure. In order to have a group in this technical sense, it is necessary first of all to have a set of somethings—in our case, transformations—and secondly to have a way of combining any two of the somethings—in our case the operation of combining or "multiplying" transformations by performing them one after another. Let the somethings, the elements of the system, be designated by lower case letters a, b, c, . . ., and let the combining operation be designated by a circle: o. Such a system will constitute a group if four conditions hold:

(1) The result of combining two elements of the set is itself a member of the set; in symbols: a o b = c. This particular condition is so important that it is sometimes referred to as the *group property*. In our example it is clear that it applies: the result of performing any two of the four transformations in succession is itself one of the four transformations.

(2) There must exist an identity element—I shall designate it by "I"—which is such that when it is combined with any given element then the result of the combination is the given element again; in symbols: a o I = I o a = a. In our example it is dear that there is an identity transformation fulfilling this condition.

(3) For each element there must be an inverse element, such that when any element is combined with its inverse, the result is the identity element. To symbolize this I shall designate the inverse of any element a by a^{-1}; then the condition stated in symbols is: a o a^{-1} = a^{-1} o a = I. In our case it turns out that all four transformations are their own inverses. This is a somewhat special situation, but it is still true that the condition is satisfied.

(4) Finally, there is a condition called "associativity": when three elements a, b, c are to be combined, it makes no difference whether I first combine a and b and then combine the result with c, or first combine b and c and then combine a with the result. In symbols: a o (b o c) = (a o b) o c. It can be shown that the transformations of our example, taken three at a time, obey this associative law.

I give a few more illustrations of groups, to help fix in mind these four conditions, especially the first three, which are most important for what follows. Countless other examples can be found in any textbook on group theory.

The positive and negative integers, including zero, form a group under the operation of addition. Clearly the sum of any two integers is an integer. The identity element is zero, and the inverse of each integer a is -a. This group is actually a subgroup of a larger group, in which the elements are all the positive and negative integers and fractions—all the rational numbers, we say—and in which the operation is again addition, and the identity element zero.

The positive and negative rational numbers, excluding zero, form a group under the operation of multiplication. Thus the product of any two rational numbers is a rational number. The identity element is 1, since 1 multiplied by any given rational number yields as a result the given number again. The inverse of any rational number a is what is usually called its reciprocal, that is $1/a$; for the product of a and $1/a$ is always the identity element 1. Zero has to be excluded here because it has no reciprocal.

If we consider all positive and negative rational numbers under the operations *both* of addition and multiplication, we obtain an example of a type of structure called a *field*. A field, in turn, is a special instance of a type of structure called a *ring*.

Now I am proposing group structure as a kind of paradigm of the structures involved in rationality, in the exercise of intelligence. Let me try to justify the proposal in terms of the characteristics of a group.

In the first place, the presence of group structure means that we have a *closed* system. To say that a system is closed means that the combination of two elements of the system always yields an element of the system, something of the same kind as the somethings I began with. If I combine two musical tones, in the sense of sounding them either in succession or simultaneously, the result is not a musical tone, but something different, a melodic move or one of the intervals of harmony. But two colors, in the form of two pigments or colored lights, can always be mixed to form a color, if you include white and black and gray as colors. And two sentences when combined by such connectives as "and," "or," "but," "although," form sentences. And so on. The importance of the group property, that is, the property of a system in virtue of which it is closed, is that it provides a criterion of relevance and a means of distinguishing one subject matter from another, or one level of consideration of a subject matter from another level, or one kind of invariant from another. On any particular level of discourse, the group structure provides a paradigm of completeness; it lies at the root of the fact that there are just five regular solids, for instance, or just 32 types of crystals. When more than one level is involved, the group property may provide a criterion for distinguishing them.

In the second place, because of the fact that every element in a group has an inverse, any given application of the combining operation can be reversed. If I add 7 to 5 and so obtain 12, I can always proceed to add the inverse of 7, namely -7, to 12 and so obtain 5 again, returning to the starting point. In other words, operations are reversible. This reversibility is an essential characteristic of an operation. All operations are actions, whether carried out on physical objects, images, or symbols; but not all actions are reversible. If I write down a sentence, going from the left to the right side of the paper, the action is irreversible, in the sense that I cannot write the sentence from right to left without first going through the labor of acquiring a totally new habit. Or if I act out of a passion, for instance wrath or love, the actions entails consequences which cannot simply be reversed, and I cannot return exactly to the original starting-point. Reversibility is, I believe, an essential condition of rationality. I would hypothesize that the possibility of progression and return, with reference to an invariant point of origin, lies at the basis of every exercise of the intelligence.

Thirdly, I return to the notion of invariance under a group of transformations. The reversibility of operations, also perhaps their associativity, the possibility of performing then in different orders without changing the end-result, give them a certain neutrality. If I cut down a tree, the action is irreversible, and the world has been irrevocably changed. But if I take ten steps eastward, I can reverse the operation, taking ten steps westward; the first operation is annulled by the second. That which remains invariant under a set of such operations may, as the operations are carried out, emerge as an organizing principle. For instance, taking ten steps east and then again ten steps west, I experience a kaleidoscopic flow or series of shifting views of colors and shapes, which I organize in terms of the notions of physical object and spatial framework. There is a physical object over there, another beside it to the left, another behind it, and so on.

I would propose, then, that invariants such as physical objects do not swim ready-made into our ken, as complete, pre-given wholes. The discovery of invariants goes hand in hand with the development of sets or groups of reversible operations. We attempt to assimilate new experience in terms of previously developed operational structures; sometimes everything does not go smoothly, we are thwarted in our attempt, and have to accommodate the operations to what is presented. The progress of knowledge proceeds in two complementary directions, in the direction of increasing articulation of experience as organized in terms of invariants, and in the direction of increasing awareness of operations as being in equilibrium and as forming groups. This twofold progress enables the individual subject to place himself within the world as part of a coherent whole.

The progressive emergence of the physical object as an invariant involves the development of a number of groups of operations. These have been traced out in some detail in the experiments of Jean Piaget on the child's conception of space. There is, for instance, the group of Euclidean transformations, also the group of projective transformations and the group of topological transformations. The Euclidean transformations are those which carry every point of space into another point, either by the translation of every point by a given amount in a given direction and sense, or by a rotation of all space, thought of as a rigid body, about a point or line, or by a combination of a translation with a rotation, that is, a twist. All these transformations, which are infinitely many, form a group. What remains invariant under this group of transformation is the congruence of geometrical figures, that is equality of size and sameness of shape. Euclidean geometry is largely the study of what remains invariant under this group of transformations.

Secondly, there are the projective transformations. Suppose I project a plane figure from one plane to another by means of straight lines running from my eye through the first plane to the second. Any given point in the original figure will have a corresponding point in the second plane, located by means of the line which passes through my eye and the given point of the original figure. Straight lines in the original figure are carried into straight lines in the new figure, but angles and lengths in the original figure may differ markedly from angles and lengths in the new figure; thus neither size nor shape is conserved. The projective transformations, however, form a group and projective geometry is the study of what remains invariant under this group. What remains invariant, principally the straightness of lines, the incidence of points on lines, and something called the cross-ratio of four points on a line, is what enables us to correlate different perspective views of an object as being views of the same object.

Finally, the topological transformations. Topological transformations are those which carry one figure into another in such a way that to every point of the one figure there corresponds a single point of the other, and points that are nearby or neighboring in the one figure remain nearby or neighboring in the other. Figures can be stretched and shrunk and enormously distorted under topological transformations. A circle can be transformed into a square or an ellipse, but not into a straight line; a sphere into a cube but not into the figure of a doughnut. The relations which remain invariant under such transformations, relations like contiguity, surrounding or being surrounded, closure or lack of closure, constitute the subject matter of topology.

I have listed these three branches of geometry in the historical order in which they have been made subjects of mathematical study. It is Piaget's conclusion that the order of development in the child is just the reverse. The child grasps topological invariants before projective invariants, and projective invariants before Euclidean invariants. Each of these acquisitions presupposes the ability to carry out, in part physically but more importantly in the imagination, a corresponding group of operations. Topological invariants can be established before there is a clear demarcation between the operating subject and the objects operated on. Projective invariants involve the correlation of different visual points of view, and hence presuppose the child's realization that he is seeing from a point of view. Here the polar opposition between self and world, subject and object, begins to crystallize out. Only with the attainment of equilibrium of the Euclidean operations does the physical object appear in its final character, as an ideal unity within a spatial framework, over against the observer, accidentally where it is and capable of being somewhere else, capable of presenting constantly different aspects, but never of presenting itself all-at-once as a whole.

There is one other sort of invariant, radically different from the physical object, which I wish to take up, namely *idea*. But first it is necessary that I consider yet another kind of thing, another sort of invariant if you wish, namely *word*. Words are in general the medium through which the result of any piece of intellectual work becomes the possession of more than one man; and operations with words, as well as with images and physical objects, may be importantly involved in the original carrying out of the work. What is a word?

As I talk to you now, we in this room are aware of sounds coming from my mouth. The sounds, considered just as sounds, are part of the world of nature, the world of physical objects and processes occurring in space and time. As sounds, they do not differ from the sounds of nature, the sounds produced by brooks and breezes; each one is a temporal event, occurring during a certain time, and apprehended at certain places.

35

We sometimes speak of a sound being repeated; we say that we have heard the same sound, for instance the sound of a whistle, several times repeated. If we were strict, we would say, not that the sounds are the same, but that they are similar, in the same way that we say that giraffes are similar, or objects produced on an assembly line are similar. Each sound is an *individual* temporal process; everything belonging to the world of nature is *individuated* in space and time. No two individuals belonging to the world of nature, whether physical object or temporal event, can be precisely the same; but two such individuals may be similar.

I submit, however, that I can repeat the same word twice. A word is identifiable as identically the same word in any number of real individual embodiments, whether sounds or concatenations of written characters. I can repeat *exactly* what I said before, if you failed to catch the words. Therefore the words are not the sounds, but are rather embodied in the sounds. This no doubt seems a bit strange.

We have a similar problem—perhaps the same problem, I am not quite sure—when we ask what a symphony or a folk dance is, independent of its individual performances; or what the *Iliad* is, independent of a thousand copies or a thousand recitations of it.

I do not believe that the word can be defined as a class of similar sounds or marks. When I grasp a word as embodied in a sound, the action is different from grasping the fact, for instance, that an individual animal belongs to the class of giraffes. One has to try to ascertain precisely what it means to speak or to listen to speech.

Listening now to the sounds coming from my mouth, we are aware of these sounds as having functions which merely natural processes lack. Seeing my body and hearing my voice, we are aware of me as willing the sounds and giving expression to thoughts which I wish to communicate. We apprehend the sound of my voice as manifesting certain psychic processes and as embodying expressions of certain meanings or ideas.

The word, then, is intrinsically relative to psychic or subjective activities. Within the objective world, there are many things which are thus relative to psychic activities. The qualification of anything as familiar or strange, useful or useless, as a tool or monument, as a slave or a king, presupposes the existence of psycho-physical beings who use things for intended ends or goals. We encounter here a new stratum of the world, over and above the stratum of the individual objects and processes of nature; let us call it the *cultural* stratum. The cultural stratum includes some things which are individuated in space and time, such as ashtrays, but it also includes objects like words and symphonies which are *not* individuated in space and time, but which can be repeatedly *embodied* in particular spatio-temporal individuals, physical objects or processes.

The fact that words and symphonies are not spatiotemporal individuals does not make them any the less objective. Through their embodiments, they are observable, distinguishable, repeatedly identifiable somethings; one can ask questions and make verifiable judgments about them.

In the grasping of such cultural invariants, there is presupposed the fact that I belong to a community, with established customs and traditions. I must have recognized myself as one among other psychophysical beings, as one man among other men, who can *intend* the same things as I. This recognition presupposes in turn the establishment of certain modes of transaction, reversible operations in which I put myself imaginatively in the place of another man, in which I identify myself, in certain respects, with every man.

Words, embodied in sounds or written characters, not only manifest psychic activities; they express meanings or ideas. An idea is not itself a psychic activity; it is not, so to speak, *within* consciousness. Consciousness is always consciousness of something; this something is an object of some kind; and an idea is one such kind of object. When I grasp an idea, I grasp it with the sense that it is something I can return to and see or grasp on another occasion; also, that it is possible for someone else to grasp it, too. We can ask questions about ideas, and make testable statements about them; they fulfill the criteria for objectivity, although they are not physical objects. On the other hand, they may be *exemplified* or *illustrated* in physical objects or processes or relations, as the idea of giraffe is exemplified by a particular giraffe, or the idea of equality is illustrated in the relation with respect to weight or size of two particular bodies. But it belongs to the idea of idea that an idea can be entertained in the *absence* of any physical exemplification of it.

The grasping of an idea presupposes certain judgments, and prefigures other judgments. We cannot sharply separate the making of judgments and the grasping of ideas, as though the ideas were first grasped, and then judgments were then built out of conjunctions of ideas the way a house is built out of bricks. Every judgment, every grasping of a relation, presupposes operations such as comparison, classification, seriation, and the ideas can emerge clearly only as these operations reach equilibrium.

In large part, ideas are presented to us through the medium of speech and writing, as *presumptive invariants*. Most words are encountered as familiar; we understand them in the sense of being vaguely, passively conscious of their meaning. An intellectual conscience will propose the task of making this meaning distinct, of articulating, one by one, the parts of this vaguely unitary sense.

One of the most important procedures here is just the free variation of the image which comes with the word. As the operations become established whereby I pass from one form of the image to another, varying one factor at a time, they tend to assume the equilibrium of group structure. Suppose, for instance, that I uncover four variants of the image (see figure); once I can pass from any one form to any other by carrying out a specific variation, then the operations form a group; any two of them are equivalent to a third, and each is its own inverse.

Besides this free but disciplined use of fantasy, there are the operations on symbols employed in hypothetical or deductive reasoning. Like fantasy, hypothetical reasoning is a way of dealing with what is physically absent; it enables us to draw out the implications of *possible* statements; and to discover invariants which fall outside the range of empirical verification, for instance the law of inertia.

The diagram I have just drawn will remind some of you of the traditional square of opposition, which is involved in what are called immediate inferences. The grasping of the relations involved in the square presupposes the establishment of operations whereby anyone of the four propositional forms is transformed into any other, and the six operations or transformations form a group.

Again, a group of operations is involved in the part of logic which deals with the propositional connectives, such as "and," and "or," "if-then," "neither-nor," "unless," together with the sign of negation of a proposition, namely "not." Given any two propositions, p and q, we can construct sixteen fundamental propositional forms by means of these connectives, for instance, p and q, p or q, if p then q, if q then p, and so on. All sixteen forms can be expressed by means of the three logical words "and," "for," and "not." The sixteen forms, thus expressed, constitute the elements of a lattice. More important, the operations whereby one of these forms can be transformed into another constitute a group.

Let me try to recapitulate what I have been saying. The work of the mind—and the mind simply is its work—is a process of attending to and grasping situations, affairs, complexes of elements that are somehow presented, and explicating them with respect to such of their determinations and relations as are likewise presented. There is no zero starting-point for inquiry, but in any inquiry something is accepted initially as pre-given. Then either of two directions may be taken; either further invariant relations are sought by tracing out the connections between the elements, which continue to be taken as pre-given, or else the pre-given elements themselves become the problem, and it is necessary to reactivate the embedded relations and connections upon which the grasping of them depends. But whether the work proceeds progressively or retrogressively, it always involves operations, that is, actions which are reversible and repeatable. To say that I understand a situation is to say that I have performed the operations of tracing out the relations in the situation, and have the dear sense that I can perform them again. Completeness of understanding on any level implies that the operations have attained the equilibrium of group structure, so that the transformations taking us from anyone element of the situation to any other have become explicit. The structure of operations thus becomes cyclical and transparent to what is operated on; new invariants may now come into view, or what has previously been taken as invariant may appear for the first time with clear boundaries.

We are prone to ask ultimate questions, and inevitably we ask how far this process goes and where it may arrive. Does there exist, at the summit of the hierarchy of structures, a pure form, the form of forms, of which every other form or structure is a refracted image? Can the edifice of forms traced out in intellectual work have completeness or is there inevitably something left out, something which is inaccessible to the intellect? Plato and Aristotle recognized some such thing called Necessity or Matter, which kept the world from being completely intelligible. Jerusalem, in opposition to Athens, has insisted that the highest principle is not accessible to the human intellect, and that our chief duty is loving submission rather than inquiry. Contemporary existentialism has claimed that the intellect itself is corrupt; that the freedom from engulfment in time which it pretends to give us is an illusion; and that the identification which it brings about between the self and every-man is a bar to direct knowledge of our existence and its mortality. But wherever the truth lies here, it is clear that the explication of these opposing positions, the drawing out of their consequences in every detail, is itself a task for the intellect at work.

Logos and the Underground
(1960)

AUTHOR'S PREFATORY NOTE

The lecture here printed was delivered in September 1960 as the dean's "opener." It is largely based on Edmund Husserl's *Erfahrung und Urteil,* which I worked my way through in the summer of that year. When Walter Sterling recently proposed [in 1984] printing the lecture in *The St. John's Review,* conscience told me I should review the text, to determine whether I could still endorse the propositions that I put forward with such somber earnestness 25 years ago. My conclusion has been both a Yes and a No.

For the heroism of Husserl's repeatedly renewed efforts to achieve a presuppositionless "beginning" in philosophy, my admiration must always remain. And the attempt to carry out phenomenological description—the delineation of how things (tables and chairs, words and sonnets and symphonies, universals like 'justice,' fictional characters like Sancho Panza, beings like my cat, persons like the reader) present themselves in awareness—has a value. In 1960 I considered the Husserlian descriptions as an antidote to the self-defeating relativism that so many freshman brought to the college: the pervasive disbelief in the possibility of improving one's opinions, the bland assurance that your opinion is as good (or as bad?) as mine. Still today I see as desirable an attentiveness to the describable character of the things that present themselves in awareness, just *as* they present themselves—to echo the Husserlian phrase. It is a mode of thoughtfulness that, in an age of reductive slogans, needs to be encouraged.

But concerning the Husserlian enterprise I today have doubts that I had not quite formulated 25 years ago. The descriptions no longer appear to me securely presuppositionless or self-explanatory; and the claim that phenomenological description constitutes "the correct method" in philosophy seems to me far too grand. "Man," says Claude Bernard, "is by nature metaphysical and proud"; and the presumption of certainty seems to me more often illusionary than not. Methods are useful or necessary; but of method that claims to have an exclusive right we must be wary, for any method presupposes more than we are likely ever to know. In short, in have long known that we must begin *in medias res,* I am no longer prepared to suppose that the mind's improvement or the advance of knowledge will consist in coming to an absolute starting-point. The very process whereby we successively pronounce the words of a sentence while intending a meaning seems to me utterly mysterious, and I think it is a miracle that we can begin at all.

This is not the place to pursue these thoughts. (Let me only mention that today I would look to linguistics and behavioral biology to throw new light on the 'underground' of the liberal arts; and I see it as a task for the future liberal artists to explore with sensitivity the intricate dialectic between genotype and phenotype, between the deep or hidden structures and what appears. This investigation would not presume to avoid hypothesis; but insofar as hypothesizing necessarily involves reduction, it would be cognizant of the dangers thereof. The human spirit is a "tangled wing," to use Melvin Konner's figure for it, and I look to linguistics and biology, as to the Bible and all deep literature, for the further elucidation of what we are and how we do what we do.)

And what of the poor freshmen, for whom the opening lecture of the college year is supposed to be a kind of exhortation? I tremble to think how widely my efforts must have missed the mark; years afterward I was informed that it was a standard bit of "put-down" on the part of upperclassmen to tell the freshmen that they could not expect to understand my lecture. But even today I know not what verbal gestures might count as useful, amidst the profusion and confusion of aims and ideas that freshmen arrive with. How can I say, in one breath: (1) work patiently and hard, for the value of what you acquire will, in general, be proportional to the care that goes into the acquiring; and (2) think! be inventive! for what is in front of you can appear in a new light, and discoveries are possible! but (3) do not expect certainty? If I should say such things, some of the brightest of my auditors would find my sayings impossibly contradictory in tendency, and the only response I could make would be that I hope and believe it is not so. In what puts itself forward as human knowledge, it is by the care and thoroughness, and by the inventiveness and the unexpectedness that throws a new light, that I attempt to distinguish the better from the worse. I know no other way.

In Plato's dialogue *Phaedo*, Socrates speaks of having, at a crucial turn-ing-point of his life, fled to the *logoi*. Previously, he says, he had pur-sued the investigation of nature, seeking the efficient and final causes of the things of the visible world. But this investigation having led to nothing that he could trust, he took flight to the *logoi*. What is charac-teristic of Socrates, the Socratic questioning, takes its start from this flight to the *logoi*.

The Greek word *logos* (plural: *logoi*) has a variety of meanings, but according to Liddell and Scott, its primary meanings are, first, the word, or that by which the inward thought is expressed, and second, the inward thought itself. Additional and related meanings are: state-ment, assertion, definition, speech, discourse, reason.

Now I am not going to give a commentary on this passage in the *Phaedo*; but I wish to take a start from the observation that there is such a thing as *logos*, meaningful speech, speech which expresses the inward thought. And I am going to explore the question: What does this fact presuppose? What underlies it?

I may as well warn you that I shall be attempting the most pedes-trian, prosaic, dry sort of description and explication. I shall try to avoid *introducing*, or *constructing*, hypotheses or theories, however attractive, which would account for what is described. I shall try, on the contrary, to describe certain kinds or types of things which are recognizably in-volved in our speaking, and my effort will be to delineate them just *as* they present themselves to us, just *as* we are aware of them. If there is an assumption in my procedure, I think it is the conviction that the "I" or self on the one hand, and the world on the other, cannot be *thought* of separately. Accurate description of my experience is description of the experience of an "I" or self *in* a situation, of a presence which is es-sentially *in* the world and bound to the world. I shall have to analyze this experience into certain strata or levels, and because of limitations of time, concentrate on certain fundamental strata which may, unfortu-nately, seem to you the least interesting.

In one respect I shall imitate the Socratic flight: I shall leave out of account all results of natural science —physics, chemistry, and bi-ology. Over the past 350 years scientists have developed imposing structures of thought which seem to reveal to us a previously hidden world, alongside of or somehow behind or underneath the world in which we live from day to day. Arthur Eddington would say, for in-stance, that besides the apparent lectern behind which I stand, there is another lectern, the real one, consisting of electrons and protons. I would maintain, on the contrary, that this is an incorrect way of speaking and thinking: there is only one lectern, the one that is before me. What is meant by the electrons and protons can only be understood by considering certain procedures and experiments and the theories built up around them. In seeking the roots of these theories I shall be led back to the world of my everyday experience, and to the language in which I formulate this experience. To ignore the layered or storied structure within and underlying scientific theories, to regard the electron as somehow on a par with and alongside the table, is to commit what Whitehead calls "the fallacy of misplaced concreteness." So I shall begin with the analysis of everyday speech and experience.

Even here I must make a reservation. I am not trying to take account of all aspects of everyday speech. We use speech to praise and to blame, to command and to pray, or even for "whistling in the dark." I shall be concerned only with the rather ordinary and colorless fact that in our speaking we make statements, assertions, which signify states of affairs, the "way things are," as we say.

The statement or assertion is the unit of fully meaningful speech. A single word, outside an assertion, does not have a fully determinate meaning. If I were to look and point in a certain direction, and to shout "Fire!", you would probably recognize that I was asserting something. But the same word "fire" in another context may have a quite different meaning, for instance in the sentence, "The captain ordered his men to fire." There are even subtler differences due to context. The meaning of the word "fire" is not quite the same in the sentence "Civilization depends on fire," and in the sentence, "The fire was burning brightly in the hearth." Precisely what a word refers to depends on the context in which it is used, which may be verbal or non-verbal or both. But in any case, nothing is really *said* until we have an assertion or statement—what traditional logic called the predicative judgment. What is predicative judgment?

The word "predication" comes from the Latin "praedicare," originally meaning to speak out, to enunciate publicly. The word was later preempted by logicians in order to translate Aristotle's term *katagorein*. The Greek word *katagorein* had originally meant to denounce, to accuse in the marketplace or assembly (the root *agora* means marketplace or assembly). Aristotle then appropriated the term to express the meaning: to say something of a subject. What is spoken of, that about which something is said, Aristotle calls the *hypokeimenon*, that is, the underlying; that which is said about or of the *hypokeimenon* is called the *kategoroumenon*; it is, one might say, what the *hypokeimenon* has been accused of. The corresponding English words, which derive from the Latin, are "subject" and "predicate." Whenever a predicate is attributed to a subject, then we have a statement, an assertion, which expresses a decision regarding the validity of the attribution, or, the justness of the "accusation"—for example, when "this" is "accused" of being a man in the statement, *This is a man.*

Doubts about the universality of the subject-predicate analysis of assertions have sometimes been raised. Consider for instance, the statement "It is raining." It might be suggested, in Aristophantic vein, that the pronoun "it" stands for Zeus. But this is surely not what we mean when we say it is raining. Where is the logical subject —or is there one?

I think this is a case in which we are fairly clear as to what we mean or intend, while the structure of the language fails to reflect the structure of the meaning. I do not believe it is possible to find an assertion so simple as not to involve at least *two* mental signs. One is an *index,* a sign which so to speak points to something; the other will be a sign signifying a characteristic or situation or action which somehow belongs or pertains to that which is pointed to. The assertion as a whole asserts something of something, and therefore necessarily involves a two-foldness. Language may fail to mirror this two-foldness. In the present case, I should say, we have a kind of idea of a rainy day. The indexical or pointed sign is that whereby I distinguish *this day* or *time,* as it is placed in my experience. The assertion "It is raining" asserts that the present time is characterized by raining-going-on.

There is another objection to the usual subject/predicate analysis. When I say "'Alcibiades is taller than Socrates," it may be argued that I am talking about two subjects, Alcibiades and Socrates. When I say, "A sells B to C for the price of D," there are four indexical signs A, B, C, and D, which are here connected by the relational predicate: ". . . sells . . . to . . . for the price . . . ." The logician may claim that there are four logical subjects here, four *hypokeimena.* The objection does not deny the distinction between subject and predicate, but points to cases in which there is not a single axis running from subject to predicate, but rather a relation which relates two or more different things.

Let me pass this objection by for the moment. Because of its greater simplicity, the assertion in which a predicate is attributed to a single subject would appear to require consideration before relations are considered. I shall return to relations later on.

An assertion, I said, expresses a decision regarding the validity of the attribution of predicate to subject, the justness of the accusation. It presents itself as knowledge; it pretends, so to speak, to be the truth. It may, of course, turn out to be false. For instance, I may have pointed at something and said, "That is a man," and then it may have turned out to be a showcase mannikin. Or the statement may become and remain doubtful or problematic. Nevertheless, I should say that it belongs to the very meaning of any assertion to make the claim to being knowledge. Negation, doubtfulness, probability, or improbability are meaningful only as modifications of this original claim. Even the statements which are used in presenting to us a world of fantasy, say the fantasy world of a novel or of the *Iliad,* make this claim *within* the context of the unity of the particular fantasy world. The truthfulness of such a work of fantasy or imagination as a whole is a rather more difficult matter, and lies in the ways in which the fantasy-world imitates, either directly or by analogy, certain features of the world in which we live.

How do we determine whether an assertion is true? Certainly we do this, day in and day out; but how? What we encounter, in asking this question, is the problem of *evidence.* What is an *evident judgment?*

The word "evidence" derives from the Latin word *evidens,* meaning *visible.* The word "evidence" when used in connection with judgments does not always mean visibility, but visibility appears to be its most primitive meaning.

I think I should digress for a moment to point out that most of the terms which we use in talking about thinking depend on visual images. We speak, for instance, of "definition," which means *setting bounds or limits;* of "synthesis" or "composition," which means *putting together;* of "analysis," which means *breaking up;* of "implication;" which means *folding back upon.* All these terms exploit, more or less evidently, an analogy between thinking and certain motor activities which we can perform, which we apprehend visually, and which in turn affect or change what we see.

The assertion itself is something which is apprehended, not visually, but by means of hearing; although, especially in a post-Gutenburg

46

era, we may tend to think of assertions as written out, visually. Now there appears to be a fundamental difference between what is perceived by hearing, and what is perceived by sight. What is perceived by hearing is something that comes to be successively, in time. What is perceived by sight *can* present itself as being there all at once, as a whole. A tone or noise or statement comes to be successively, so that its different parts exist in different times; it is a temporal event. When I see a table, on the contrary, I take all of its parts to exist simultaneously, even although what I see at any one time is only one or two sides of the table. I never see all parts and sides of the table at once, I can only come to see all the parts in the sense that, by moving about, I am able to examine them one by one in succession. But the table is not a temporal event.

This is an important difference, which may have important consequences; but the point for the moment is this: A statement or assertion, coming to be in time, makes a *prima facie* claim to knowledge. Knowledge of what? We have to say, I think: knowledge of what is, and of how it is. The judgment has a subject or *hypokeimenon,* about which it is. This *hypokeimenon* must somehow be pre-given, evidently given, prior to any asserting, if the assertion is to be what it claims to be, namely knowledge of what is. But what is evidently given? Many things, perhaps, but first and foremost, what we can all agree upon, the individual, visible objects which are presented to us in the world. The object or thing presents itself to us as being *there,* as a whole, with all its parts, within the visible world. A temporal event, say a sound or a motion, seems, on the contrary, to demand further analysis: we want to know *what* is moving, or what is the source of the sound or other temporal event. The world as it presents itself to us is first and foremost a world of individual objects.

Therefore, I am going to start the discussion of the problem of evidence by discussing the kind of given-ness which a visible object has. Then I shall go on to discuss other kinds of objects of awareness, which can also be made subjects of predicative judgments, and which may have their own modes of being evidently given. These other objectivities, potential subjects of judgments, are in a certain way founded upon our experience of the visible world; they arise for us in connection with our experience, but as Kant would put it, they do not simply arise *from* experience—I think that will be apparent.

How, then, are the individual objects of experience given or presented to us? As I stated previously, I am leaving out of account all that the physical and biological sciences can tell me of the processes involved in sensation and of the objects of experience. I wish to make, in addition, certain further simplifications.

In sense experience I am confronted with individual objects which present themselves as *bodies,* as *corporeal.* But there are many individual objects of experience which do not present themselves *simply* as corporeal. Animals, men, and man-made objects, products of art, are indeed perceived as bodies within the spatio-temporal world, but they differ from rocky crags, rivers, and lakes, in expressing the presence or activity of what I shall call "soul." An ash-tray is not simply a natural body; what it is can only be understood by a reference to human beings who indulge in a certain vice. A human being is not perceived as such in quite the same way as a rock is perceived as a rock; there is involved an *interpretation* of what is perceived, as expressing the presence of soul, the psychic, the subjective, the "I" or self of this other who is before me. The soul of the other is not simply perceptible in the manner of a corporeal object; but it is understood, through interpretation of the simple perceptions, as being *in* and *with* what is simply perceptible. Now this whole stratum of experience, involving as it does the interpretation of what is bodily as *expressing* the psychic, I wish to leave out of account, so as to attend entirely to what all such experience presupposes, the experience of individual objects as corporeal.

Finally, as a further simplification, you must permit me to imagine that I am a purely contemplative being, examining the individual objects out of a pure interest in finding out about them. It is probably a rather rare occurrence for such a pure interest to govern our activity. Ordinarily we pass over the perceptions to go on to manipulating objects, or valuing them in relation to certain practical aims. The "I" or self, living concretely in its surroundings, and among other selves or persons, is by no means primarily contemplative. A pure contemplation of a particular object can occasionally occur; this involves a stopping of normal activity; it need not be especially important. As subordinate to a philosophic reflection which seeks to discover the structure of the world, such contemplation can become serious. My supposition here of a purely contemplative interest may be regarded as a fiction, designed to enable me to uncover a basic stratum of experience.

The object does not present itself to me in isolation, all by itself, nor does it present itself as something completely novel. With the awakening of my interest, it comes forth from a background, in which I take it to have been existing already, along with other objects. Suppose, for instance, that the object which I am about to examine is this lectern; I grasp it as something already existing, something which was already there, in the auditorium, even before I was looking at it. Similarly the auditorium, with its stage and curtain and rows of seats, including the part I do not see because it does not come within my field of vision, was already there, was within the bounds of the familiar St. John's campus, within the familiar town of Annapolis, within the farther and less familiar reaches of Anne Arundel County, and so on, till I say: within the world. This pre-given-ness of objects and of the world in which they are is prior to every inquiry which seeks knowledge; it is presupposed in inquiry. The presupposition is a passively held belief, a belief which I hold with unshakable certainty. It is *doxa,* the Greeks would say. There is a passive doxic certainty in the being of the world and its objects; I cannot imagine it possible *earnestly* to doubt this belief. Every inquiry into an object proceeds on the basis of the believed-in-world. Belief precedes inquiry which in turn aims at knowledge.

The object itself is never completely novel, it never presents itself as something completely indeterminate, about which I can then proceed to learn. The world, for us, is always a world in which inquiry has already gone on; it is a familiar world the objects of which belong to more or less familiar types, with more or less familiar kinds of characteristics. When I examine an object for the first time, I already know, in a sense, something about it. Not only do I perceive the side which is presented to me, but I anticipate, in an indeterminate way, certain of the characteristics of the unseen side. The other side of the curtain here I imagine at this moment as being grey; it is quite possible that it will turn out to be of another color, but I am confident that it will have some color. At the very least, the object is pre-given as a spatial object, with such *necessarily* accompanying characteristics as color and shape; probably also as a spatial object of a more particular type, belonging to a more specific category. The progress of the inquiry takes the form of correcting anticipations, or replacing vague anticipations by definite, perceived characteristics. Every advance of the inquiry has the form "Yes, it is as I expected," or "Not so, but otherwise"; in the latter case, the correction is always a correction within a range of possibilities

49

which is not limitless. For instance, I may expect "red"; it will not turn out to be Middle C. To each single perception of the object there thus adheres a *transcendence* of perception, because of the anticipation of the possibility of further determinations. In the succession of perceptions of the object, I am aware of it as an identical something which presents itself in and through its characteristics and relations, but I am also aware of it as a unity of possible experience, a substratum about which I can always acquire further information.

As I turn to the object for the first time, there is a moment in which my attention is directed to the object as a whole, before I go on to note particular characteristics, parts not quite perceived of the obvious whole. This moment has short duration; the attempt to make it last turns into a blank stare. But even as I go on to examine particular aspects of the whole, there remains an effect, a precipitate, so to speak, of this first mental grasping of the object, this taking-it-in-as-a-whole. As long as the object remains the theme of my inquiry, the characteristics and aspects are not viewed separately, by themselves, but always as aspects *of* the object. If S is the object, and p, q, and r are characteristics, then my perceptions of p, q, and r are not isolated, but each perception of a characteristic adds to, enriches the meaning for me, of the substrate S; first S is seen to be p, then S which is P is seen to be q, and so on. And always in the background there is the sense of the object S as a temporally enduring something which *has* these characteristics. The persistence of S as an identity in time presents itself passively, in the harmonious succession of perceptions, as though I had nothing to do with it. Yet I must note at least in passing that this grasp of the object as an enduring thing is complex, and presupposes a structure in my inner time, in the flux of changing awareness, whereby the object presented at any moment is grasped as having been and as yet to be.

What I am seeking to describe here is a receptive experience of the object in which I am first aware of the object and then examine it, noting characteristics, without actually making judgments or assertions; passing from perception to perception, without attempting to fix once for all the results of perception in the form of assertions. But it is apparent that even in this receptive experience of one particular object, prior to all judging, there emerges the basis of the distinction between subject and predicate, namely, in the distinction between substrate and characteristic.

I can of course make anything which presents itself into the theme of an examination or inquiry—the color of the curtain, for instance, or the aggregate of seats in the auditorium, which presents itself in a particular spatial configuration. That with which the inquiry is concerned as its theme then comes to be a substratum or substrate, of which I proceed to ascertain the characteristics. The distinction between substrate and characteristic would thus seem to be relative to the theme of the inquiry. Some of the things I perceive and attend to, however, are of such a kind as to exist *only* as determinations or characteristics of something else—for instance the color here which I take as the color of the curtain. Other things I perceive and attend to are not essentially dependent in this way. The curtain, for instance, is not a characteristic of the auditorium in the same sense as its color is its characteristic. That the curtain is where it is is in a sense accidental; it could be somewhere else, and if it were, its color would have gone with it. In grasping the curtain as an object of perception, I grasp something which has a certain independence of everything else, which does not present itself always and necessarily as an aspect or characteristic of anything else.

I have been using the word "characteristic" in a vague sense; and some further distinctions will be in order.

An individual object of perception, a body, has parts, into which it could be divided by some process; one part could be severed from another. Such parts are to be distinguished from characteristics which qualify the thing as a whole, for instance the color of the whole, if it is of a single color; its shape or form; its extendedness; its roughness or smoothness. Characteristics of the latter kind may be called *immediate properties* of the whole. The parts, too, may have properties, which are not immediately properties of the whole, but first and foremost properties of the parts: *their* shape, color, and so on. Moreover, there are aspects of the thing which are properties of properties; for instance, the surface of the thing is not an immediate property of the whole, but is essentially the *limit* of its extendedness, and hence a property of its extendedness.

Some characteristics or determinations of a thing involve an essential reference to other things. The other things may be actually nearby and therefore perceivable along with the thing I am examining, or they may be presented in memory or in the imagination. I have already said that we perceive an object as being of a more or less determinate type—it is a kind of tree, or rock, and so on. The recognition of type depends upon a precipitate of past experience. I do not necessarily remember particular objects which were previously experienced and which are similar to the one before me; I do not make an explicit comparison; but past experience, now apparently forgotten, has somehow produced a precipitate of habitual familiarity which operates without my being explicitly aware of it as such.

But comparisons of objects with respect to likeness or similarity can also become *explicit*. The comparison then involves a mental going-back-and-forth between one of the objects and the other, with at the same time a holding-in-grasp of the one I am not at the moment attending to. The object with which I am comparing the one in front of me may be present or else absent; in the latter case it is either remembered or imagined. The similarity may amount to complete alike-ness or sameness, or it may have to do with the whole of each object but still involve difference, as the large, bright red ball is similar to but not completely like the small, dark red ball. Or again, the similarity may relate only to particular aspects of the objects, as the table and chair may be alike with respect to color or ornamentation.

The relations I have mentioned thus far—relations of similarity and difference—are to be distinguished from relations which presuppose that the things related are actually present and co-existing, and not given in imagination or memory. For instance, the distance of one object from another is a relation which requires both objects to be given as present. Again, in order that an object be perceived as part of a configuration of objects, say a constellation of stars, it is necessary that all objects of the configuration be present in a perceptually grasped whole. Such relations I think I shall call *reality* relations, because they require the real, simultaneous presence of the objects related.

All relations, whether comparison relations or reality relations, presuppose that the objects related are taken together as a plurality. The awareness of the objects as forming a plurality is, however, only a *precondition* for the grasping of a relation. In order for me to grasp a relation, there must be a primary interest in one of the objects, in relation to which the other objects are seen as similar, or nearby, and so on. I see A as taller than B; the focus of interest is for the moment on A, which thus forms the substrate of the relation. The interest, of course, can shift to B, in which case I see B as shorter than A—in a sense the same relation. All relational facts are thus reversible. The general fact that in relating objects I go from one to the other would be my reason for regarding the simple subject-predicate analysis of assertions as fundamental.

The grasping of a relation presupposes that the plurality of objects related is given; but the given-ness of the plurality can be of different kinds. In a comparison relation one or more of the objects compared may be an imagined or fictive object rather than a perceived object; and in this case the togetherness of the objects is brought about only in my own awareness, in my own inner time, but not in the visible world. The objects related in a reality-relation, on the other hand, stand next to one another in a real duration, in an objective time valid for all objects of the visible world. This objective time is also valid for other persons besides myself. If someone tells me of his past experience, what he remembers has its fixed place in the same public time as does my own past experience. Objective or public time is a form in accordance with which everything perceivable is ordered in succession. Just how such a form comes about is a difficult problem. But the point I am making here is that this objective time, which is presupposed in reality-relations, binds together my own experience and the experience of others, so that it is experience of *one* world.

All the distinctions I have been making—between substrate and characteristic, immediate property and mediate property, part and whole, comparison relation and reality relation—are, I am claiming, recognizable as involved in our experience of the visible world, the world of broad daylight, independently of the forms of our speech. The forms of speech, I am claiming, are *rooted* in these distinctions. In our actual lives, the receptive experiencing of the perceivable world, on the one hand, and our speaking, our predicative judging, on the other, are not separate but interlaced. I have separated them in analysis because they *are* separable, and because in separating them I find it possible to discern the ways in which objects present themselves in experience. It is a very simple and obvious thing I am saying. Speech, logos, presupposes a world, the world, in which it is a fundamental fact that there are distinguishable, relatively independent objects which present themselves in and through their characteristics. The world, on the contrary, does not presuppose speech or language.

In calling our experience of the world "receptive," I do not mean to imply that the "I" or self is altogether passive in such experience. Every awareness is an awareness *of* something; there is a polarity here, with the "I" or self at one pole and the object of awareness at the other. The "I" is affected by the object; it attends to or grasps it. Activity and passivity are interlaced in each awareness.

If we turn now to the predicative judgment, we encounter a new kind of interest and activity on the part of the "I" or self. Let us suppose that I have perceived a certain object or substrate S, and then noted a characteristic of it, p. For instance, I may have isolated the curtain as an object and then noted its color. These activities—the grasping of the substrate, the holding of the substrate in grasp while I note the color, which is thus grasped as *belonging* to the substrate—these activities are bound to what is immediately given. The result of such activities, if I do not fix it once for all in a predicative judgment, is not really my *possession.* Perhaps it is not altogether lost, but sinks into the background of awareness and there works to build my general familiarity with the perceivable world. But it is not yet knowledge. We have some way yet to go before we reach anything which can be called, in a strict sense, knowledge.

The predicative judgment presupposes an active will to knowledge. I return to the substrate S, and now grasp *actively* and *explicitly* the fact that it is determined by the characteristic p. The transition from S to p no longer occurs passively, but is guided by an active will to hold S fast by fixing its characteristics. The substrate becomes the subject of a predication. Fixing the gaze on the hidden unity of S and p, I now grasp actively the synthesis of the two which was previously given only in a passive way. I say: "S is p"; or, "The curtain is beige" (if that is the right name for this color).

Having uttered or thought a judgment, my fictional contemplative fellow has for the first time used words. Now what does this involve? Let me first distinguish two kinds of words in the sentence "The curtain is beige." First, words like "curtain" and "beige," which could by themselves constitute assertions in certain contexts, for instance as answers to questions. These have been traditionally known as *categorematic* terms. Secondly, there are words which influence within an assertion the way in which the categorematic terms signify what they signify; these have been traditionally known as *syncategorematic* terms. For instance, the word "the" before "curtain" is a demonstrative which makes the word "curtain" refer to this curtain; the copula "is" a sign indicating the synthesis of subject and predicate in judgment. But it is the categorematic term "curtain" which tells me what I am talking about, and the categorematic term "beige" which tells me what I am saying about it. These words are *common* nouns and adjectives; verbs are also categorematic. The meanings of such terms are what we call *universals* because the words in virtue of their meanings are able to refer to many particular instances. All predication involves such universals. This fact points back to the fact that every perceived object or characteristic in the perceivable world is perceived as of a more or less known or familiar *type*. The common nouns and adjectives used in predication refer to such types. When I say, "This object is beige," there is implicit in this predication a relation to the general essence *beige*. The relation to the general essence or universal is not yet explicit here, as it would begin to be if I said, "This is *a* beige object," where the indefinite article a points to generality. Later on I shall try to discuss the problem of the given-ness of universals. But in assertions about individual objects of the perceivable world, the explicit grasping of universals is not involved; the use of common nouns and adjectives is based on our passive, doxic familiarity with types of things and characteristics.

Assertions about individual perceivable objects run parallel to our receptive experience of such objects. I have already mentioned judgments of the type "S is p," "The curtain is beige." Such judgments express the fact that a substrate is characterized by the *immediate property* p. If the focus of interest passes to a second immediate property q, we get an assertion of the form, "S which is p is also q"; or the subordinate clause "which is p" may be replaced simply by the attributive adjective p modifying S. To take another case: if the property p of S is itself characterized by a property α, we get an assertion of the form, "S is p which is α"; and again the subordinate clause "which is α" may be replaced by an adverb modifying p. The use of adverbs and attributive adjectives thus presupposes prior assertions.

There are assertions of the form "S has T," which express the fact that an individual object S contains a certain part; for example, "The house has or contains an attic." Assertions of this kind refer back to experiences in which an object is perceived as being a whole made up of parts. These assertions *cannot* be converted into assertions of the form "S is p"; the part which is separable cannot lose its independence and become a property. On the other hand, a statement of the form "S is p" *can* be turned into a statement of the form "S has T"; for instance,

55

the assertion "This object is red" can become "This object has redness," or reversely, "Redness belongs to this object." This shift involves a *substantifying* of the property designated by the word "red." *Substantivity* means standing as something which can have characteristics, and which can therefore become the subject of a predication; it is opposed to *adjectivity,* which means *being in* or *on* something else. Substantivity and adjectivity are not merely a matter of grammatical forms; the difference in the two depends on a difference in the manner of grasping something, either as *for itself,* or as *on* or *in something.* Any characteristic of a thing, although initially presented as in or on a substrate, can be substantified. This freedom in substantifying rests on the fact that already in the receptive experience of the world everything that presents itself, whether substrate or characteristic, can be made the theme of inquiry; it has characteristics which can be ascertained, including relations of similarity and difference to other substrates or characteristics.

Again, there are assertions based on our grasping of relations in experience, for instance the assertion "A is similar to B." Once more we have a subject and a predicate, but the predicate is more complicated than in the previous cases. The word "similar" is adjectival, but its adjectivity is different from that of the word "red"; it is grasped only through the transition in awareness from A to B, from the *subject* to the *object* of the relation. Once again, what is adjectival can be substantified, and we can come to speak of the "similarity of A to B."

Now this freedom in substantifying extends further, and at this point we can take a very large step forward. Having uttered assertions, I can now substantify that which they mean, the synthesis of subject and predicate which is intended in the act of asserting. I can make statements of the form, "The fact that S is p, is q," where q can be an adjective like "just" or "pleasant." Here the subject of the sentence is itself a sentence expressing a state of affairs. As subject of the new sentence, the assertion "S is p" is no longer traversed in a two-membered, upbeat-downbeat rhythm; it is caught, so to speak, in one beam of the attention, is treated as a substrate of which I can ascertain characteristics. We here encounter a new kind of object of awareness, the unity of meaning in a completed judgment. Such objects I shall *call objects of reason,* because they presuppose the activity of reason or *logos,* the faculty of making judgments.

These new objects, constituted in the activity of reason, differ radically from the objects presented in our experience of the perceptual world. The perceptual object is indeed presented in a temporal process; further examination always enriches its meaning, adds to its ascertained characteristics. But the object is always there; the examination of the object can be broken off at any point, and yet the object is always presented as being one and the same and *there*. The activity of the "I" or self produces *presentations* of the object, but not the object itself. In the case of an object of reason, on the other hand, the synthesis of subject and predicate is required for the object to be given at all; the activity of the "I" cannot be broken off at an arbitrary point, but must be carried through to completion, in order for the object to be present.

The difference may be stated differently. The perceptual object, the individual object of the visible world, is presented in the course of my inner time, the succession of awareness, but it always stands before me as existing in an objective time, a time which is valid for the whole world of individual objects. It is an individual thing, distinguished from every other individual thing of the visible world in virtue of being localized in public space and time. An object of reason, the unity of meaning in an assertion, does not belong to the visible world in this way; we do not find meanings in the world in the same way in which we find things. The meant states of affairs are indeed constituted and grasped in my inner time. But what is grasped when I grasp the content of an assertion is not given as itself belonging to any particular stretch of the objective time of the world. I am not concerned here with the truth or falsity of the assertion, but only with the mode of given-ness of its content. That Caesar crossed the Rubicon may be true or false; but the kind of object I grasp when I grasp the content of this assertion, namely a meaning, presents itself as transtemporal, something which is identically the same every time I grasp it, that is, every time I think of it, but which is not itself individualized in the space and time of the visible world.

What I am saying here is, I believe, quite elementary, and is tacitly presupposed in every assertion I make. For in making an assertion I intend that the auditor grasp my meaning, and I am disappointed when what he says and does implies that he has failed to understand. Any particular uttering of the assertion is an event in the objective time of the world; but I act as if what is asserted in many repetitions of the assertion is self-identical, always the same, and capable as such of being communicated.

57

Now there is one more kind of object whose mode of given-ness has to be discussed; this is the universal, the idea, or in the Greek, *eidos*. The Greek word *eidos* comes from the verb "to see," and meant originally the "look" of a thing. The look of a thing, what we see on first impression, is the general type to which the thing belongs. In the sense of familiar type, the universal has been with us all along.

Up to now I have been talking about experience of individual objects of the visible world, and about assertions immediately based on such experience. All such assertions involve an *implicit* relation to generalities or universals; this is shown by the fact that in making an assertion we have to use *common* nouns and adjectives and verbs, which in virtue of their meaning are capable of referring to many individuals. Words of this kind, capable of referring to many instances, seem to be fundamental to any language. Even proper names often derive from common nouns, Smith, Brown, Klein, and so on. The implicit relation to universals rests on our typical familiarity with the world, the fact that every object presents itself as belonging to a more or less definite type.

Is there any way in which ideas or universals can be *explicitly* grasped, as evidently *given* objects of consciousness? This is a difficult question. Let me point out first that every inquiry aiming at knowledge seems to *presuppose* that the universal can be clearly and distinctly grasped, insofar as it assumes that questions of the form, What is so-and-so, for instance, What is what we call a tree, or a meson, or courage, can be inquired into, and with effort and good luck, be answered. In Greek, the question is *ti estin*—What is it? The *what* is the universal, capable by its nature of being applied to many.

You must permit me once more to proceed on the basis of the simplest example. Suppose I am confronted with two objects, S and S', each of which has the property p, say "red." Of course S has *its* redness, and S' has *its* redness; there is a separateness of the properties as well as of the substrata. But there is also a unity here, an identity, which I can grasp in shifting the attention from S to S' and back again. There is a oneness in the manyness. The comparison of objects, the focusing upon that with respect to which they are similar, can go on to further cases, and need not be limited to actually presented cases, but may include the consideration of *imagined, possible cases*. Thus through the medium of the imagination I arrive at the notion of an infinity, an endlessness of possible individuations of the same *eidos*. It may be difficult to define the limits of the possible variation of instances, but in some

58

cases I do seem to be able to do this, and to see that the universal involves definite limits, a definite structure, definite relations to other universals. For instance, I can imagine the colors of the objects to be different; there is a range of possible colors, but I seem to grasp that whatever color is, it will always be extended; an unextended color is unimaginable. Similarly, it appears clear to me that a tone or any sound must in every instance have an intensity, as well as the quality we call timbre or tone-quality.

I introduce these cases of intellectual perception, not because of any importance they might or might not have in themselves, but because of what they show. It is not enough, and not quite correct, to say that they *derive* from experience, that they are inductions or abstractions from experience. If I observe 100 swans, and find them all to be white, I may indeed guess that all swans are white; but the conclusion is not *necessary,* and is in fact false, since there are black swans in Central Africa. It is not the same with the connection between color and extension; color involves extension *essentially,* and I see this not just by observing particular instances of color, but by a variation of instances in the imagination, which allows me to "see" what must be involved in any case of color. And the idea or *eidos,* which thus appears as an identity running through the imaginable instances, presents itself, like the objects of reason previously described, as something transtemporal, something not in the objective world with its objective time, not even immanent in the acts of consciousness, but as an identity which can be repeatedly *intended* by consciousness.

Permit me to summarize what I have been saying. I have been aiming, not to make hypotheses, but to describe and to explicate; what I have been attempting to describe and explicate is that which is involved or presupposed in the making of assertions, judgments, predications. The description has proceeded by stages; at each stage I seek to delineate precisely what the I or self grasps, as being somehow presented to it.

The making of predications presupposes, in the first place, my prepredicative, pre-reflective experience of the world. My pre-predicative experience of the world can be separated, in analysis, from speech; our speaking, on the contrary, appears when analyzed always to point back to the pre-predicative experience of the world. Prepredicative experience is first and foremost experience of perceivable objects. The objects present themselves as in the world, along with other objects, in an objective time which is valid for all such objects. They present themselves as belonging to more or less familiar types. And they present themselves in and through their properties, parts, and relations. There is always a sense of "and so on" attaching to my experience of a perceivable object, in that I can always make further determinations, both of the internal characteristics of the object and of its relations to other objects. But it remains throughout an identity, a locus of possible experience, a substratum of possible determinations.

Predication, on the simplest level, involves an active repetition of the passage in pre-predicative experience from substratum to characteristic. The flow of perceptions in our pre-predicative experience goes on harmoniously almost of itself. Predication, on the other hand, presupposes an active will to fix, once and for all, that which is given in experience, to make it my possession. The predication is embodied in a temporal event, in a succession of sounds, the spoken sentence; but it is not itself this temporal event. The sound emerges from silence and falls back into silence; it passes like an arrow, leaving no trace in the air. But that which the sound expresses, the predication, is a unity or identity of meaning which can be repeatedly intended and repeatedly expressed; speaking quite strictly, it is not in the objective time of the world, but is grasped as trans-temporal. It is constituted in the activity of the I or self, but it is nonetheless an objectivity; it can be substantified, and itself made the subject of predication.

Finally, I have described one further and essential condition of predicative speech, namely the universal. Every assertion I make involves categorematic terms, universals, which in their nature are capable of referring to many instances. The use of the universal in speech is based, to begin with, on the typical character of my experience of the world, the fact that objects and characteristics present themselves as belonging to more or less familiar types. The universal first enters the assertion so to speak tacitly, without its range of meaning being explicitly grasped. But the will to knowledge can be satisfied only if the universal can itself be made the subject of predication. The empirical sciences approach such universal predications by means of statistical inference; their results are always open-ended, subject to revision. But it also appears that there is such a thing as intellectual perception, eidetic insight, by which one can grasp the range of a universal, define it, and make necessary predications about it, on the basis of a variation of instances in the imagination. I may note that, on a rough count, nearly half the assertions I have made this evening are such universal assertions.

My effort at description has to end here, although the stopping-point is arbitrary; there is a vast range of possible explications of this kind, which would have the aim of delineating each objectivity or kind of objectivity presented to awareness just as it presents itself. I regard such description as important, because I believe the correct method of philosophy is that of attending to and grasping states of affairs just as they are given or presented, and explicating them with respect to such of their connections and relations as are likewise presented and grasped. Only by a repetition of this process can philosophically primitive ideas and propositions receive adequate confirmation. Principles should not be just postulated or constructed, accepted merely on faith, whether animal or spiritual, or justified by the emotional comfort or practical success they may bring. That is part of the meaning, I think of the Socratic return to the *logoi*.

What, finally, about the Underground, since the announced title of this lecture included that term? The German word "Underground" can mean anything which either in a direct way or analogically underlies something else. So in talking about *hypokeimena,* or subjects of predication, and of the way in which they present themselves, I can claim to have been talking about the underground of speech. But as everyone knows, there is a more subversive and indeed altogether more interesting sort of underground, the one which, Dostoevski intends in *Notes from the Underground.* This underground is the location, so to speak, of certain writers of the present and of the last hundred years who throw to us, and in fact to the whole tradition of philosophy, a certain challenge. There are really many challenges which they throw; the challenges are difficult to characterize as a whole, but they might perhaps be subsumed under the formula of the old myth of Prometheus, according to which all the gifts which make man man, including speech, are based upon, and therefore infected by a fraud. So Camus and Jaspers and Heidegger speak of man as a castaway, shipwrecked on an island of everyday-ness. And Heidegger above all has sought to pull the tradition of philosophy up by the roots, and to show that our awareness of the world and of ideas as constituted in inner time involves a fraud. Then wisdom can only lie in the destruction, the total dismantling, of what is fraudulent in our awareness. And perhaps the four revolver shots of Meersault, the hero of Camus's novel *The Stranger*, are more efficacious in this respect than the discipline of listening to and following the *logos*. On the other hand, it might just be that the staccato notes which issue from the Underground will shock us, and cause us to look once more with open eyes and with wonder at what is our most characteristically human endowment, speech.

Reflections on the Idea of Science
(1961)

We and our world stand within the unity of an encompassing history, a vast culture or set of traditions, inherited techniques and patterns of behavior, interlocking and diverging patterns of transmitted thought. And perhaps the mightiest of these traditions today is that which we call "science." Its effects are omnipresent. It has transformed, and continues to transform, at an ever accelerating rate, the visible world around us as well as the routine of our lives. It has made possible the extinction of man and his culture in a universal holocaust; and it has presented, for the first time in the history of the human species, the possibility of banishing material want from the face of the earth.

All this is journalistic commonplace, and the actual or possible material effects of science are not my concern here. These effects testify, of course, to a certain kind of success of the on-going tradition of modern science, and this success brings with it a certain claim. When I set out to think in the attitude of one who seeks to arrive at truth and to avoid falsehood, this claim appears upon the horizon of my thought, whether invited or not. It is the claim of the *objectivity* of science. Whether I choose to welcome this spectre as a friend, or to duel with him as an enemy, or even to dismiss him as irrelevant to my concerns, his appearance, I believe, will not have been without lasting effect upon my thought. For as participant in a particular culture at a particular time—and every human being is that—I do not find it possible to determine my thought just as I please; problems, and the terms in which they are couched, are *presented*. And we cannot set out to think in our time without being confronted with the claim upon our thought of scientific objectivity. It seems best, therefore, to attempt to *question* this spectre, to try to elicit the meaning of his presence.

Why do I say "spectre"? Surely there is nothing ghostly or frightening about scientific objectivity. Here is a realm of light. We are out of the dust of metaphysical disputation. Superstition and prejudice have been left behind. Rigorous standards of procedure are upheld; results are arrived at which are reproducible with a known order of precision. Should a mistaken assumption be made, it will surely be found to be such, for there is a constant reference to observation and experiment, to reality. We gain the image here of a machine which functions smoothly, dispassionately, according to rules of operation which are

63

clearly set forth once and for all. As its product we obtain universally established, objective truths.

"Objectivity," however, is a polar term; it evokes its contrary or opposite, "subjectivity"; it has its meaning in relation to the meaning of this second term. I should mention that the use of the terms "subject" and "object" which is involved here is relatively recent. In the thirteenth century the use of these terms in philosophic discourse was almost opposite to the present-day use; "subject," from *subjicio*, to place or throw under, could mean a thing as a sustainer of properties and attributes, something so to speak "thrown under" the qualities of the thing; "object," from *objicio*, to place or throw before or opposite one, normally meant the concept intended by the mind. The present-day use appears to derive from a particular setting of problems in modern philosophic thought, beginning with Descartes's assumption that what exists must either be a thinking thing, *res cogitans*, or an extended thing, *res extensa*. Particularly since Kant, the words "subject" and "object" have been the key terms of a complicated and sometimes acrimonious dialectic. Kierkegaard says: "Subjectivity is the truth." The theologian Berdyaev says: "The self-alienation of spirit in objectivity is a fall." The psychoanalyst Theodore Reik speaks of the bitchgoddess objectivity. He suggests that there is an historical connection between the most elevated passion or thirst for knowledge and the desire to devour something; this hypothesis, he adds, will explain why people open their mouths when surprised.

What is at issue here? Clearly the terms "subjectivity" and "objectivity" are being used as firearms; they are, as we say, "loaded terms." Is not this opposition basically the one of which C.P. Snow complains?— the splitting of the intellectual life of western society into two polar groups, between which there is a gulf of mutual incomprehension and even hostility—those typified as literary intellectuals on the one hand, those called scientists on the other. "The non-scientists," Snow says,

> have a rooted impression that the scientists are shallowly optimistic, unaware of man's condition. The scientists believe that the literary intellectuals are totally lacking foresight, peculiarly unconcerned with their brother man, in a deep sense anti-intellectual, anxious to restrict art and thought to the existential moment.

It is not hard, Snow adds, to produce plenty of this kind of subterranean backchat.

It is my purpose here, not to rehearse the debate, but to try to specify more exactly the notion of scientific objectivity, which I have called
a ghost, and then to compare this notion with what I believe to be the
actualities of scientific practice.

When we think or speak of scientific objectivity, I believe we are
evoking a kind of mosaic of meanings, different components of which
will be in the foreground of our attention in different contexts of discussion. I believe I can distinguish the following components. First,
we are thinking of science as *empirical*, as based on observation of
what we call facts. We are thinking of science as a set of statements
which is "objective" in the sense that its substance, its essential content,
is entirely determined by observation, even though its presentation may
be shaped by convention. Second, we are thinking of science as a set
of techniques, exact methods for establishing control over experience.
We are thinking that there is something called *scientific* method, a set
of precise rules for proceeding which can be formally set out and empirically tested. The scientist has only to follow the rules faithfully in
order to arrive at reliable results. Third, we are thinking of science as
proposing a certain type of explanation as the only proper and final
kind of explanation; roughly speaking, a physico-chemical explanation
of all things, including living and thinking beings.

I shall take up the three components one by one. It will be clear
enough that I regard these assertions embedded in the notions of scientific objectivity as false and misleading; but there is a central and
important element in the notion which I shall seek to disengage in the
end. I begin with the notion of science as *empirical*, as determined in
its essential content by observation.

Kirchhoff, the nineteenth-century physicist, said that science is ultimately concerned with nothing else than a precise and conscientious
description of what has been perceived through the senses. Suppose
now that a man devotes his entire adult life to writing down in notebooks a precise and conscientious description of what he perceives
through the senses; when his life approaches its close, he forwards
these notebooks to the National Academy of Sciences. It will not be
merely because of bureaucratic inefficiency that these notebooks are
never read by anyone, and end in the discard.

Yes, of course, you say, a selection has to be made; the observations
have to be sorted. In order that knowledge should arise from sense experience, you must abstract or separate certain aspects from the differ-

ent perceptions, associate similar aspects in order to form general ideas, and correlate those aspects which are constantly conjoined.

I counter with a question: What do you mean here by "observation" or "sense-perception"? Am I supposed to be thinking of myself as a blind computer harnessed to a brainless photoplate? Consider the following passage from Duhem's *La theorie physique* (p. 218):

> Enter a laboratory; approach the table crowded with an assort-
> ment of apparatus, an electric cell, silk-covered copper wire,
> small cups of mercury, spools of wire, a mirror mounted on an
> iron bar; the experimenter is inserting into small openings the
> metal ends of ebony-headed pins; the iron oscillates, and the
> mirror attached to it throws a luminous band upon a celluloid
> scale; the forward-backward motion of this spot enables the
> physicist to observe the minute oscillations of the iron bar. But
> ask him what he is doing. Will he answer "I am studying the
> oscillations of the iron bar which carries a mirror"? No, he will
> say that he is measuring the electric resistance of the spools. If
> you are astonished, if you ask him what his words mean, what
> relation they have with the phenomena he has been observing
> and which you have noted at the same time as he, he will an-
> swer that your question requires a long explanation and that
> you should take a course in electricity.

Is it altogether clear that the visitor sees the same thing as the physicist? Consider the following cases. A musician listening to a quintet *hears* that the oboe is out of tune; the non-musician does not. Or a Westerner and a FarEasterner listen, both for the first time, to a Mozart concerto. For the one there is a perception of form, for the other there is confusing, unadulterated pure sound. For the experienced listener, the interpretation is there in the music; it is not something taking up a time of its own, subsequent to the hearing.

The same goes for seeing. I see a bird in the air; my seeing takes him in as a being that has just been flying and that will continue in the arc of his flight. Or compare the way in which the freshmen see the College campus with the way it appears to his ancient Tutor.

Seeing, hearing, perceiving are through-and-through interpretative; only in a limiting case, maybe in the case of a newborn babe, or a person fainting, does observation become an encounter with unfamiliar and un-connected flashes, spots of color, sounds, bumps. A physicist confronted with observations which he could describe only in terms of color patches, shapes, oscillations, pointer readings, would feel himself to be in a conceptually confused situation; he would try to get his observations to cohere against a background of established knowledge. It is in terms of a perceiving in which the elements already cohere in a pattern and interpretation that new inquiry proceeds, and not in terms of an encounter with pure flashes, sounds and bumps. Physicists observe new data as physicists, and not as cameras.

What I have been saying argues against the view that observation is simply opening one's eyes and seeing, or that facts are plain and unvarnished, and not laden with theory. But I want to go further. Observation, measurement, experimental result—all these have had their roles in the development of what we call modern science. These roles have not been the same as the role of experience in, say, theological speculation—where, by the way, experience does have a role; for I do not think that we can make any statement at all which does not have roots of some kind in our experience of the visible, sounding world. But the question is how we are to describe the relation of theory and fact in the modern scientific development; and I maintain that we do not describe that relation correctly if we say that theories are merely convenient summaries of experience, or economic adaptations of thought to facts, or logical constructions whose sole purpose is to predict what will be observed. I do not even believe that it is correct to say that validity or rightness of a theory can be simply judged by the degree to which it is confirmed by experiment, or that a theory is automatically discarded when experiment fails to confirm it. The rightness or wrongness of these statements that I make can only be determined through a close examination of the ways in which scientific speculation and experimentation have proceeded in actual, particular cases; I can only give a few indications here. I take physical science as paradigmatic, and avoid the social sciences where the maxim often seems to be: If you cannot measure, measure anyhow.

In the development of modern physics one has to distinguish the great theoretical achievements, or times of theoretical break-through, from the mopping-up exercises which follow. Textbooks of physical science tend to give the following picture. First a theory is proposed. Then there is a certain amount of logical and mathematical equipment, a kind of machine, used in manipulating the theory; the theoretical assumptions are fed into this machine along with certain initial conditions specifying the situation to which the theory is to be applied, the crank is turned; logical and mathematical operations are internally performed, and numerical predictions emerge from the chute at the front of the machine. These predictions are arranged in the left-hand column of a table; in the right-hand column appear the results of actual measurements. If the numbers in the two columns are in reasonable agreement, the theory is said to be confirmed; otherwise, it is disconfirmed.

We must first ask: what is meant by "reasonable agreement"? Is an average deviation of three per cent acceptable, is ten per cent good enough, or should we insist on 0.001 per cent? These questions cannot be decided *a priori*, independently of the total theoretical background within which the physicist works in each case. In effect, the tables in the textbooks define what is meant by reasonable agreement in the case which is being described.

In the second place, we must examine what physicists do in the period following the proposal of a new and encompassing theory. Are they engaged in attempting to confirm the theory by experiment? If so, then a failure to obtain agreement between theoretical prediction and actual measurement should lead to discarding the theory. Nothing of the sort happens. In general, a theory is not discarded unless there is another theory to replace it. In general, the theory has already been accepted; it has been accepted because it brings potential order to a large number of natural phenomena. Finer and finer investigations of the quantitative match between theory and observation are not attempts to confirm the theory, but attempts to make *explicit* what was previously *implicit agreement* between theory and the world. Again and again, nature's hand has to be forced. If the physicist succeeds in achieving reasonable agreement between theory and observation, he achieves a result already anticipated by the general community of physicists. If he fails, his failure counts not against the theory, but against himself; his talents have not proved equal to the task.

In 1638 Galileo published his description of the famous inclined plane experiment. He claimed that comparisons of the times required by a sphere to roll different measured distances down the plane confirmed his prior thesis that the motion was uniformly accelerated; he did not report his measurements. Subsequently a group of well-known scientists in France announced their total failure to get comparable results, and publicly doubted that Galileo ever tried the experiment. Presumably

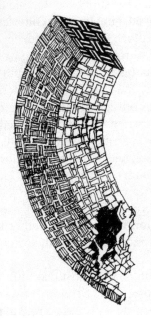

Galileo did perform the experiment; presumably he got results which appeared to him to be reasonable agreement with his hypothesis. Anyone who has performed this experiment with a present-day electric timer or stop-watch may doubt that Galileo's results were in anything like *unequivocal* agreement with the hypothesis. But for the development of physics this did not particularly matter. What mattered was that there should appear a detective like Newton, who, taking as clues such apparently unnoticed unrelated items as Galileo's thesis, Kepler's laws for the planets, observations of the lunar tides, and precession of the equinoxes, and so on, could produce a coherent pattern of intelligibility, a rational structure of potential explanatory power. The coherence and rationality of such a pattern is recognized precisely when one understands the theory, and not otherwise; there is a quality of wholeness there, an inter locking of parts in the theoretical structure, which commands the assent of the mind. And there is also an indeterminate range of yet unknown implications which later investigators will be years in ferreting out and trying to realize in experimental situations.

In the years between 1902 and 1926, D. C. Miller repeated the Michelson-Morley experiment many thousands of times, in an effort to disconfirm the theory of relativity. The Michelson-Morley experiment, you know, is generally described as having shown that the velocity of the earth with respect to the ether is indetectible. (The ether is, or was, the medium hypothesized as carrying electromagnetic vibration—radio waves, light waves, and so on.) Actually the experiment detected a positive effect, corresponding to a velocity of about eight or nine kilometers per second. This is considerably less than a pre-Einsteinian physicist, believing in absolute space, would have predicted; but it is not zero. In 1925, Miller announced that the whole series of his experiments confirmed overwhelmingly the existence of a positive effect of about eight or nine kilometers per second. Miller was known to be a careful experimentalist. One would have supposed that the theory of relativity would be instantly abandoned, or at least that physicists would have withheld judgment until Miller's results could be accounted for without impairing the theory of relativity. Nothing of the kind. Only in Russia were Miller's results taken as casting doubt on the theory, and there the theory of relativity had not been accepted as yet anyway, since it was believed to be in conflict with the dialectical materialism of Engels and Lenin, and no material benefits seemed to flow from it. For physicists elsewhere, however, the theory of relativity continued to command belief for the reasons which had led to its original acceptance: it provided a coherent vision of laws, theories, facts, which had previously appeared disparate, rationally unconnected. There is still no generally agreed-upon interpretation of Miller's result; but the indetectibility of the earth's motion relative to the ether has been shown experimentally in ways quite independent of the Michelson-Morley apparatus.

According to Einstein himself, and contrary to most textbook accounts, contrary even to certain implications in the St. John's manual on this subject, the Michelson-Morley experiment played no role in the formulation of relativity theory. Einstein was concerned fundamentally with certain anomalies in the theory of electrodynamics. For instance, he felt that when a magnet is moved relative to an electrical conductor, or a conductor relative to a magnet, the situation was fundamentally the same, and should be determined by the relative velocity alone; whereas Maxwell's electrodynamics gave different accounts of the two cases.

It is sometimes alleged that Einstein's motive was to eliminate untestable conceptions from theory, for instance, the notion of absolute space. Such conceptions, it is urged, are meaningless. This assertion is an attempt to assimilate Einstein's work to the notion of science as economic description. Actually, the Newtonian conception of space was not untestable; Einstein showed not that it was meaningless but that it was false.

What I am urging here is that the great and revolutionary theories of physics—and the number of these has not been large—have all possessed qualities of wholeness and coherence, intellectual beauties and harmonies and profundities, and that it is by these qualities that the theories have laid claim to truth. Observations and experimental results function as clues; but the theories transcend such experience by embracing a vision of the world. This vision speaks for itself and as such becomes accredited with prophetic powers. This view of the nature of physics will be confirmed, I believe, by a study of the major theoretical achievements from Copernicus to Einstein. Nor does the newer quantum mechanics deny it, so far as I can tell from a slight acquaintance. Unlike previous theories, it is peculiarly concerned with the processes involved in observation itself; but I do not find that it is a convenient or economical summary of experimental results. It requires of the physicist startlingly new ways of thinking about the world. Every one of the major theories has done just that—changed the framework of interpretation. And it is just for this reason that such major discoveries cannot be arrived at by continued application of a previously accepted framework of interpretation.

This brings me to the second component of the notion of scientific objectivity which I have distinguished: the conception of science as method. Let me begin by considering the subject of methods generally. We are all able to do many things. We walk, talk, and eat with fork and spoon. All these actions involve sets of skills, or arts. What is an art or skill?

Consider a simple case like this: the use of a hammer to drive a nail. The carpenter is aware of both the nail and the hammer, but it is the nail which occupies the focus of his attention. He watches the effect of his strokes, and wields the hammer in such a way as to hit the nail effectively. He is aware in a *subsidiary* way of the feeling in his fingers and hand: even more dimly, he may be aware of the contractions in the muscles of his arm and shoulder, and of his whole bodily posture. But these feelings are not the *object* of his attention; they are not watched in themselves. The subsidiary awareness is merged into a focal awareness of driving the nail. The adjustments in hand and arm and body are *instrumental* in achieving an end; the hammer is used as a tool, an extension of the body. By the effort of concentration on the operation to be performed, the successful nail-driver *absorbs,* one might say, the elements of the situation of which he might otherwise be aware in themselves; he is aware of them only in terms of the operational results achieved through their use. He is no doubt following here a complex set of rules; but he is not aware of these rules as such.

The same thing, I believe, is true of every skilled performance. The process of bicycling can be analyzed in accordance with the theory of mechanics. It is found that when the cyclist starts falling to the right, he turns the handlebars to the right, so that the bicycle moves along a curve to the right. This action results in a centrifugal force which pushes the cyclist to the left and offsets the gravitational force which is dragging him down to the right. This maneuver soon has the effect of throwing the cyclist out of balance to the left, a lack of balance which he counteracts by turning the handlebars to the left; and so he keeps in balance by winding along a series of appropriate curves. An analysis in terms of Proposition IV of Book I of Newton's *Principia* shows that for a given angle of unbalance the radius of the curve must be inversely proportional to the square of the cyclist's speed. The cyclist, of course, knows nothing of all this; nor would such information be useful in learning to ride. In any skilled performance there are countless rules which are observed but of which the performer is unaware.

An art or skill is a set of potentialities which is brought into play in the accomplishment of an end. The elements of the successful performance are merged in a focal awareness of the end. Bringing one of these elements into awareness may be occasionally helpful in improving performance; but focal awareness of the elements, if maintained, is destructive of the skill, and leads to a paralysis like stage-fright.

It may indeed be possible to analyze every aspect of a skilled performance; and it will then be possible to replace the performer by a machine. The studies of the industrial arts made in preparation for mechanization have repeatedly shown that such analysis is enormously difficult. The resources of microscopy, chemistry, mathematics, and electronics have as yet failed pathetically to produce a single violin of the quality which Stradivarius achieved as a matter of routine 200 years ago. Even in modern industries based on the discoveries of pure science there is a considerable amount of undefined knowledge or knowhow which forms an essential part of the technology. Hence the importance of imitation in learning a skill; the apprentice has to submit himself to the authority and example of the master; he thus learns to obey rules of which he is not focally aware and which may not even be explicitly known to the master himself. An unbroken tradition, from generation to generation, is essential here.

I want to apply these conclusions now to the cognitive or intellectual arts, the arts not of doing but of knowing. Let me begin with the arts of language; it is the possession of these arts which distinguishes man from the other animals; and it is their exercise which has made possible the constant extension of human knowledge on the basis of previously achieved results.

A language is no doubt a construction, the product of the activity of generations of human beings belonging to a given society. The ready use of nouns, verbs, adjectives, which have been invented and endowed with meaning by unknown men of the past, expresses a theory of the nature of things. Every child who learns to speak accepts unwittingly this theory or framework as the basis for all further efforts of understanding.

There is a prevalent view that language is a set of convenient symbols used according to conventional rules of a "language game." Likewise, the nominalistic doctrine which was put forward in medieval times and is still with us, maintains that general terms are merely names designating certain collections of objects. The implication is that a language is essentially arbitrary and unrelated to the way things are. This view is adopted in abhorrence of its metaphysical alternatives.

On the contrary, I think that we can appraise skill and lack of skill in the use of words. I am sometimes aware of groping for words and phrases, and I recognize that something is awry when I get the wrong one. A skilled artist in speech is thoroughly conscious of the figurative and metaphorical elements in speech. He continues to correct and supplement one metaphor by another, even allowing contradictions to enter at times, but always attending focally to the unity of his thought. In skillful speech, there is only a subsidiary awareness of words; one sees through the words to things; attention is focused on the object of the thought. This characteristic has been called the transparency of language.

Every situation to which speech is applied is to some extent unprecedented. In the adaptation of speech to new situations, there is a focal awareness of the situation of which we wish to make sense, and a subsidiary awareness of the words we are using as instruments. In this process the meanings of words become modified, but we are not focally aware of this change. Our framework of interpretation thus changes, and the words for which we grope become invested with a fund of unspecifiable connotations.

What I have just said applies not only to the education of a single person, but to the development of the language of a whole society. The efforts of men to adapt language to situations and things have the result, after years and centuries and millennia, of modifying the instrument of interpretation itself. Some of the changes are degenerative; but words which have great human significance tend to accumulate a wealth of connotations adapted to the situations in which they have been meaningfully applied. It is because of this fact that when we speak we say more than we know; that language seems to have a wisdom of its own. And it is also because of this fact that inquiries of the Socratic type are worthwhile. We have the power to take cognizance of a subsidiary element in the comprehension of a term, say "justice" or "courage"; we can try to define the term. Such an enterprise presupposes an understanding of the subject-matter to which the term refers. Only if we are confident that we can identify what is just or courageous in particular cases, can we reasonably undertake to define the term. If we want to analyze the meaning of the term, we must be using it as thoughtfully as we can, and at the same time watch ourselves doing this. We must look, with all the discrimination we can bring to bear, *through* the term "justice" at justice itself; for this is the use we are trying to define.

I am urging, then, that the skills of using a language are like other skills; that the employment of linguistic skill involves the merging of a subsidiary awareness of words and grammar in a focal awareness of an end, say persuading someone of something, or expressing a truth. The process depends on a fund of unspecified connections and connotations which constitute a framework or instrument of potential explanation. Any attempt to step altogether outside this framework and to criticize the structure of language as wholly arbitrary and conventional, is lacking in frankness; for the attempt employs and appeals to the very instrument whose validity it denies. Language commits us, far beyond our comprehension, to a vision of the world. It is a shirt of flame in which we are garmented; the responsibility of wearing it we cannot avoid.

The view I am advocating would deny that there is a single scientific method, or method of achieving truth, the rules of which can be set down once and for all. I do not deny that there are methods (plural) which have been developed in the particular sciences, and which continue to be applied effectively in a variety of situations. But formulations of these techniques, even by a competent scientist, tend to be inadequate because the scientist automatically supplements the explicit formulation by a tacit knowledge of how the techniques are applied in particular cases.

As an illustration of the way in which tacit appraisals are involved in the use of a given technique, let me mention the application of probability theory. This theory is applicable to systems of objects and events which have a characteristic called *randomness,* and also to significantly ordered systems which interact with random systems. "Randomness" means the absence of significant order or pattern; "significant order," of course, means absence of randomness. The randomness of a system cannot be specified in terms of the particular elements of the system; such specification, if it were possible, would in fact destroy the randomness. The appraisal of a system as random or as orderly depends

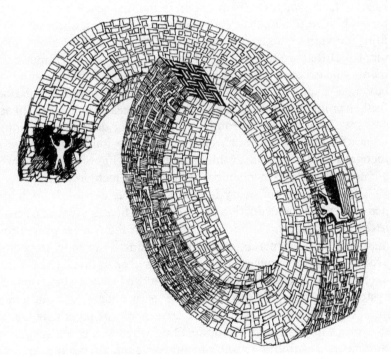

on tenuous criteria peculiar to the system under consideration, and cannot be reduced to universal rules. This becomes evident when probability theory is applied to a live scientific issue; in such cases there may be intense controversy over the proper experimental design and statistical technique.

If we turn to the deductive or mathematical aspects of the science, I believe we shall find again the same supremacy of art over mechanical procedure. The teaching of all the mathematical sciences such as mechanics or electromagnetic theory relies to a large extent on practice in solving problems. The skill striven for in all these cases is that of converting a language which one has assimilated only receptively into an active tool for answering new questions. The rules for problem-solving that can be specified are but vague maxims: Polya, in his book, *How To Solve It*, says "Look at the end. Remember your aim. Do not lose sight of what is required. . . . *Look at the unknown*." In attempting to solve a problem we use the known particulars as clues, and try to feel our way toward an understanding of the manner in which these known particulars relate with each other and with the unknown which is sought. To recognize the problem in the first place means to anticipate a hidden potentiality. As we proceed in tracing out relations and in trying various transpositions of symbols, we may at a certain moment feel that we are getting close; we sense—not without excitement—the accessibility of a hidden inference. Finding the solution is having a "happy thought"; it is crossing a logical gap. The solution of further problems of the same kind may increase our facility. In some cases we may discover a routine technique for dealing with all problems of a given class; such problems are no longer problems. The recognition of a *genuine* problem, and the solving of it, are acts which are not reducible to mechanical or systematic technique.

That genuine discovery is not in principle capable of being dispensed with in the deductive disciplines has been shown as a result of research in *metamathematics,* the study of the formal properties of mathematical systems. In some such systems, a decision procedure is available, a finite sequence of predetermined operations which suffices to resolve every question or problem that can be set in the terms of the system. Reckoning of sums, differences, products, and quotients of numbers is of this character. There are other systems in which a decision procedure is available for deciding whether a given sequence of statements constitutes a *proof* of a given statement, but in which there is no decision procedure which would enable one to decide, in a finite number of steps, whether a given statement couched in the terms of the system is *provable* or not—is a theorem or not. In these cases a machine can be built which, by operating on the axioms according to specified rules of inference, will churn out theorems, one after another; but there can be no guarantee that it will turn out in a given finite time the answer to any particular question. The determination as to whether a given statement in the language of science is deducible or not is contingent on time, luck, ingenuity, and intelligence directed toward a goal. Euclid's geometry is of this character. We may speak of a primordial darkness of reason here; we are unable to envisage the total outcome of a series of acts whose generating principle we can envisage with complete clarity.

There is still a further kind of situation in the deductive sciences, discovered by Goedel in 1931. Within any deductive science of sufficient scope to include arithmetic, it is possible to formulate sentences which cannot be proved within the science, that is, starting from the stated axioms of the science and employing the stipulated rules of inference, but which can nevertheless be shown to be true by reflections on the science as a whole. In 1949 Turing showed that a machine could be devised which would construct and assert as new axioms an indefinite sequence of these Goedelian sentences, as they are called. It nevertheless remains true that any given set of mathematical inference machines can only cut a swathe out of the total field of mathematical truths. Mathematics cannot be formalized, once and for all, in a single linear deductive development; the methods of procedure and inference in mathematics are, in principle, inexhaustible.

What I am saying here, in sum, is that no fixed, impersonal, and fully specified technique can be laid down for attaining all and only the truth. The knowledge we have or gain is shaped within a framework of tacit acceptance and incompletely specificable arts which are logically prior to any particular assertion we may make. Such a framework can be altered, or as I would say, improved, in the very process of examining a topic in its light. Either this is so, or liberal education is nonsensical. My acceptance of one of these alternatives rather than the other is no doubt a passionate act, a commitment.

I turn now to the third claim associated with the notion of scientific objectivity, roughly speaking, the claim that only explanations based on physics and chemistry can be accredited as final in the sciences. Huyghens, in his *Treatise on Light* of 1678, speaks of "the true Philosophy, in which one conceives the causes of all natural effects in terms of mechanical motions." Laplace, in his *Treatise on Probability* of 1814, writes than an intelligence which knew at one moment of time

> all the forces by which nature is animated and the respective
> positions of the entities which compose it . . . would embrace
> in the same formula the movements of the largest bodies in the

universe and those of the lightest atom: nothing would be uncertain for it, and the future, like the past, would be present to
its eyes.

Such a mind, Laplace claims, would possess a complete scientific knowledge of the universe. K.S. Lashley, speaking in 1948 at a
symposium on cerebral mechanisms in behavior, states:

Our common meeting ground is the faith to which we all sub-
scribe, I believe, that the phenomena of behavior and of mind
are ultimately describable in the concepts of the mathematical
and physical sciences.

With these views I believe it is necessary to do battle. For they
present us with a picture of the universe in which we ourselves are absent, in which there are no scientists and hence no science. This is a
simple-minded objection. I would support it and amplify it by the following considerations, which will have to be brief.

Suppose, first, that the universal knowledge of which Laplace
dreamed were possible. Then from the positions and velocities of the
n atoms or particles of the world at a given time t_1, it would be possible
to compute the positions and velocities of all these particles at any later
or earlier time t_2. As it turns out, if n is greater than two, no exact and
general solution of the computational problem is possible. But even if
it were, it would remain true that this knowledge would not constitute
knowledge of all past and future events, unless "event" be defined in
so narrow a fashion as to exclude the events of which I have experience. The Laplacean picture supplies no clue as to how the data of experience are to be accounted for, how I am to pass from information
about atoms to data of experience. It merely claims, wrongly as it turns
out, that an answer is possible to a question raised by the theory of mechanics itself. The change from Newtonian to quantum mechanics
makes no difference in this argument. The wave equation of the world
in quantum mechanics represents our ultimate knowledge of all the
particles in the world, leaving open within this framework only variations which are strictly random. There is no accounting here for living
beings or for intelligent behavior.

Computing machines and feedback mechanisms, as is well known
today, can simulate or improve upon the behavior of living and thinking
beings. But here again I must point to a very obvious fact, which is
nevertheless often forgotten: No knowledge of physics or chemistry

would ever suffice to enable us to recognize or account for a machine. A machine, for instance a clock, a steam engine, or a digital computer, is an instrument or implement which is operated in accordance with certain more or less specifiable rules for the sake of a certain advantage. Of instruments, operational rules, and advantages, physics and chemistry can tell us nothing. Suppose you are confronted with a problematic object and try to explore its nature by a physical and chemical analysis of its parts. A complete physical and chemical account of the object, and of all its future possible transformations, would still not enable you to discover that it is a machine, if it is one, and if so, how it operates. Such a discovery could only be made by testing the object as a possible instance of known or conceivable machines. For you must know that the problematic object embodies a rule of rightness, or operational rule; that it can succeed or fail, depending on whether it operates in accordance with this rule or not. In the subject matter of physics and chem-

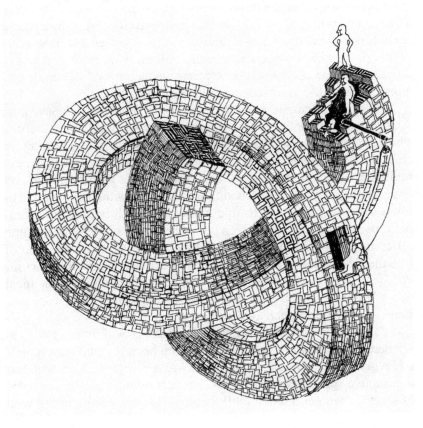

istry proper, the notions of success and failure do not occur. Given the rule of rightness, the physicist and chemist may be able to find the causes of a failure of the machine, or the material *conditions* under which it will operate successfully; but the *reasons* for the consecutive stage of operations of the machine, and for the ways in which its parts are coordinated, are not specifiable in physico-chemical terms. The relation between *reasons* and physico-chemical causes or *conditions* is like the relation between *logical rules* and *psychological explanations* as applied to processes of thought. A given sequence of thought may be started by appetite or intellectual passion; it may depend on memory, visual imagination, and verbal or other symbolism. But a psychological analysis of these conditions will never reveal whether the sequence of thought embodies a correct inference or not.

The thesis that all living beings are physico-chemical automata, the operations of which are in principle totally specifiable in terms of spatio-temporal determinations, is not strictly inconceivable. It forms a closed interpretative system which is passionately pursued by a whole school of geneticists and neurologists today. The fact that most non-psychopathic persons become morally indignant when treated as automata might be said to be due to primitive patterns of mentality that a perfect scientific knowledge would eliminate.

There are, of course, many biological phenomena which have thus far resisted reduction to physico-chemical or spatio-chemical terms. These include the powers of improvisation discovered by Driesch in embryonic fragments, and the powers of adaptive reorganization exhibited by many animals in the achievement of a predetermined end under profoundly modified conditions. Thus a rat which has learned to run a maze will continue to find his way through it after the neural paths used in learning have been cut, although he has to employ quite different patterns of locomotion. All along the evolutionary scale of life, there is evidence for the presence of active centers which act inventively in ways which are not fully specifiable in physico-chemical terms. It can always be claimed, of course, that further knowledge will enable us to explain such evidence away.

These particular and no doubt intricate issues within the biological sciences I cannot follow up here. My central point is that if man is himself regarded only in his factuality, only as a complex object which is in principle specifiable in physico-chemical terms, then the very idea of science becomes unintelligible. I can no longer accredit myself with

the responsibility for drawing an ever indeterminate knowledge from unspecifiable clues with an aim to universal validity; nor can I acknowledge other persons as responsible centers of equally unspecifiable operations, aiming likewise at universal validity. In the resulting image of the world and of man there is no longer room for the norms and ideas of reason. Thus western man, who since the sixth century B.C., has *defined* himself by the idea of reason, as *animal rationale,* loses sight of himself. Reason, through a partial realization of its goal in modern science, appears to betray itself.

This is the primary root, I believe, of the intellectual crisis of our time.

Let me recur here to the fact with which I began. What we call science is at any moment, and for anyone, a part of an inherited culture, a set of techniques and patterns of thought which have been *cultivated* and *transmitted,* and which as such *must* have arisen, *must* have had origins, in human activity. Thus science has a *history.*

This fact would be of little import if the past of science or of any cultural configuration were merely behind and extraneous to its present. But surely this supposition is false. For do we not know that the historical present comes out of the historical past, and contains this derivation implicitly in itself? Is it not so in the case of languages, customs, laws, and indeed of every cultural achievement? And does not this fact point to a possibility—the possibility that tradition may allow itself to be questioned; that it is not necessary merely to live within a tradition, accepting it as a matter of course, or to set oneself up blindly as nihilist, rejecting traditions which have formed us?

To follow up this line of inquiry in the case of the sciences is to refer scientific knowledge to the generating and producing activities of the mind; it is to attempt to discern how objective science came to be, how it must have stepped into history as a human production. The history which is in question here is not primarily factual history; and indeed the factual origins of the idea of objective science are irrecoverably lost. The concern is rather with general and necessary truths about the way in which this idea came to be *there* for human beings; and these truths are implicit in the mode of being of science in the living present.

Do we not know, for instance, this simple truth that science is transmitted from generation to generation, from teacher to learner, and that at the same time it is continually broadened, with the achievement of new results? And in this process is it not manifest that there is a continual synthesis, an incorporation of new results with prior results to form a totality; and that at any time the entire achievement becomes the total premise for further results? Do we not know, further, that this history must have had a beginning; that there must have been a moment in time when a man of the past for the first time grasped, in full awareness, a truth as being *there* in its own right, as being *evident,* as constituting knowledge which was self-justifying and capable of indefinite expansion?

Each science is thus related to an open chain of generations of investigators, working with one another and for one another. For the later investigators, the earlier acquisitions or results are not, in general, grasped in the same way as they were by the original discoverer in the original act of discovering. They have become embodied in speech and writing; and indeed in no other way could they be objectively and self-identically *there,* as ideal objects, for Everyman, for every real or possible investigator of every place and of every future time. Also, it is their embodiment in speech and writing which allows them to be used as stepping-stones to further results, stepping-stones in a deductive development. But these advantages bring with them a seduction. The sounds of speech and the signs of writing, these indefinitely repeatable sensible forms of the embodiment of the ideal objects of knowledge, are, for the most part, taken in passively, in an unreal way, because they are given in the sphere of trust which is language. This seduction is a kind of forgetting, a lapsing of the originally active grasp of evidence, which nevertheless permits the deductive process to proceed. Thus Galileo and Kepler, and we also, can accept the geometry of Euclid as a self-contained science, with no roots or foundations outside itself; we have lost sight of its beginnings, of the idealizing activities which, starting from the vaguely typical objects of experience, and the rules of thumb in the practical arts, produced a universe of ideal entities among which exact and necessary relations hold.

Can the original evidence be regained, reactivated? Can we, for instance, rediscover the original meaning of objective knowledge, science, as present to the mind of the man who first envisaged its possibility?

I would claim that this unidentifiable man is not totally unknown to us. He lived, no doubt, in an already developed cultural world. Like his contemporaries and like all men, he had first lived naively within that world, which he had then taken for granted and accepted unquestionably as reality. It was a magico-religious world, thoroughly imbued with meaning, with a traditional meaning bestowed on it by the members of the community whose world it was. All activities there were traditional, and were undertaken for the sake of living and making one's way about in this limitedly meaningful world. Even cognitive activities would there be motivated by, and essentially related to practical human interests. The notions of knowledge, truth, and error, would be understood only in relation to the specific world belonging to the community. Speculative activities would occur only within a finite horizon.

Some moment, for some man, was a moment of emergence from this limited world. It was, if we are to believe Plato and Aristotle, a moment characterized by wonder. And indeed, to wonder is to suspend practical activity, it is to adopt the attitude of the detached, onlooking observer. And in the attitude of wonder there arises the conception of Being, Being-as-it-really-is-in-itself, a conception standing in contrast to the limited world belonging to a specific human community. Also, and correlative to this conception, there arises the idea of objective science, or knowledge of being, *epistēmē,* standing in contrast to the opinion, *doxa*, by which men relate themselves to their traditional and everyday world. This moment is a disclosure of unlimited horizons. For the idea of *epistēmē* is an ideal norm, an ideal limit with respect to every cognitive endeavor. The grasping of any single truth will henceforth be regarded as a transitional phase within an infinite process oriented towards this ideal limit—*epistēmē* as finally accomplished.

The rise of this idea in Greek Antiquity marks the appearance in history of a new type of man who in all his finitude assumes an infinite task. To belong to this tradition is less a glory than a responsibility. To seek to uncover its original meaning is an essential step toward the discovery of what we are.

On the Discovery of Deductive Science
(1973)

How did the notion of deductive science—science based on definitions, postulates, and axioms, science consisting of a sequence of propositions, each of which is deduced, either from previously deduced propositions, or from the definitions, postulates, and axioms initially set out—how did this notion first come to be thought of, and then realized? For there seems to have been a particular moment in which this idea was first conceived; so far as we can tell, it did not make its appearance at different times and places, independently. Can we learn anything about the original conception? I am going to pursue this question, although as you will at once realize, it is not the sort of question that is likely to receive a non-conjectural answer. The ground here has been worked into a deep and slippery mud by the trampling feet of contending scholars; mere non-classicists or not yet classicists like myself are liable to stumble over the mouldering carcasses of defunct theories, not yet decently interred. Certain questions of historical fact that are material to this discussion I am able to answer only conjecturally. At the same time, I wish to affirm that my primary aim is not to establish historical facts, nor yet to hypothesize possible causes for those facts, but rather to locate the meaning of facts that, it seems to me, come nearest to being reliable.[1]

---

1. This lecture owes everything, or nearly everything, to a number of studies by historians of mathematics, particularly: O. Becker, "Die Lehre vom Geraden und Ungeraden im neunten Buch der euklidischen Elemente," *Quellen und Studien zur Geschichte der Mathematik,* Abt. B, Band 3 (Berlin: Springer-Verlag, 1936), 125-145; B. L. van der Waerden "Die Arithmetk der Pythagoreer," *Mathematische Annalen* 120 (1947-1949): 127-153 and *Science Awakening* (New York: Oxford University Press, 1971); G. Vlastos, "Zeno of Elea" in *Encyclopedia of Philosophy,* VII, 370; O. Neugebauer, *The Exact Sciences in Antiquity,* (New York: Dover, 1969); and above all articles by Arpad Szabo: "Zur Geschichte der Dialektik des Denkens," in *Acta Antiqua Academiae Scientiarun Hungaricae* 2 (1954): 17-62, and "The Transformation of Mathematics into Deductive Science and the Beginnings of its Foundation on Definitions and Axioms," in *Scipta Mathematica* 27 (1964): 28-48 and 113-139. For the Proclus text I depended on *Procli Diadochi in Primum Euclidis Elementorum Commentarii*, ed. G. Friedlein (Leipzig: Teubner, 1873), and the recent translation by the late Glenn R. Morrow, *A Commentary on the First Book of Euclid's Elements*, (Princeton: Princton University Press, 1970). In the section on Parmenides there may be recognized a certain inspiration, much diluted, of Martin Heidegger's *What is Called Thinking?*, tr. F. D. Wieck and J. G. Graz New York:Harper, 1954).

I want to begin by saying something about pre-Greek mathematics. The oldest mathematical documents known from any place on this earth are Egyptian papyri stemming from the Middle Kingdom, 2000-1800 B.C., and clay tablets dug out of the sands of Mesopotamia, and stemming from about 1800-1600 B.C.

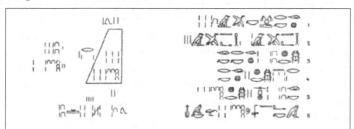

Figure 1. This illustration is taken from B. L. van der Waerden, *Science Awakening* I (1971), Plate 5, facing p. 44; also W. W. Struve, *Mathematical Papyrus des Museums in Moskau, Quellen und Studien* A1 (Berlin: Springer, 1930).

In Figure 1 you see a transcription from a papyrus now in Moscow, showing the computation of the volume of a truncated pyramid with square base and top. The base is four cubits on a side, the top two cubits on a side, and the height or distance between base and top is six cubits. The text says: "Add together this 16 with this 8 and this 4. [16 is the area of the base, 4 the area of the top, and 8 the product of the side of the base by the side of the top.] You get 28. Compute one-third of 6 [the height]; you get 2. Multiply 28 by 2. You get 56. Behold: it is 56. You have found right."

Now the result *is* right. It is something you might want to know if you were building pyramids; but by the time of the Middle Kingdom the Egyptians had ceased building pyramids, enjoyable though that occupation seems to have been, as we gather from the inscriptions of rival work gangs. It is not clear that there was any immediate practical reason for anyone in the Middle Kingdom to know the rule for computing the volume of a truncated pyramid. But the real puzzle is how this rule was discovered in the first place. It is a complicated rule, and there is no plausible empirical way of arriving at it by, say, weighing certain objects; therefore reasoning was involved. But on the other hand, the Egyptian mathematicians would not fall back on algebraic transformations in the modern manner, since their mathematics dealt explicitly only with particular numbers. There are a number of hypotheses as to how the Egyptians' procedure could have been arrived at, the most plausible, I think, involving a slicing of the pyramid into parts.

Let us take another example.

> "A square and a second square whose side is $\bar{2} + \bar{4}$ of the first square, have together an area of 100. Show me how to calculate this."[2]
>
> "Take a square of side 1, and take $\bar{2} + \bar{4}$ (3/4) of 1 as the side of the other square.
>
> "Multiply $\bar{2} + \bar{4}$ by itself; this gives $\bar{2} + \overline{16}$.
>
> "Hence, if the side of one of the areas is taken to be 1, and that of the other is $\bar{2} + \bar{4}$, then the addition of the areas gives $1 + \bar{2} + \overline{16}$.
>
> "Take the square root of this; it is $1 + \bar{4}$.
>
> "Take the square root of the given number 100; it is 10.
>
> "How many times is $1 + \bar{4}$ contained in 10? Answer 8."
>
> The two squares then have sides $8 \times 1 = 8$ and $8 \times 3/4 = 6$, the sum of their squares being 100.

Now Egyptian mathematics has certain general characteristics. First, Egyptian mathematics, whatever it is dealing with—areas, volumes, numbers of bricks or loaves of bread or jugs of beer—is always a matter of numerical calculation. The mathematician is a computer who uses both integers and fractions. Second, there are no explicit proofs whatever, but reasonings *have* to have been employed in the solution of problems. Finally, while the problems presented in the papyri seldom appear to be actual practical problems, they give the general impression of being the sort of problems that an instructor might think up for his students, in order to prepare them for solving practical problems. Instructors seldom succeed in being strictly practical, but the Egyptian ones appear to have understood their activity as occurring within the horizon of the practical.

Aristotle claimed that the mathematical arts had been founded in Egypt, because there the priestly class was allowed leisure; but this is incorrect. The Egyptian calculative art was the possession not of a priestly class, but of scribes who had practical functions in the state,

---

2. I note that Egyptian fractions, with one exception, are unit fractions, fractions we would write with 1 as numerator. They are written by putting a line above the number we call the denominator. The exception was 2/3, written by putting two of these lines above the numeral 3.

and among whom there was rivalry. So we find one scribe ridiculing another:

> You come to me to inquire concerning the rations for the soldiers, and you say "reckon it out." You are deserting your office! . . . I cause you to be abashed when I bring you a command of your lord, you who are his Royal Scribe. A building ramp is to be constructed, 730 cubits long, 55 cubits wide, 55 cubits high at its summit. . . . The quantity of bricks needed for it is asked of the generals, and the scribes are all asked together, without one of them knowing anything. They all put their trust in you. . . . Behold your name is famous. . . . Answer us how may bricks are needed for it?[3]

It seems likely, then, that the mathematical papyri were textbooks used in the school for scribes.

In Babylonia, the mathematical texts appear to have been produced by a similar class of scribes. The texts give problems with their solutions; proofs are entirely absent; the procedures are always numerical. Are the problems *practical* problems? Once again, yes and no. Here is an example from the time of Hammurabi, 1700 B.C.:

> "I have multiplied length and width, thus obtaining the area.
> Then to the area I added the excess of the length over the width.
> The total result is 183. I have also added the length and width,
> with the result 27. Required: length, width, and area."[4]

I omit the solution. For us it would involve the solution of a quadratic equation. This Babylonian problem does not strike me as a practical problem, or a near neighbor to one. The adding of a length to an area seems to me decidedly impractical, perhaps even nonsensical. This is mathematics gone a bit haywire: a pedagogue might invent it to bemuse his pupils, always understanding, of course, that calculating is a good thing.

Babylonian mathematics, however, is a good deal more powerful than Egyptian mathematics. When the Babylonian scribe wrote:

$$\sqrt{2} = 1; 24,51,10$$

---

3. Quoted in O. Neugebauer, *The Exact Sciences in Antiquity,* 79.
4. Paraphrased from van der Waerden, *Science Awakening,* 109.

(I am using the Indian numerals in place of the Babylonian), he meant

$$1 + 24/60 + 51/60^2 + 10/60^3$$

This is the Babylonian approximation to the square root of 2, or diagonal of a square of unit side. The Babylonians definitely knew and used the proposition we call the theorem of Pythagoras, which is involved in getting this approximation, but nowhere do any of the clay tablets that have been deciphered give a proof of this or any other theorem. The approximation, which is probably the result of a series of successively closer approximations, is good to one-millionth. Ptolemy will still be using it, having acquired it probably indirectly from the Babylonians, when he computes his table of chords in the second century A.D.

Now if we turn to other civilizations besides the Egyptian and the Babylonian, but still uninfluenced by Greek thought—the civilization of the Yellow River valley, say, or Mayan civilization—I think we shall once again find a computational art, often highly developed, but not explicit deductions. You may on occasion find the contrary asserted. Joseph Needham in his *Science and Civilization in China* (vol. 2 [Cambridge: Cambridge University Press, 1956], 22) gives a passage from a Chinese mathematical text which perhaps originated as early as the 4th century B.C.; it is accompanied by a diagram which he labels "proof of the Pythagoras Theorem" (See Figure 2).

I quote from the text:

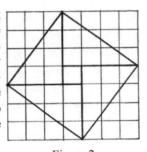

Of old, Chou Kung addressed Shang Kao, saying, "I have heard that the Grand Prefect [that is Shang Kao] is versed in the art of numbering. May I venture to inquire how Fu-Hai anciently established the degrees of the celestial sphere? . . . I should like to ask you what was the origin of these numbers?"

Figure 2

[In the course of his reply Shang Kao says:]

"Let us cut a rectangle diagonally, and make the width 3 units, and the length 4 units. The diagonal between the corners will then be 5 units long. Now after drawing a square on this diagonal, circumscribe it by half rectangles like that which has been left outside, so as to form a square plate. Thus the outer half rectangles of width 3, length 4, and diagonal 5, together make two rectangles [of total area 24]; then the remainder [that

89

is, of the square of area 49] is of area 25. This is called 'piling up the rectangles.'

"The methods used by Yu the Great in governing the world were derived from these numbers. . . . He who understands the earth is a wise man, and he who understands the heavens is a sage. Knowledge is derived from a straight line. The straight line is derived from the right angle. And the combination of the right angle with numbers is what guides and rules the ten thousand things."

Chou Kung exclaimed, "Excellent indeed!"

Nothing here, I would urge, has really been proven, certainly not the theorem of Pythagoras, so-called. Needham has shown in overwhelming detail that between the fifth century B.C. and the fifteenth century A.D. no people on earth exercised more technical ingenuity than the Chinese. Long in advance of the West, they possessed cast iron, an escapement clock, the navigational compass, gunpowder, printing by movable type, the segmental arch bridge. But as for deductive science, the Chinese would not encounter it until the Jesuits came to China in the late 16th century, bringing the textbooks of their fellow Jesuit, Christopher Clavius. It is a curious fact that, for some centuries thereafter, Chinese students reciting their Euclidean theorems out of Clavius would finish not with our Q.E.D. but with the Chinese word for "nail." Apparently they were citing their authority: "clavus" being the Latin word for "nail."

It is conceivable that some day, in the investigation of early civilizations uninfluenced by Greece, evidence will turn up for the existence of some pieces of deductive mathematics. On the basis of what is known today, the prospects for such a find are dim. Deductive mathematics is a rare bird, which first settled, so far as we know, in Greece.

How did it happen? What did it mean that it happened? Seeking an answer, I turn to a document of late antiquity, a commentary on the first book of Euclid's *Elements* written by Proclus in the middle of the fifth century A.D. Proclus was a member of the Platonic Academy in Athens during the last century of its 900-year existence. The commentary includes a kind of catalogue of ancient geometers which is based on an earlier history of geometry, now lost, by Eudemus, a disciple of Aristotle writing in the late fourth century B.C. The account begins by saying that geometry was first discovered among the Egyptians, and originated in the remeasuring of their lands necessitated by the annual flooding of the Nile. Proclus then proceeds as follows:

> Thales, having travelled in Egypt, first introduced this theory into Hellas. He discovered many things himself, and pointed the road to the principles of many others, to those who came after him, attacking some questions in a more general way, and others in a way more dependent on sense perception.

After mentioning the names of two other ancient geometers, Proclus continues:

> After these, Pythagoras transformed the philosophy of this (geometry) into a scheme of liberal education. He surveyed its principles from the highest on down, and investigated its theorems separately from matter and intellectually. He it was who discovered the doctrine of irrationals and the construction of the cosmic figures.

A little farther on we read:

> Hippocrates of Chios, who invented the method of squaring lunules (crescents formed from arcs of circles) and Theodorus of Cyrene became eminent in geometry. For Hippocrates wrote a book on elements, the first of whom we have any record who did so.

With respect to Hippocrates of Chios, there is no reason to doubt what Proclus says. A fragment of Hippocrates's work on lunules still exists, and it shows a high level of rigor. Thus Hippocrates may very well have written a book on the elements of geometry. At the time Hippocrates was teaching geometry in Athens, around 430 B.C., the process of turning geometry into a deductive science was in all probability well advanced.

Thales, who was active about a century and a half before Hippocrates of Chios, is a much more shadowy figure, and it is unclear how we should interpret what Proclus says about him. Proclus attributes to Thales the discovery and proof of five propositions:

(1) A circle is bisected by any diameter.
(2) Vertical angles of intersecting straight lines are equal.
(3) The base angles of an isosceles triangle are equal.
(4) Two triangles such that two angles and the included side of one are equal to two angles and the included side of the other, are themselves equal.
(5) The angle at the periphery of a semicircle is right.

Now these are general, theoretical propositions, *theorems,* propositions to be contemplated rather than mere rules for solution of problems. The enunciation of them may therefore mark a decisive step in the emergence of theoretical science. But how were they proved? The usual guess is that it was by superposition, the visual showing that one figure or part of a figure would coincide with another. If this is right, then it is unlikely that we have here the notion of a logically constructed theory which begins with expressly enunciated premises and advances step by step. Thales need not have enunciated any premises explicitly. He pointed the road to the principles, as Proclus says; the extent to which he laid out principles is totally unclear.

As for Pythagoras, whole books have been devoted in recent times to showing that the ancient accounts of his mathematical exploits are unworthy of trust.[5] These accounts stem from members of the Platonic Academy from the fourth century and later, men who saw in the 6th century Pythagoras a forerunner of Plato, and who tended to attribute to him discoveries that had been made later on in the Pythagorean tradition. Pythagoras cannot have known all the five cosmic figures, because two of them, the octahedron and the icosahedron, were first discovered by Theaetetus, a contemporary of Plato. Contrary to what Proclus says, there is no good evidence that Pythagoras knew anything about the doctrine of irrational lines. The old verse quoted by Plutarch, according to which Pythagoras, on making a certain geometrical discovery, sacrificed an ox, cannot be true, because it is well attested that Pythagoras was a vegetarian, who believed in transmigration of souls and was opposed to the killing of animals. What we can be fairly sure of, with regard to Pythagoras, aside of course from his having had a golden thigh, is that he had made the flight to the Beyond and had become the leader of a cult, a medicine man, a shaman. He can well have taught that odd numbers are male, even numbers female; that five is the marriage number; that ten is perfect, being the sum of 1, 2, 3, and 4. Somewhat similar beliefs have been found all over the world, in connection with rituals and creation myths, and have not led to deductive mathematics. Pythagoras's thought seems to have been cosmogonic, concerned with the corning-to-be of our world out of something prior and more fundamental. There is no trustworthy evidence that Pythagoras ever carried out an explicit proof.

---

5. In particular, see Walter Burkert, *Lore and Science in Ancient Pythagoreanism,* trans. E. L. Minar (Cambridge, Mass.: Harvard University Press, 1972).

On the other hand, the transformation in the character of mathematics that Proclus attributes to *Pythagoras* may well have been brought about by *Pythagoreans*. The old accounts speak of a split within the Pythagorean tradition; the Mathematikoi, those who wished to discuss and teach openly the mathematical disciplines, separated off from the secret cult, the Akousmatikoi, the hearers of the sacred and secret sayings. Reliable fourth-century sources speak of the arithmetical studies of the fifth-century Pythagoreans. Aristotle says that the so-called Pythagoreans were the first to deal with *mathemata,* mathematical disciplines. According to the *Epinomis,* a dialogue written either by Plato or a follower of Plato, the first and primary discipline or *mathema* of the Pythagoreans was arithmetic. Now it is possible to make a plausible reconstruction of some of this early Pythagorean arithmetic. When this is done, we find ourselves before a piece of deductive science, quite possibly the earliest that ever was; it is the science of the odd and the even. It is a science in which the principles are explicit, and in which the theorems are, to use Proclus's terms, investigated independently of matter and intellectually.

The reconstruction necessarily starts from Euclid's text, which appears to be, to a certain extent, a compilation from earlier texts which it drove out of circulation, and which are now wholly lost, so that we know of them only from certain references by Aristotle or Plato or other ancient authors. The reconstruction proceeds by a kind of literary archaeology.

Permit me to give here a set of not very reliable dates.

Flourishings

Thales	flor. 585 B.C.
Pythagoras	flor. 550 B.C.
Parmenides	flor. 475 B.C.
Hippocrates of Chios	flor, 430 B.C.
Archytas of Tarentum	flor. 400 B.C.
Theaetetus	c. 415-369 B.C.
Plato	c. 428-348 B.C.
Aristotle	384-322 B.C.
Euclid	flor. 300 B.C.

Flourishing was something Greeks did as a rule at age 40, just as they often died at 80, to suit the taste for symmetry of a certain second-century B.C. chronographer named Apollodorus. Euclid wrote about 300 B.C. There are good grounds to believe that much of geometry had been organized as a deductive science by the time of Hippocrates of Chios, about 430 B.C.; and there are plausibilities in assuming that portions of arithmetic had been organized deductively even earlier. In discussing this development, I shall want to refer to Parmenides, who lived in the first half of the fifth century; to Archytas of Tarentum, a Pythagorean and friend of Plato living around the turn of the fifth and fourth centuries; and to Theaetetus, another friend of Plato, who died as a result of battle wounds in 369 B.C., and was one of the great mathematicians of antiquity, being the author, in all probability, of nearly all of books X and XIII of Euclid's *Elements*.

In 1936, Oskar Becker pointed out a number of peculiar facts concerning Propositions 21-34 of Book IX of Euclid. These theorems are for the most part so obvious that it is hard to imagine why anyone would be so fussy as to want them proved. "If as many even numbers as we please be added together, the whole is even." Certainly. "If from an even number an even number is subtracted, the remainder will be even." Who will doubt it?

The proofs, with one exception, do not depend on any previous theorems in Euclid's *Elements:* they depend rather on certain *definitions* given at the start of Book VII, the first of the arithmetical books. The one exception, IX.32, depends on IX.13. But Becker suspects the proof as we now have it to be Euclid's emendation of the original proof; he shows that IX.32 follows quite straightforwardly from IX.31. Thus Propositions IX.21 to IX.34 are a self-sufficient set of propositions dependent only on certain definitions. Moreover, with one curious exception nothing else in Euclid's *Elements* depends on these propositions. The exception is the last proposition of Book X, which modern editors delete as not fitting into the argument of Book X. It is the ancient proof of the incommensurability of the side and diagonal of the square, and what it depends on is the doctrine of the even and the odd, and more specifically, Propositions 32-34 of Book IX.

Becker believed that, originally, before incorporation in Euclid's *Elements,* the doctrine of the even and the odd had led to another consequence, the traces of which have been left in Euclid. Propositions 21-34 of Book IX are followed by two final propositions, 35 and 36;

35 is used for the proof of 36, and 36 shows how to construct a perfect number—perhaps all perfect numbers, but that I believe is not yet known. Euclid's proofs for these two propositions depend on propositions in Book VII having to do with *ratios* of numbers. Becker shows that 35 and 36 can be proved on the basis of the immediately preceding propositions of Book IX, independently of any reference to ratios. Thus Becker's conjecture is that, long before Euclid, there existed a treatise on the even and the odd, including Propositions 21-36 of Book IX and the last proposition of Book X; that out of piety Euclid or some ancient editor added this treatise to the *Elements,* then, in an effort to integrate this addition with the whole, changed some of the proofs (32, 35 and 36), making use of propositions on numerical ratios from Book VII. This hypothesis at least accounts for the peculiarities of Book IX that I have cited: the fact that, with an easy revision of the proof of IX.32, propositions IX.21 to IX.34 form a treatise independent of the rest of Book IX, to which IX.35 and 36, with revised proofs, can also be added.

Such, then, is Becker's reconstruction of the doctrine of the even and the odd. That such a doctrine existed in the fifth century is supported by the fact that Plato *defines* arithmetic as the doctrine of the even and the odd, and refers to this doctrine as a familiar discipline.

Following Becker, van der Waerden has argued that most of Book VII of Euclid had also been worked out in the fifth century. One of his arguments is that Archytas of Tarentum, in a work on musical theory written about 400 B.C., depends on propositions found in Book VIII, and these propositions depend in turn on propositions found in Book VII. Since Archytas is punctilious in working out the most trivial syllogisms, it is unlikely that he merely assumed the propositions he needed; he must have known them to be already proved. If the propositions of Book VII existed in any form in Archytas's time, then van der Waerden concludes that they must have been in almost exactly their present form and thus in apple-pie order; for Book VII is worked out with great care and in such a strictly logical fashion that no step can be removed without the whole collapsing. There are other clues that lead van der Waerden to believe that most of Book VII was complete before Hippocrates of Chios wrote on lunules about 430 B.C.

These two pieces of deductive arithmetic—the doctrine of the even and the odd, and what became Euclid's Book VII—along with Hippocrates's quadrature of lunules, constitute the available presumptive evidence for the character of fifth century deductive mathematics. Can we learn anything from them, which might throw light on the question of what it meant for them to come to be? I want to take up, first, the demonstrations, then, the premises on which they are based.

Every Euclidean proposition ends with the stereotyped formula, *hoper edei deixai*, meaning: the very thing that it was necessary to show. The infinitive here, *deixai*, seems to have had the original meaning of *showing visually*. Thus in Plato's dialogue *Cratylus* Socrates says:

> Can I not step up to a man and say to him, "This is your portrait," and show him perhaps his own likeness or, perhaps, that of a woman? And by "show" *(deixai)* I mean, bring before the sense of sight (430e).

Early geometry must have been primarily a kind of visual showing, the pointing out of a symmetry, or the possibility of the coincidence of two figures, superposition. But in Euclid's text every effort is made to reduce the dependence on superposition to a minimum. We come to suspect that there was present a kind of anti-illustrative, anti-empirical tendency in mathematics, as it was being transformed into deductive science. This tendency is detectible in arithmetic as well as geometry.

Pythagorean arithmetical doctrines seem to have been originally worked out and taught with the aid of calculating pebbles. There is a fragment of the comic poet Epicharmus, written probably before 500 B.C., that runs as follows:

> "When there is an even number present, or, for all I care, an odd number, and someone wants to add a pebble or to take one away, do you think that the number remains unchanged?"
> "Not me!"
> "Well, then, look at people: one grows, another one perhaps gets shorter, and they are constantly subject to change. But whatever is changeable in character and does not remain the same, that is certainly different from what is changed. You and I are also different people from what we were yesterday, and we will still be different in the future, so that by the same argument we are never the same."[6]

---

6. Quoted in van der Waerden, *Science Awakening*, 109.

Presumably the sly rogue goes on to argue that he need not pay the debt he contracted the day before.

Aristotle, too, speaks of the Pythagorean pebble figures, the triangles, squares, and rectangles formed of pebbles with which the Pythagoreans taught arithmetical truths. We can easily see how they could have satisfied themselves, with their pebble figures, of the propositions concerning the even and the odd. Take Proposition IX.30: if an odd number is the divisor of an even number, then this same odd number is also the divisor of half the even number. (See Figure 3).

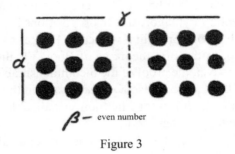

Figure 3

The number will be a rectangular number, with our odd number, the divisor, represented by the pebbles forming one of the sides. But the number as a whole is even, hence divisible in half, as by the vertical line. We see at once, then, that our odd number is a side of the half rectangle, hence a divisor of the half.

The proof of this in Euclid is quite different. The numbers are not represented by points, but rather by lines. We know that Archytas represents numbers in this way, by lines, as a matter of course, and presumably, therefore, this mode of representation had become conventional before his time, that is, already in the fifth century. Now by looking at a line which represents a number, one cannot tell whether the number is even or odd, since any line can be halved; consequently, Euclid's visual representation of the numbers does not help us at all to see *why* the proposition is true. A pebble configuration could only represent visually a *particular* number; the new representation has the advantage of generality, but it also has the disadvantage that it forces one to look for an entirely new proof. The new proof that Euclid gives us involves the famous reduction to the absurd, or *indirect* demonstration. The important step is to show that the odd number *α*, the divisor, measures the even number β an even number of times, or in other words, that the quotient, γ is even. Euclid's argument runs as follows: I say that γ is *not* odd, for *if* possible, let it be so. Now α multiplying γ makes β, and α was taken at the start to be odd, and an odd number multiplying an odd number yields only an odd number, as Euclid has previously shown. Therefore it would be odd, which is impossible (*adunaton*), because it was taken at the start to be even.

The anti-illustrative tendency brings with it the *reductio ad absurdum* proof. It is surprising how many *reductio* proofs occur in the arithmetical treatises that, according to Becker and van der Waerden, stem from the fifth century Pythagoreans. In propositions 21-36 of Book IX there are six such proofs, or eight if we accept Becker's reconstructions of 32, 35 and 36. In the first theorems of Book VII there are fifteen such proofs. Moreover, the proof of the incommensurability of the side and diagonal of the square is also a *reductio,* and in this case we have to do with a truth which is altogether non-visualizable. Let me pause to review the strategy of that proof.

Suppose, if possible, that the side and diagonal of a square *are* commensurable. Then there would be a length that measured both, and also a largest such length. Let this largest such length measure the diagonal α times, and the side β times, where α and β are integers or whole numbers. Now α and β cannot both be even, for otherwise our unit length could have been doubled, and the numbers halved, contrary to the assumption that the unit length was the largest possible; so at least one of the numbers must be odd. The sequence of the proof then shows that both must be even, or as Aristotle says in referring to this proof, that the same number must be both even and odd. For $\alpha^2 = 2\beta^2$; therefore α^2 is even, whence α is even. Let $\alpha = 2\gamma$. Then $\alpha^2 = 4\gamma^2 = 2\beta^2$, or $\beta^2 = 2\gamma^2$. Therefore β^2 is an even number, and so β is an even number. *Adunaton!* The only alternative left is to relinquish the original assumption that α and β exist, or that side and diagonal are commensurable. In this demonstration human reason exhibits a rather astonishing power, the power to discover what eyesight could never in any way disclose. This discovery would encourage the anti-illustrative tendency, and the recourse to indirect proofs. It also implies that geometry cannot be subsumed under arithmetic, and needs therefore to be built up as an independent science on its own right, for magnitudes if incommensurable do not have to one another the ratio of a number to a number, while surely having a relation with respect to size. But the relevant point at this moment is that the emergence of deductive science appears to be connected with an anti-illustrative tendency, and with the closely-connected introduction of reductio proofs.

What about the *principles* or *premises* of Pythagorean arithmetic? I have already mentioned that the premises of the doctrine of the even and the odd are to be found among the definitions of Euclid's Book VII, and only there. The same thing goes for the doctrine concerning divisibility and proportionality found in Book VII itself. And fundamentally, all the definitions of Book VII rest on the first two definitions, the definition of number—a number is a multitude composed of units or monads—and then the definition of monad: monad is that according to which each of the things that are, each of the beings, is called one. What these definitions do, above all, is to limit the following discussion to *whole* numbers. Comparing this Greek arithmetical theory with Egyptian and Babylonian numerical work, we see that the Greek theory is sharply distinguished by its careful avoidance of fractions; and the first definition, whatever else it is doing, is expressing this prohibition against fractions, this insistence on the indivisibility of the one or unit. This insistence was already traditional in Plato's time.

In the *Republic* Socrates speaks of "the teaching concerning the one" *(hē peri to hen mathēsis)*, and explains what he means by it:

> You are doubtless aware that experts in this study, if anyone attempts to cut up the "one" in argument, laugh at him and refuse to allow it; but if you mince it up, they multiply, always on guard lest the one should appear to be not one but a multiplicity of parts. . . . Suppose now . . . someone were to ask them, "My good friends, what numbers are these you are talking about, in which the one is such as you postulate, each unit equal to every other without the slightest difference and admitting no division into parts?" What do you think would be their answer? This, I think—that they are speaking of units which can only be conceived by thought, and which it is not possible to deal with in any other way.

Socrates's explanation tells us *why* the indivisibility of the one had to be insisted upon; if the one were divisible, then it would be a multiplicity of parts, hence many, *not* one. In other words, the thought that the one is divisible is self-contradictory, Thus the insistence on the indivisibility of the one, which is Euclidean and also, according to Plato's Socrates in the *Republic,* pre-Platonic, is the conclusion of an indirect demonstration, a reduction to the absurd: The one is indivisible; for if not, then it is divisible, and therefore a multiplicity of parts, hence not one; therefore, etc.

I have not yet taken up the principles used in early deductive geometry, but let me recapitulate what I have said, and consider what it suggests. The earliest deductive science, as far as we can tell, was the arithmetical theory of the so-called Pythagoreans of the fifth century. Their science differs from all earlier mathematics, first, in exhibiting an anti-empirical tendency, which sought to eliminate mere visual showing, as with the pebble figures; secondly, in making use of indirect demonstration, or proof of something by reduction of its opposite to absurdity; thirdly, in insisting upon the indivisibility of the one, on the ground that admission of its divisibility would contradict the very meaning of the word "one." Now these features call to mind certain lines that remain of a poem written early in the fifth century, the poem by Parmenides of Elea; and to no other author of this time can these features be related. I must try to say some words about the poem of Parmenides.

Only fragments of it remain. Their interpretation is thoroughly controversial. There is widespread assurance that, whatever it was that Parmenides meant, he was wrong. On the other hand, it will be little contested, I believe, if I say that Parmenides was the founder of Dialectic. Aristotle says that Zeno of Elea, Parmenides's pupil, was the founder of dialectic; but I think that may be because Zeno wrote out arguments in prose, whereas Parmenides wrote a poem in epic verse, while dialectic has essentially nothing to do with verse. I believe there is also rather general agreement that Parmenides, in composing his poem, was responding to, and attacking, earlier cosmogonies, which sought to derive all the variety and diversity of the world out of some underlying stuff, understood to be the real stuff of the world.

The poet begins by describing his journey in a chariot, drawn by mares that know the way, and escorted by the Daughters of the Sun. They arrive, high in the sky, before the gates of Night and Day. The Sun Maidens persuade the Goddess Justice to open the gates, and Parmenides is welcomed by the goddess who takes his hand and assures him that it is right and just that he, a mortal, should have taken this road. He must now learn both the unshaken heart of well-rounded truth, and the unreliable beliefs of mortals. The goddess describes three ways of inquiry: First, "That it is (*esti*), and cannot not be; this is the way of

Persuasion, for she is the attendant of Truth." Second, "That it is not (*ouk esti*), and must necessarily not be; this I tell you is a way of total ignorance." Third, "That it is, and it is not, the same and not the same; this is the way that ignorant mortals wander, bemused."

An initial difficulty that we face is that, although the pronoun "it" is not expressed in Greek, we can hardly resist the impression that there is something that is being talked about, and we should like to know what it is. The next fragments may be helpful.

> "It is the same thing that can be thought and can be." "What can be spoken of and thought must be; for it is possible for it to be, but it is not possible for *nothing* to be. These things I bid thee ponder."

In a preliminary and superficial way I think I can conclude that the subject of the verb *estin* is: that which is *intended* in thought, what we call the object of thought. The goddess is presenting an argument: that which thought intends *can* exist; but *nothing* cannot exist; therefore that which thought intends cannot be nothing; hence it *must* exist.

The syllogism holds, I believe, although at that point in time logic had not been invented. But what does it mean? That which in no way is cannot be entertained in thought. Thought always is of something, it is intentional in character. Hence I must accept the Goddess's rejection of the second way, or non-way, of inquiry.

But the Goddess means something more. Some of this "more" emerges as she proceeds to dispose of the third way of inquiry. This is the way whereon, she says, mortals who know nothing wander two-headed; perplexity guides the wandering thought in their breasts; they are borne along, both deaf and blind, bemused, as undiscerning hordes, who have decided to believe that it is, and it is not, the same and not the same, and for whom there is a way of all things that turns back upon itself. "Never," says the goddess, "shall this be proved: that things that are not, are; but do thou hold back thy thought from this way of inquiry, nor let custom that comes of much experience force thee to cast along this way an aimless eye and a noise-cluttered ear and tongue, but judge through logos (through reasoning) the hard-hitting refutation I have uttered."

"It is necessary," adds the Goddess, "to say and to think that Being is."

Now in one way, this is all simple and undeniable. When I entertain an idea, when I use a word to signify some idea, I intend what I am thinking of as a constant, invariable. Never mind that my thought, my intending of what I am thinking about, is a shifting and not very controllable process. What is thought and named is intended as having a certain fixity. Otherwise, as Aristotle puts it, to seek truth would be to follow flying game. We would be reduced to the level of Cratylus, who did not think it right to say anything, and instead only moved his finger, and who criticized Heracleitus for saying that it is impossible to step twice into the same river, for he, Cratylus, said that one could not do it even once. On one level, the words of the Goddess are simply telling us what the prerequisites and necessities are for speech and thought that will be free of contradiction. Parmenides' poem is the earliest document preserved from the past which speaks explicitly of the logical necessities of thought.

Yet the discourse of the Goddess is more strange and frightening, or insane, or as Whitehead might say, important, than I have been making it out to be. The Goddess is not concerned with just anything that might be thought; she is concerned—she says so again and again—with Being.

What is all this silly talk about Being? What else is there for Being to do but be? "It is necessary," says the Goddess, "to say and to think that Being is." Is it? Then is it necessary to say and to think that rain rains, that thunder thunders, or that lightning lightnings, and are not these parallel cases? I shall later come back, very briefly, to this question. It is just here that the poem becomes exasperating and impossible, prompting Aristotle to say more than once: the premises are false, and the conclusions do not follow. From the fact that Being just is, the Goddess proceeds to conclude that Being is precisely One, and contains no plurality, no multiplicity or differentiation within it, and no motion. In particular, and to Aristotle's great disgust, the Eleatics claim to have discovered the self-contradictory character of motion. It is Zeno, Parmenides's pupil, who formulates this discovery in the most memorable way. The flying arrow is in every instant exactly where it is, is at rest in the space equal to itself, and since this is true of every moment of its flight, it is always at rest, it does not move. Never mind the mortal wound we think it can inflict; this does not answer the argument, it does not tell us how motion can be consistently thought. The question is not whether Zeno is wrong but how. It is still being debated in the philosophical journals.

In the case of Parmenides, a more insistent question is what he can have meant by his poem. There is a second part to it, called the Way of

Seeming or Opinion, of which forty lines remain, and this speaks of the coming-to-be of the visible things of our ordinary world out of Fire and Night. Did Parmenides intend the Way of Opinion to have any validity at all, or only to present the bemused and erring beliefs of mortals? Plutarch remarks that

> Parmenides had taken away neither fire nor water nor rocks nor precipices, nor yet cities . . . for he has written very largely of the earth, heaven, sun, moon and stars, and has spoken of the generation of man.

Traditions credit Parmenides with having given laws to the city of Elea, and with having been the first to say that the Earth is round, that the Moon shines by reflected light, and that the morning star is identical with the evening star—momentous discoveries everyone of them. But such actions and discoveries do not seem easily compatible with the teaching about Being that the goddess has set forth, with such emphasis, such imperial absolutism. The heart of well-rounded truth, Being which is precisely and only One appears to have no place in it for human law, for the earth's rotundity and its conical shadow, for Venus and her irregularities, or for Parmenides or you or me.

The speech of the Goddess is nevertheless fateful. With Parmenides, as I have said, dialectic takes its start. The age of those called sophists begins. One of the earliest of them, Protagoras, is clearly reacting to Parmenides when he makes his famous statement: man, he says, is the measure of all things, of the things that are, that they are, and of the things that are not, that they are not. Who but Parmenides had raised these questions about Being and not-Being? Protagoras has concluded that the Parmenidean standard of truth, that is, freedom from contradiction, is unreachable; thought, he thinks, inevitably involves contradiction. Therefore he turns to sense-experience, and asserts his right to say that the same thing can at one time be, and at another time not be, according as he, Protagoras, holds it to be or not to be. In Protagoras's time and later, there will be other objectors with other formulations, rejecting the speech of the Parmenidean Goddess in other ways. Gorgias, for instance, argues, first, that nothing is; second, that if anything is, it cannot be known; third, that if anything is and can be known, it cannot be expressed in speech.

Among the Parmenidean sequels, I want to suggest, was deductive arithmetic. For according to Aristotle, Parmenides was the first to speak of the One according to *logos,* according to definition; and arithmetic seems to have become deductive just when the Pythagoreans set out to found the doctrine of the even and the odd on the definition of the One, on its essential indivisibility, and proceeded in Parmenidean style to formulate proofs which relied no longer on visualization but rather on non-contradiction of the *logos.*

Of course—and this is a crucial qualification—no arithmetician could follow the teaching of the Parmenidean Goddess strictly. When the deductive arithmetician took his start from the indivisibility of the One, he was proceeding in accordance with a Parmenidean necessity of thought. When he went on to multiply the One, in order that arithmetic might be, he was violating the Parmenidean Way of Truth. Parmenidean-wise, how could there be many ones, each exactly the same as every other, and yet each retaining its identity to the extent of remaining separate from the others? The way in which these many ones can be, or are in being, is a question not for arithmetic, but for meta-arithmetic, but apparently the arithmeticians recognized that their discipline depended on the question about being. The Euclidean definition of Monas, One, reads: Monas is that in accordance with which *each of the beings* is called one. A plurality of beings—what they are remains unclear—is here presupposed.

As for geometry, the violations of the Parmenidean logos that are necessary in order for it to become deductive are more drastic. The definitions of *point* and *line* with which Euclid begins Book I were no doubt modelled on the definitions of *One* and *Number,* but there is a world of difference between the cases. The definitions tell us that a point is without parts, that a line is without breadth, but we cannot go on to derive any geometrical proposition from these rather problematic denials. It was the questionable character of the geometrical things that led Protagoras to reject the possibility of geometry altogether: a wheel, he said, does not touch a straight pole in one point only; therefore geometry is impossible, Q.E.D. But even if the geometrical definitions are granted, they do not provide a sufficient basis for the organization of geometry as a deductive science.

At the beginning of Euclid's *Elements* three kinds of principles are set out. First, definitions or *horoi*; second, postulates or *aitēmata*; third, common notions or *koinai ennoiai*. About a century ago, it was argued that the term *koinai ennoiai* had to be of late Stoic origin, and therefore not due to Euclid. Was there an earlier Greek term? It may well have been *axiōmata;* this is the term that Proclus constantly uses instead of *koinai ennoiai,* and it may have been the term in front of him in his Euclidean text. Instead of *horoi* for definitions, Proclus commonly uses *hupotheseis;* this usage is found earlier in Archimedes, and earlier still in Plato's *Republic*, where the odd and the even, and the various kinds of figures and angles are said to be treated in the sciences that deal with them as *hupotheseis*. All three of these terms, *hupotheseis, aitēmata,* and *axiōmata,* were connected at one time with the practice of dialectic.

The term *aitēmata,* postulates, comes from the verb *aiteō,* to require, to ask. "Whenever," Proclus tells us, "the statement is unknown and nevertheless is taken as true without the student's conceding it, then, Aristotle says, we call it an *aitēma*." The term *axiōmata* comes from *axioō,* which can also mean to require, to demand; it is often so used in the Platonic dialogues. To be sure, Proclus says of the axioms that they are deemed by everybody to be true and no one disputes them. I believe this statement reflects an Aristotelian and post-Aristotelian usage. Aristotle himself refers to the earlier, dialectical usage when he says: *axioō* is used of a proposition which the questioner *hopes* the questioned person will concede.

As for the term *hupotheseis,* there is perhaps little need to mention its dialectical use. At a certain point in Plato's *Republic*, Socrates speaks of the principle of non-contradiction, the presumably unshakeable principle according to which it is not possible for the same thing at the same time in the same respect and same relation to suffer, be, or do opposite things. And having enunciated the principle, he says, "Let us proceed on the hypothesis that this is so, with the understanding that, if it ever appear otherwise, everything that results from the assumption shall be invalidated" (437a). And as even the not-so-dialectical Aristotle recognizes, this principle can only be established controversially, that is to say, dialectically, against an adversary who offers to say something.

My general point is a simple one. The first book of the elements of geometry of which we have record was written in the middle of the fifth century, by Hippocrates of Chios. An anti-visual, anti-illustrative tendency that had first emerged, so far as I know, in Parmenidean dialectic, is already present in the geometrical proofs of Hippocrates of Chios that have come down to us, e.g., proofs of inequalities that would be obvious to visual inspection. The fact that at an early stage the terms adopted for the premises of geometry were terms of dialectic, terms referring to assumptions or concessions that do not entirely lose their provisional character but are required in order that a discussion might proceed, reinforces the impression that the transformation of geometry into a deductive science was carried out in a context determined by the practice of dialectic.

It is also important to realize here that the premises of geometry had to be *concessions:* propositions needed to derive what not only geometers but even surveyors and carpenters knew, yet propositions which violated, in the most obvious way, the canons of the Parmenidean logos. Two things equal to the same thing, Euclid tells us, are equal to each other. But what is equality but sameness, and how can three things that are exactly the same be three? How, moreover, are we to perform the absolutely impossible feats that the *aitēmata* require—to draw a straight line from point to point, to extend a line, to describe a circle? Part of the paradox here is described by Socrates in the *Republic*: "The science (of geometry)," he says, "is in direct contradiction to the language spoken by its practitioners. They speak in a ludicrous way, although they cannot help it; for they speak as if they were doing something and as if all their words were directed towards action. For all their talk is of squaring and applying and adding and the like, whereas the entire discipline is directed towards knowledge" (527a-b). It is probably this peculiar mixture that *Timaeus* is referring to when he speaks of geometry as apprehending what it deals with by a bastard kind of reasoning.

I should like to conclude with a short summary of and comment on what I have been saying, followed by a brief epilogue.

Deductive science appears to have been first discovered by a few Greeks; so far as I know, this discovery remained unique. Knowledge of it fell into oblivion during certain times; at whatever later times the possibility of deductive science has been recognized, the recognition has come through the recovery of Greek deductive science.

What did the original discovery involve, what did it mean, for those who made it? That is the question I have sought to examine. From a plausible reconstruction of Pythagorean deductive arithmetic, I am led to conclude that the essential moves were (1) the turning away from visualization and taking recourse in *logos;* (2) the application of a negative test, the method of indirect proof or reduction to the absurd. Now these two steps are dialectical steps, they are the steps of the method that Socrates in the *Phaedo* describes as his own: "I was afraid," he says, "that my soul might be blinded altogether if I looked at things with my eyes or tried to apprehend them only by the help of the senses. And I thought I had better have recourse to the *logos*. . . . This was the method I adopted. I first assumed some principle, which I judged to be the strongest, and then I affirmed as true whatever seemed to agree with this, and that which disagreed I regarded as untrue." But in all the features that Socrates mentions, Socratic method is essentially Eleatic, Parmenidean dialectic. The search for the sources of Pythagorean deductive arithmetic thus leads us back to Parmenides, or to someone else, who lived about the same time, and whose utterances had the same effect.

What was so special, so peculiar, about the discourse of the Parmenidean Goddess, that it could precipitate what followed?

"Thinking and the thought that *it is,*" says the Goddess, "are one and the same. For you will not find thought apart from that which is . . . ; for there is and shall be no other besides what is, since Destiny has fettered it so as to be whole and immovable."

"It is necessary to say and to think," the Goddess adds, "that Being is." These words are spoken not *by* Parmenides but *to* Parmenides. He is being called upon to say and to think, and the saying and thinking are not separated, although the order in which the Goddess names them is worth noticing, being the opposite of that which we moderns tend to choose. The thinking, we had better remind ourselves, is Greek thinking; the verb is *noein,* which once meant: to perceive by the eyes, to observe, to notice. It is not to conceive, to analyze, to grasp, to attack in our thinking. And that which Parmenides is asked to say and to notice, what will it do for him to say and to notice it? The sentence, "Being is," does indeed offer nothing to grasp, nothing to conceptualize, nothing to attack in our thinking, nothing to analyze. Except— there is a twoness there. There is the noun and the verb, essentially, of course, the same word. Yet, there is that which is present, and there is its presence. To say and to notice not only what is present but its presence is to be arrested in front of something. It is to be, at least a little bit, astonished. It is to respect what lies before us. It is to think appropriately, as befits the matter. At some point Greek thought ceased asking: Out of what do the many things come to be? and began to ask instead: What is the Being of that which is in front of us? *Ti to on* is the Greek: what is the being? In this question, there are implicit the so-

107

called laws of logic: A is A, A is not not-A. Deductive science, I am proposing, takes it start here. What seems to have been important, for these beginnings, was not answering the question but pursuing it. Even Aristotle, from whom we have received more answers than questions, nevertheless says:

> Both formerly and now and forever it remains something to be
> sought and something forever darting away: *Ti to on?*

Suppose, if you will, that the account I propose is something like the truth. Then deductive science came to be and perhaps still comes to be as a result both of a logos from beyond the gates of Night and Day, and of the fracturing of Being and of the Motion going on in the Realm of Fire and Night. Or can deductive science proceed on its own way, simply leaving behind what triggered its coming-to-be? It has sometimes attempted to do this: to become, for instance, purely formal, with the specification of every element and every rule of operation, and the exclusion of every bit of explicit or implicit ontology, with the intent of insuring logical completeness and consistency. The effort has led to many refinements; but the odd result of modern metamathematical study is that the effort cannot succeed in its original intention. Mathematics does not succeed in being completely in itself and for itself. Its triumph lies not in isolated grandeur, but in coping as best it can with necessities that appear.

Deductive mathematics, not quite a century after coming to be, underwent a crisis with respect to its foundations, The discovery of incommensurability can well have been early in the fifth century. It implies, rather obviously one would think, the falsity of the old Pythagorean doctrine that all is number, whatever that doctrine may have meant. But if the discovery was early, an important consequence of it was somewhat slow in being realized. The teaching concerning ratios of magnitudes was originally conceived in a numerical fashion: four magnitudes are proportional when the first is the same part, parts or multiple of the second that the third is of the fourth. That definition was still being used by Hippocrates of Chios. Archytas, around 400 B.C., was saying that logistic, the doctrine of ratios of numbers, has the highest rank among the arts, and in particular it is superior to geometry, "since it can treat more clearly than the latter whatever it will." Archytas thus fails to notice that the fact of incommensurability sets a new task for mathematics, the formulation of a new definition of proportionality, one which will apply to magnitudes that may be incommensurable.

The problem is solved by the early fourth century, possibly by Theaetetus; at least he is the first we know to have used the new definition, and he did so extensively. The new definition of same ratio or proportionality is not the one embodied in Euclid, the definition due to Eudoxus, but a precursor of the latter, one which we can argue Euclid excised from Book X as it came down to him from Theaetetus. The manner of the new definition is worth noting. At the beginning of Book VII of Euclid, a method is given of determining the greatest common divisor of two numbers; it has come to be called the Euclidean algorithm.

What is the greatest common divisor of 65, 39?

$65 - 39 = 26$

$39 - 26 = 13$

But 13 measures 26.

Ans: 13 is g.c.d. (65, 39).

The lesser of the two numbers is subtracted from the greater until a yet smaller number remains. This smaller remainder is subtracted from the preceding subtrahend in the same manner, and so one continues, obtaining a series of decreasing remainders, until one arrives at a remainder that measures the preceding remainder. In the case shown, this number is 13, which is the greatest common divisor of 65 and 39.

This same procedure of successive, in-turn, subtractions—its Greek name was *antanairesis*—can be applied to magnitudes, in order to determine their common measure. But suppose they are incommensurable; then the subtractions would go on forever, without any remainder being found that measured the preceding remainder. A particular such situation is shown in the following diagram, which shows the side and diagonal of a square (See Figure 4):

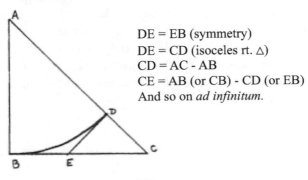

DE = EB (symmetry)
DE = CD (isoceles rt. △)
CD = AC - AB
CE = AB (or CB) - CD (or EB)
And so on *ad infinitum*.

First the side is subtracted from the diagonal, leaving CD; CD is then subtracted from the side CB twice, and so on; I will not go into the proof of incommensurability here, which necessarily involves a reduction to the absurd; but one can get a hint from the diagram as to why the process would be infinite. Nevertheless, this infinite process of *antanairesis* would go on in a determinate way, for a given pair of original magnitudes. The *n*th remainder, for example, might subtract two times from remainder *n*-1; and remainder *n*-1 might subtract three times from remainder *n*. The two and three, along with the corresponding numbers for all the other subtractions, would characterize and define the antanairesis as a whole. Then *same antanairesis* could be the definition of *same ratio:* a first magnitude would have to a second magnitude the same ratio as a third to a fourth if the first and second magnitude had the same *antanairesis* as the third and the fourth. With this definition, it is possible to prove, for example, that rectangles under the same height are to one another as their bases, because one sees that the antanairesis will go on in the same way with the rectangles as with the bases, even though the antanairesis be infinite.

There are other mathematical exploits of Theaetetus, embodied in books X and XIII, and they are of a kind with the formulation of the definition of proportionality that I have just described. Using theorems about numbers in new ways, Theaetetus succeeds in rendering what was inexpressible expressible. Such achievement, I would suggest, should be put down under the rubric of Pascal's *esprit de finesse*, rather than under his *esprit de geometrie*, the geometrical turn of mind, which Pascal so berates for its blindness to the problem of the principles. The Pythagorean mathemata, arithmetic, geometry, and the rest, are not liberal arts merely or primarily in being deductive, in proceeding stepwise in accordance with certain rules. Their liberality, it seems to me, has an essential relation to the awareness not merely of logical necessity, but of that necessity with which they are designed to cope: we are free men when we are aware of that necessity and can begin to cope with it. The liberal arts become fully liberal only as we turn toward the problem of the principles, toward the matrix of necessity in which those principles are embedded, toward the question of being from which those arts take their rise.

Kepler and the Mode of Vision
(1974)

How do we see?

Opening our eyes, looking outward from the eye sockets in our heads, we perceive—unless perhaps we are being subjected to an experiment by a psychologist, or are about to faint away—we usually perceive objects in a world: chairs, the floor, walls, windows, trees, houses, automobiles, dogs, cats, birds, people; each object situated at any moment in some place in an environment that spreads out from here, that is, from wherever we happen at the moment to be. Visual perception, almost always, is of persisting, stable, space-occupying things or objects, things not only extended and shaped in length and width but modelled in depth, and located with respect to certain background surfaces, like the floor or walls of the room, or the terrain outside, or the surface of the bay. The solid objects look solid, the square objects look square, the horizontal surfaces look horizontal, and a person who approaches me from 100 feet away does not grow to ten times his previous size. The visual world with its stable, meaningful objects remains patiently there for my inspection, in all its meaningful known-ness and unknown-ness.

It is all so familiar. And yet the performance of seeing, all of you surely know, is a complex affair. It depends on certain organs or instruments and conditions. In order for anyone to see, there must be light to see by; the eyes must be open, and must focus and point properly; certain sensitive cells of the retina of the eye must react to light in certain ways, nerve fibrils must transmit impulses to certain nerve ganglia, and these must transmit impulses on to what has been called the enchanted loom, the brain. Let anyone of these conditions not be fulfilled, and seeing will not occur. Yet, seeing does not "feel like" a complex, physical process. Rather, it "feels as if" things are simply there when we open our eyes. The fact that, on the basis of physiological processes occurring in my head, I should see a world *out there,* a world that I believe I share with you, though at any moment I acknowledge—and I think you acknowledge—that we are seeing it under different perspectives: this, when I start to reflect on it, seems little short of miracle.

Please do not expect me to explain away that miracle. The world that I perceive before me, into which I enter perforce, and where I meet

you on a basis of essential equality—that perceived world I take to be
a primary datum. Its existence seems to me to be presupposed, in one
way or another, in whatever I may say, on any subject whatsoever, even
when, seeking to philosophize, I pretend or imagine that it is not so.
My language and hence my thought are deeply rooted in the human
experience of the humanly perceived world. My seeing and my knowing seem to be from *within* that world. True enough, I can be persuaded
that the humanly perceived world is not the world simply. It differs
from a cat's world or a rabbit's world, painted only in shades of gray;
it differs from the world of an arthropod, a bee, say, with its compound,
movement-sensitive eyes. The human eye, we are told, responds only
to a narrow band in the spectrum of radiation frequencies. Does this
mean that our seeing and hence our knowing are fundamentally perspectival and partial? Is there some act of prestidigitation, whereby I
can gain a perspective on my seeing and my knowing as though from
outside my world?

Such questions were raised in antiquity. One is reminded, for instance, of Protagoras's assertion that "each thing is as it appears to him
who perceives it." Protagoras is identifying perception with what is.
He is attempting to surmount the paradox presented by earlier cosmol-
ogists, for example the atomists, who on the one hand insisted on a
sharp disjunction between appearance and reality, between what appears and what is, while on the other claiming sensation or perception
as the basis of their theories. Protagoras proposes, in contrast, that there
are no abiding elements underlying the world, but that the world consists of motions. Such motions, encountering one another, produce both
the thing perceived and the perception. Neither the perception nor the
thing perceived exists of itself, but only each for and with the other.
Where nothing is perceived, nothing exists, and conversely, whatever
is perceived, exists. Protagoras concludes, "my perception is true for
me, since its object at any moment is what is there for me, and I am
judge of what is for me, that it is, and of what is not, that it is not." Protagoras is asserting that it is impossible to err. The implication, unfortunately, is that the notion of truth is empty, no statement being
controvertible. Protagoras attempts to encapsulate his doctrine in the
famous formula, "Man is the measure of all things." But it seems to be
human beings in the plural he is referring to, and they disagree. Moreover,
it is not clear why other creatures, say cats or crustaceans, should not be
the measure of all things. How can we maintain the Protagorean thesis

against one who denies it, since he, too, is a "measure of all things"?

Issues of this kind were being talked about and argued over at the time modern science came to be. Skepticism with regard to the possibility of true knowledge arose out of the religious conflicts of the sixteenth century; out of the discovery and exploration of the New World, which revealed the existence of plants, animals, and peoples unknown to Aristotle; out of certain challenges to traditional medicine and astronomy posed by Paracelsus and Copernicus. The men most notably responsible for setting modern science on its way, Kepler, Galileo, Descartes, were each of them aiming to make a new beginning, to lay new foundations for human knowledge. The attempt at a new beginning becomes most radical and far-reaching in Descartes.

It is with the theory of visual perception as it relates to the emergence of Cartesian thought that I am here concerned. My account will begin with the development of Kepler's theory of the eye as an optical instrument, including his discovery in 1603 of the formation of the retinal image. I shall go into some detail as to how this discovery came about, as an example of learning and discovery. Then I shall turn to the Cartesian interpretation of the Keplerian result. For Descartes, Kepler's theory of the optics of the eye becomes an inspiration and vindication of Cartesian physics and philosophy. In the final section, I shall undertake a brief critique of the Cartesian interpretation.

Prior to the publication of Kepler's theory, from the thirteenth through the sixteenth centuries, the generally accepted theory of vision in the European universities was that due to an Arab optician of the eleventh century, Ibn al-Haytham—the Europeans referred to him as Alhazen. Without describing this theory in much detail, let me say, first, that, following Galen, the Greek physician, Alhazen supposed the sensitive organ of the eye to be the crystalline humor, what we now call the lens. This crystalline humor was supposed to be right in the center of the eye, a notion not challenged till the second half of the sixteenth century. The ciliary muscles and ligaments, connecting the lens to the coatings of the eye, were thought by Galen to be nerves, and this was a reason for supposing the crystalline humor to be the sensitive element. Secondly, Alhazen assumed that not all the light rays entering the eye were effectual in producing vision, but only those rays which started from the surface of the object, crossed the surface of the eye perpendicularly, and came to an apex in the center of the crystalline humor, thus forming a pyramid. If more rays were to take part in the production of vision, Alhazen thought, the result would be indistinctness and confusion. He had no notion of the bringing to a focus of a bundle or *pencil* of rays within the eye, nor did anyone before Kepler; it was Kepler who introduced these terms *pencil* and *focus* into optics. Thirdly, as to what the soul senses, Alhazen essentially followed the Aristotelian account. The soul senses the sensible form of the thing seen, without its matter. This is not a copy theory; no images are involved; it is an identity theory. To perceive the sensible form of a thing is to perceive the object as it is. The soul is passive; it receives the sensible form. What is received in the eye must travel along the optic nerve to the brain, according to Alhazen, in order for sensation to be completed. But already in the eye, the motion of light has somehow brought about a sensation, that is, the reception of the sensible form.

How did Kepler come to break with this optical tradition of the universities? Certain anatomical discoveries played a role. In the latter half of the sixteenth century, it was discovered that the crystalline humor was far forward in the eye, and that the cutting of the ciliary processes does not prevent vision. But at a crucial point, I believe, Kepler was dependent here on a tradition that was not an academic one, the tradition of *perspectiva pingendi*. In this term, *perspectiva* means not what we mean by "Perspective," but rather the same as optics, the science of vision. Optics as studied in the universities went under the name "perspectiva"; thus the standard university textbook of optics for 300 years up to Kepler's time was Johannes Pecham's *Perspectiva communis*, where the adjective "communis" or "common" apparently meant merely that the text was the standard one. *Perspectiva pingendi*, in contrast, was the optics of painting, the science that forms the basis of the art of drawing *in* perspective, to use the term in the sense that it has now come to have. *Perspectiva pingendi* was sometimes referred to as a "secret art," with its rules passed orally from one painter or draftsman to another in the fifteenth century. But Kepler was able to read about it in a book by Albrecht Dürer published in 1538, and entitled *Instruction in Mensuration*. I shall be pursuing this connection shortly.

The starting point of the train of thought that led Kepler into an intensive study of optics was an astronomical anomaly. In April of 1598, Tycho Brahe, the famous observational astronomer, wrote to the professor of mathematics at Tübingen, to report his observation of a solar eclipse earlier that year. What was especially remarkable was the apparent size of the moon when eclipsing the sun.

> Truly [Tycho wrote], it must be recognized that the moon when it is on the ecliptic and when it is new [this means it is going to eclipse the sun] does not appear to be the size which it is at other times at full moon, even though it is then at the same distance from the earth; but it is, as it were, constricted by about one part in five, from certain causes to be disclosed elsewhere.

The causes were to be disclosed elsewhere, but not by Tycho, who was totally ignorant of what they were. The only thing he did conclude for sure was that there could never be a total eclipse of the sun, in which the sun was completely covered by the moon, and in this he was wrong.

Kepler read of the anomaly in a letter from Maestlin, his former teacher at Tübingen, in July. He was at this time twenty-seven years old, and employed as a schoolteacher in the Duchy of Styria in Lower Austria. The report of a twenty percent shrinkage of the moon during solar eclipses perplexed and intrigued him. He imagined various hypotheses to explain the appearance, for instance that the moon had a transparent atmosphere. He also studied medieval books on the *perspectiva* of the schools, seeking an optical explanation. On July 10, 1600, there occurred another eclipse of the sun, and Kepler observed and measured it with special apparatus set up in the marketplace of Graz, in Styria. By the end of the month he had correctly resolved the Tychonic paradox in terms of optics. As he reported a little later to Maestlin:

> I have been fully occupied in calculating and observing the solar eclipse. While I was involved in preparing the special instrument, setting up the boards under the sky, some fellow took the opportunity to observe another shadow and it produced not an eclipse of the sun but of my purse, costing me 30 Gulden. A costly eclipse, by God! But from it I have deduced the explanation why the moon shows so small a diameter on the ecliptic at new moon. And so in what was left of July, I have written a "Paralipomena" to the Second Book of the Optics of Witelo.

Witelo's *Optics* had been written about A.D. 1270, in northern Italy, and was based largely on the *Optics* of Alhazen. *Paralipomena* means "Things Omitted." When Kepler's book on optics finally appeared, in 1604, it had swollen from a brief explanation of Tycho's paradox into a 450-page treatise, the foundation of modern optics, the science not of vision but of light. But it still carried the old title, *Things Omitted by Witelo*.

Now what was Tycho's paradox, really, and how did Kepler resolve it? You must first understand that in Tycho's time, eclipses were being observed by means of the *camera obscura,* a dark room with a single, small hole in the wall, through which the rays of the sun are admitted (Figure 1). The idea of using the camera for quantitative astronomy goes back at least to the thirteenth century, but its common use for the meauring of eclipses dates from the 1540s. The

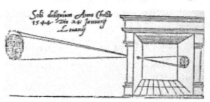

Figure 1[1]

figure is from a book on astronomical measurements published in 1545. It shows a double cone of imagined light rays passing through the tiny aperture in the wall, and it clearly depicts the inversion of the solar crescent on the wall. Let me incidentally call your attention to the fact that the depiction is in focused or linear perspective, that is, the lines that we take to be receding perpendicularly from the plane of the drawing are so oriented as to intersect at a point, called the vanishing point.

Kepler's eclipse measuring instrument in 1600 was not an obscure room, but an obscure tent (Figure 2). It consisted like the camera of a small aperture and a screen for receiving the image. The bar carrying the screens could be adjusted so that the receiving screen would be perpendicular to the rays of the sun. When an observation was in

Figure 2[2]

---

1. From Gemma Frisius, *De radio astronomico*, 312, as reproduced in Stephen Straker, *Kepler's Optics: A Study in the Development of Seventeenth-Century Natural Philosophy*, Indiana University Ph.D. Dissertation, 1971, 319.
2. From Kepler, *Ad Vitellionem Paralipomena*, Frankfurt, 1604, 338.

progress, the whole apparatus, including Kepler himself, not shown here, was covered by a black cloth.

Now Tycho's shrunken moon is actually a shrunken shadow. To understand the shrinking, we have to understand how a luminous object forms an image behind an aperture that is not a point, but has a finite size, In the *camera obscura,* it won't do to use a *very* tiny aperture, because then the image is feeble, and its boundaries become indistinct. But let the aperture be the size of a pea, as Kepler did, and the size of the image will be affected in the way shown by Figure 3. On the left imagine a luminous triangle. The rays of light emerging from any given

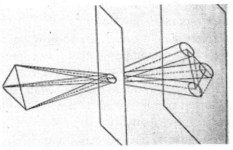

point of the triangle, and passing through the round aperture in the middle, form a cone. On the right you can see that the illuminated portion of the screen will not be triangular, but will be of such a shape as one obtains from a triangle by augmenting it on all sides by a border of uniform width; Figure 3[3]

that is, it will be a three-sided figure with rounded corners. The width of the border depends on the distance of the object and on the distance and size of the aperture. If the luminous object is distant enough so that the rays of light coming from any point of it and passing through the aperture are very nearly parallel, then, very nearly, the width of the border will be half the diameter of the aperture.

Actually, there are two ways of conceiving the formation of the image. One can think of the whole luminous object as being projected through each point of the aperture, as in Figure 4, to yield innumerable overlapping images on the screen. This was the standard medieval way

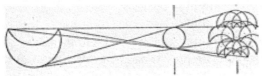

Figure 4[4]

3. From Stephen Straker, *Kepler's Optics*, Ph.D. dissertation, Indiana University, 19.
4. From Straker, *Kepler's Optics*, 24.

of analyzing the rays coming from an object: that is, to think of them as forming a pyramid with base on the object, apex in the eye, or in the case here, in the aperture of the camera. The other way of analyzing the image formation, the one Kepler used, is shown in Figure 5. Here one considers all the rays emerging from each point of the object. To the cone of rays emerging from a given point and passing through a round aperture, Kepler gave the name *pencil,* which in his day meant painter's brush or pencil. The medieval analysis takes the *object* as its starting point; Kepler's analysis takes the point

Figure 5[5]

source of light rays as its starting point. Although the pencils and the pyramids encompass exactly the same rays, the analysis in terms of pencils is probably the more helpful in letting us see what the shape of the image will be. In the case of the luminous crescent of the partially eclipsed sun, for instance, it will be a crescent with rounded horns. The quantitative analysis of Tycho's paradoxical phenomenon is now easy; see Figure 6. The crescent image of the sun has been augmented on all sides by a border of width equal to the radius of the aperture, while the shadow of the moon has been diminished by a border of the same width.

This Keplerian solution to the Tychonic paradox happens to have been very important in the history of seventeenth century astronomy, but it is so straightforward that I do not anticipate your being moved to rapture over it. It simply makes a rigorous applica-

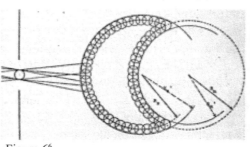

Figure 6[6]

tion of the rectilinear propagation of light to the *camera obscura.* Had this never been done before? It had, (by Alhazen), but the correct so-

---

5. From Straker, *Kepler's Optics,* 23.
6. From Straker, *Kepler's Optics,* 26.

discussed a special form of the problem, namely, why it is that, behind an *angular* aperture, when the screen is sufficiently distant, the image cast by the sun is *round*. The most common solution among the medieval opticians was to say that light has a tendency to round itself out, that it contains an active power which brings this about. Light, it was said, was the bearer of all creative and causal action, and thus had this power.

Kepler read these medieval discussions. He was even attracted to the neoplatonic theory of light that they contained. He has left us a very explicit account of the way in which he came to reject the notion that light could round itself out, in violation of rectilinear propagation.

> A certain light (he writes in the *Paralipomena*) drove me out of the shadows of Pecham several years ago. For, since I could not comprehend the obscure sense of the words from the diagram on the page, I had recourse to a personal observation in three dimensions. I placed a book on high to take the place of the shining body. Between it and the floor I set a tablet having a many-cornered aperture. Next, a thread was sent down from one corner of the book through the aperture and onto the floor; it fell on the floor in such a way that it grazed the edges of the aperture; I traced the path produced and by this method created a figure on the floor similar to the aperture. . . . [He goes on to describe the tracing of the aperture from each salient point of the book]. And so it became possible for solving the problem to bring in circularity, not of the rays of light, but of the sun itself; not because the circle is the most perfect figure, but because it is the figure of the shining body.

So it is a thread that leads Kepler out of his perplexity, and to the vindication of rectilinear propagation of light, and of the invariant size of the moon. It may seem preposterous to ask where the idea of using the thread came from, but Stephen Straker, a recent student of Kepler's optics, has made an interesting guess. It has to do with *perspectiva pingendi*.

What is *perspectiva pingendi*, focused or linear perspective, Renaissance perspective, as it is variously called? The rules of linear or focused perspective seem to have been worked out, at least in part, in ancient times, apparently in connection with the painting of scenery for dramas. It was the Greeks who first entered upon the path of trying to produce, on a flat surface, the illusion of three-dimensional figures and scenes; in the painting of other ancient peoples before they come under Greek influence, there is almost no evidence of an interest in such illusion. The difference seems to be connected with the Greek interest in *fiction,* epic and drama freed from ritualistic constraints. So on countless Greek vases and

Figure 7[7]

mixing bowls, one finds painted scenes in which something wicked and interesting is going on. The photograph of Figure 7 shows a scene from the ambush of Dolon in the *Iliad.*

The mastery of techniques for producing the three-dimensional illusion was progressive. Foreshortening, consistent handling of light and shade, the increasing mistiness with distance—the various tricks of illusionist painting were mastered over a period of some centuries. Ancient authors who write of the history of painting, like Pliny and Quintilian, record it as a series of triumphs in the production of progressively more persuasive illusions. In the final stages of the Greek progress, the paintings were painted not on vases, which tend to be preserved, but on walls which tend to crumble. However, the accident of the eruption of Vesuvius in A.D. 79 has preserved for us some of those paintings in the ruins of Pompeii and Herculaneum.

Figure 8[8]

Such signatures as one finds on the paintings are Greek. A large part of the art lay in learning to rely on the imagination of the beholder, which is very obliging, and assures us that the young woman in Figure 8 is a

7. From the British Museum.
8. From Museo Nazionale, Naples.

creature of grace and beauty; we can never know whether it is good luck or bad that she can never turn around. In Figure 9 we have an imaginary, sacred landscape, in which spatial recession is suggested by an adroit handling of light and linear per-

Figure 9[9]

spective. Also present is what Leonardo will call *aerial perspective,* the increasing indistinctness as the scene recedes into the distance. But linear perspective emerges most clearly in wall decorations from about the middle of the first century B.C., that represent theatre sets, apparently a common form of wall decoration for homes. In Figure 10 the lines of the colonnaded court, seen above the facade wall, recede to a single vanishing point, as they should because of their parallelism.

That a mathematical theory of linear perspective had been worked out in antiquity is suggested by certain passages in

Figure 10[10]

the book on architecture written by Vitruvius about 25 B.C. He describes "scenography" as the sketching of the front and of the retreating sides of buildings and the correspondence of all lines to a fixed center, the 'vanishing point' of the later theory of linear perspective. "It is necessary," he says, "that, a fixed centre being established, the lines correspond by natural law to the sight of the eyes and the extension of the

9. From Museo Nazionale, Naples.
10. From Museo Nazionale, Naples.

rays, so that certain images may render the appearance of buildings in the painting of stages, and things which are drawn upon certain surfaces may seem in one case to be receding, and in another to be projecting."

During medieval times the interest in visual illusion faded, to be revived in the late 13th and 14th centuries by the artists that Dante praises, Cimabue and Giotto. But it was not till the early 15th century that a mathematical theory of perspective reappeared. It was the work of Brunelleschi, the architect of the great dome of the cathedral in Florence, and a reader of Vitruvius. Brunelleschi's procedure started from the architectural ground-plan and elevation; lines were drawn from every salient point of these plans to the position of the eye as projected onto the same plane; the intersection of these connecting lines with the picture plane gave the dimensions to use in the perspective construction. Brunelleschi's procedure came to be called the *costruzione legittima*. Other, less time-consuming procedures for producing a similar result were later introduced. The first treatise on painter's perspective was written by Leon Battista Alberti in 1435, and Alberti makes much of the *sottilissimo velo,* shown in Figure 11 in a somewhat uncomfort-

Figure 11[11]

able representation by Dürer. It is a grid of threads through which the object to be drawn is looked at from a fixed point. It has sometimes been asserted that a *curvilinear* system of perspective would be more correct than the linear perspective used in the Renaissance. Panofsky, the late historian of art, endorsed this notion, claiming that Renaissance perspective was a mere convention, comparable to the conventions of versification in poetry. This is surely wrong. Linear perspective is based on certain constraining assumptions, namely that only one eye is used

---

11. Dürer in E.H. Gombrich, *Art and Illusion*, New York, 1960, 306.

and that the head is kept immobile. But given these assumptions, it does what the Renaissance artists thought it did; it sends to the eye the same pattern of lines and points as the object itself would. It has nothing to do with the shape of the retina or with neurophysiology or psychology; it is simply a matter of rectilinear light rays and projective geometry.

Figure 12[12]

The rectilinear light ray is *materialized* in the procedure represented in the picture by Dürer in Figure 12. A perspective picture of a lute is being constructed. A needle or nail having a large eye has been fastened into the right-hand wall. A heavy thread is led through the needle, a weight being attached to the lower end of the thread. Between the needle and the lute, a frame is set up which has a little door hinged to it that is free to move in and out of the plane of the frame. Crossing the rectangular space enclosed by the frame are two other threads which the picture does not clearly show; they are free to be moved across the plane of the frame; they intersect at right angles and so define a point of the plane. The free end of weighted thread is led through the plane of the frame and held on a point of the lute by the man on the left. The threads crossing in the frame are moved till their intersection coincides with the point at which the weighted thread cuts through the plane. The weighted thread is then taken away, the little door bearing the paper is shut, and a mark is made on the paper where the movable threads intersect. The process must be repeated for other points of the object—as many as the draftsman feels are necessary for its proper portrayal. The finished picture will show

12. Dürer, *Underweysung der Messung*, in Gombrich, *Art and Illusion*, 251.

the lute as it would be seen by a single eye situated at the position of the needle in the right-hand wall. It was a standard problem for Renaissance artists to avoid the undesirably sharp foreshortening that results when the beholder's eye is only an arm's length from the plane of the picture. Dürer's apparatus solves this problem mechanically.

It is Stephen Straker's conjecture that Kepler's thread was the direct offspring of Dürer's thread. Whether this is so or not, Kepler's choice of the *pencil* of light as the element of his analysis of the *camera obscura* surely stems from *perspectiva pingendi*, the projective geometry of light rays, rather than the *perspectiva communis* of the schools, according to which the light rays had the capacity to round themselves out. Later on Kepler himself, as we learn from a letter of 1620 from the English ambassador Henry Wotton to Francis Bacon, used a *camera obscura*—in this case a black tent with an aperture for admitting light—to obtain projections of landscape scenes from the outside; ensconcing himself within the tent, Kepler proceeded to copy the landscapes in detail, producing drawings "not as a painter but as a mathematician," to Wotton's great wonder and admiration. The evident principle was rectilinear propagation.

Rectilinear propagation had resolved Tycho's paradox and accounted for the operation of the *camera obscura*. But Kepler did not stop here. His *Paralipomena* is a big book. In the fifth chapter he went on to consider the operation of the eye.

The eye evidently resembles a *camera obscura:* the substances within are transparent, and the external coverings exclude light except where the pupil or aperture is. But the eye could *not* be a *camera obscura,* because as Kepler had learned, the larger the relative size of the aperture, the fuzzier the image, and the pupil of the eye was large enough to produce a total confusion of overlapping images at the back of the eye.

Kepler turned to the accounts of the anatomists. All but one held to the standard view, according to which the crystalline humor was the sensitive element. A certain Felix Plater, however, had learned that what were supposed to be the nerves to the crystalline humor could be cut without loss of vision, and further that the crystalline humor was far forward in the eye. To him the eye looked like a *camera obscura* with a magnifying glass in front. Such instruments had been constructed in the late sixteenth century. It was left to Kepler to explain how a lens forms an image. Convex lenses had been in use as burning and magnifying glasses for centuries; eye-glasses had been in use since the thirteenth century. But Kepler was the first to account for their operation in terms of geometrical optics, the paths of light rays. He recognized that there could be a one-to-one correspondence between luminous points outside the eye, and illuminated points on the retina, provided that the cornea and crystalline humor were understood to act as lenses which refract the diverging pencils of light from each bright point outside, and bring each of them to convergence to illuminate a single point on the retina. With a globe of glass containing water, he constructed a model of the eye to show how it worked. He it was who first formulated the simple quantitative rules governing the distances and sizes of images formed by lenses. He was the first to be able to give an account of nearsightedness and farsightedness, the function of eyeglasses in correcting these defects, and why, despite the finite size of the aperture, a sharp image can be formed. And he always describes the geometrical optics of the eye in terms that remind us of painting: the image he calls a *pictura*; he speaks of its production as a process of painting the world outside on the retina of the eye; and for the conical bundles of light rays that do the painting, he introduces the term *pencil,* meaning painter's brush.

What about the upside-down-ness and left-for-right reversal of the retinal image? Kepler regards these as simply necessitated by the behavior of light. How, *given* the retinal image, vision then occurs, he does not offer to say:

> I leave it to be disputed by natural philosophers how this picture
> is put together by the visual spirits that reside in the retina and
> nerve. . . . The impression of this image on the visual spirits is
> not optical but physical and wonderful (*admirabilis*).

Kepler is saying that what happens in the retina and optic nerve and brain surpasses the powers of the mathematical optician; not being reducible to mathematics, it seems to be, in Kepler's view, ultimately inexplicable, analogous to the mystery of creation.

> Just as the eye was made to see colors, and the ear to hear
> sounds, so the human mind was made to understand, not whatever you please, but quantity. . . . It is the characteristic of the
> human understanding which seems to be such from the law of
> creation, that nothing can be known completely except quantities. And so it happens that the conclusions of mathematics are
> most certain and indubitable.

Kepler's view of the world was profoundly affected by the theory of *perspectiva pingendi*. The theorists of *perspectiva pingendi* had resurrected the famous formula of Protagoras, "man is the measure of all things." But they meant by it not the denial of the possibility of knowledge, but in a special way, the contrary. Seeing is perspectival, no doubt, but the perspective can be understood. Perspective measures space; space is known through quantities; quantities measure the permanent order of nature. At the center of every perspective system is man himself, who becomes the judge and standard for all comparisons. It is he, for instance, who apprehends and judges the beauty of things, as consisting in harmonious proportions. (This is almost a quotation from Alberti).

Kepler shared these ideas. He viewed the whole cosmos as a divinely created, three-dimensional work of art, an image of the Divine Trinity, the structure of which we can determine from the perspectives we have of it. Man, created in the image of God, is the contemplative creature, the measuring creature, as Kepler repeatedly calls him. He is placed on the midmost planet, so that by taking account of his changing position he can carry out triangulations, like a surveyor, and determine the distances of the primary cosmic bodies and their harmonious arrangement. Travel, Kepler explains, is broadening. We view the cosmos always by means of a continually changing perspective, but by calculating for our own displacement, we can use that very perspective to determine the cosmic dimensions and harmonious order. And so doing, Kepler says, the soul which is like a point, becoming contemplative, expands as it were into a circle. The sphere is reserved as the image for God himself.

I turn now to Descartes. In a letter of 1638 to an acquaintance, Descartes acknowledged Kepler as his "first master in optics." Descartes was not in the habit of admitting intellectual indebtedness; I do not know of another case where he did so. But for Descartes, the Keplerian theory of the eye as an optical instrument was both an inspiration for and vindication of Cartesian physics and philosophy. With Descartes, optics displaced astronomy as the key science for the understanding of the world.

The Keplerian theory of the eye as an optical instrument presents itself to Descartes as banishing mystery from the eye. Following Kepler, it becomes possible to construct a model of the eye, complete in just about every detail, down to the image at the rear that one may catch on a translucent piece of parchment. Of course, we do not see the inverted, reversed, perspectival images on our own retinas. For that, there would be needed an eye behind the eye, and there is no such eye. As to what happens in the optic nerve and brain, in order that vision may be completed, Kepler leaves this mystery as deep as he found it.

The Keplerian theory of image formation means, for Descartes, that the traditional Aristotelian and scholastic theory of visual perception is wrong. Sensation is not the reception of the sensible form of a thing without its matter, because, in the first place, visual sensation is perspectival. At the very beginning of the first book he completed for publication, a book entitled *The World,* Descartes wrote:

> It is commonly believed that the ideas we have in our thoughts
> entirely resemble the objects from which they proceed . . . , but
> I observe, on the contrary, several experiences that ought to
> make us doubt it.

One of the experiences Descartes has in mind here is that of looking at pictures:

> [Y]ou can see that engravings, being made of nothing but a little
> ink placed here and there on the paper, represent to us forests,
> towns, men, and even battles and storms, even though, among
> an infinity of diverse qualities which they make us conceive in
> these objects, only in shape is there actually any resemblance.
> And even this resemblance is a very imperfect one, seeing that,
> on a completely flat surface, they represent to us bodies which
> are of different heights and distances, and even that following
> the rules of perspective, circles are often better represented by
> ovals rather than by other circles, and squares by diamonds
> rather than by other squares; and so for all other shapes. So that
> often, in order to be more perfect as images and to represent an
> object better, they must not resemble it.

But Descartes's critique of the traditional account of perception goes deeper. According to Aristotle and the schoolmen, things were very much what they appeared to be. The objects perceived were themselves colored; heat and cold were what in ordinary experience we apprehend them as being; the qualities of objects were the specifications of the things that made each one of them to be what it was. But how can such perception occur? The traditional theory fails to explain how such resemblance or identity is physically achieved. The proponents of this theory cannot show us how sensations "can be formed by these objects, received by the external sense organs, and transmitted by the nerves to the brain."

Now if the traditional assumption of resemblance between sensations and their objects is questionable or without warrant, then the traditional attempt to *found the sciences* by a step-wise advance from sense perception is also questionable or without warrant. In that case, how are the sciences to be founded?

At the beginning of his book *The World*, Descartes proposes to construct a fable of a world, "feigned at pleasure." Matter in this feigned new world, Descartes proposes, should be something of which we cannot even *pretend* ignorance. This requirement is met by *extension,* for "the idea of extension is so comprised in all other things which our imagination is able to form, that it is necessary for you to conceive it, if you imagine anything whatsoever." Having seized upon this first principle, Descartes proceeds to frame in terms of it a fabulous account of the world and of man. All is to be accounted for in terms of matter, that is, figured extension, shaped portions of space, in motion. Some features of this system can be deduced from the first principle, others must be constructed. The fabulous world that results is scientific in the sense that it involves only what can be clearly conceived; it thus embodies the rigor of mathematics.

In this fable of a world, Descartes is able to provide a clear conception of the way sensations are formed by their objects, received by the external sense organs, and transmitted to the brain. First,

> All the external senses . . . serve in a purely passive way, precisely in the manner in which wax receives shape from a seal. We have to think of the external shape of the sentient body as being really altered by the object precisely in the manner in which the shape of the surface of the wax is altered by the seal.

This description applies not only to touch but also to sight, in which light, conceived as a pressure transmitted through a medium, plays the same role as a blind man's stick. Since with our two eyes we apprehend a single thing, with our two ears a single sound, with our two hands a single body, when we are touching one, Descartes concludes that there must be a center in which the incoming stimuli are coordinated; this he identifies with the single organ in the upper brain which he knew to be single and central in position, the pineal gland. The whole process of vision Descartes now describes as follows:

If we see some animal approach us, the light reflected from its body depicts two images of it, one in each of our eyes. The two images, by way of the optic nerves, form two others on the interior surface of the brain which faces its cavities. From these, by way of the spirits (or subtle fluids) which fill these cavities, the images then radiate towards the small gland which the spirits encircle, and do so in such fashion that the movement which constitutes each point of one of the images tends towards the same point of the gland as does the movement constituting that point in the other image which represents the same part of this animal; and in this way, the two brain-images form but one on the gland, which acting immediately on the soul, causes it to see the shape of the animal.

The Keplerian optics of the eye is thus interpreted as a mechanism for transmitting pressure, and this same mechanism is imagined as transmitting the patterned pressures onward from the retinas through the nerves to the brain, with final composition of a single image on the pineal gland, or *sensus communis* or imagination as Descartes calls it. The entire process, up to the final apprehension by the soul of the image on the imagination, is to be considered as consisting simply in the alteration of the spatial disposition of the parts of the body, not less, says Descartes, than the movements of a clock or other automaton.

Readers of Descartes' *Discourse on Method* and *Meditations* are familiar with the course of reflection whereby he undertakes to justify, and to lay unshakeable foundations for, this fabulous physics and physiology. According to this physics and physiology, we, that is, our Souls, have immediate knowledge only of our ideas, including in this term, as Descartes does, sensations. Between our sensations, on the one hand, and that which provides the occasions for these sensations, namely the patterned pressures on the pineal gland, there opens an abyss. On one side is a qualityless world of matter in motion; on the other, worldless qualities in a consciousness that is out of the world, is extra mundane. Between world and self, Descartes attempts to construct a metaphysical bridge, taking for starting point his certainty that the isolated, extra-mundane self, even when it doubts everything it can, at least exists as the source of the doubt. He calls it a *thinking thing*. The central arch of the construction is theological; it finally enables Descartes to conclude that what can be clearly and distinctly conceived, provided it fits all the facts, is true. I shall not pursue this argument here, my concern being rather with our perception of the visual world.

Nor shall I deal further with the detailed mechanisms of perception in Descartes' fabulous physics. Action by pressure is not regarded today either as the mode of action of light, or as that of the impulses in the nervous system. Among neurophysiologists today there is considerable doubt that anything is to be gained by supposing that the neural processes copy or are isomorphic with the contents of consciousness, are for example triangular in some way when one sees a triangle. Nevertheless, we still hear from neurophysiologists such statements as the following one by Lashley:

> All phenomena of behavior and of mind are ultimately describable in the concepts of the mathematical and physical sciences.[13]

The problem about how we are *in* or *related to* the world, so insistently posed by the Cartesian theory, here recurs. Descartes's great achievement, it would appear, was to make the world safe for mathematical physics; but this achievement has left us outside, puzzled and questioning as to what we are.

Once more, in this final section as in the beginning, I must ask you not to expect me either to do miracles or to explain them away. What I can attempt is to point in directions in which non-Cartesian perspectives open up. Let me begin by citing a certain number of results of recent studies of human visual perception.

(1) To begin with, it is worth noting that the eye is in constant motion. It has a tremor with a frequency of between 30 and 80 cycles per second, and with an amplitude such as to shift a focused pencil of light from one retinal cell to the adjacent one. There are also wider flicks of up to a third of a degree, coming at irregular intervals of up to five seconds, with slow drifts in between the flicks. The apparatus shown in Figure 13 is attached to a contact lens, and because it moves with the eye, it produces a stabilized retinal image. The effect of this is, first, within a few seconds, distorted vision, and shortly afterward, the complete breakdown of vision, in the sense that the viewer can no longer see anything at all, although a clear image is being focused on his retina. Vision is thus an active process, and fails altogether when it ceases to be so.

---

13. K. S. Lashley, in *Cerebral Mechanisms in Behavior,* ed. L. A. Jeffress (New York: Wiley, 1951), 112.

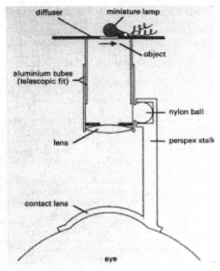

Figure 13[14]

(2) Next, let me point out that the clues to depth are multiple, more, in fact, than can be reviewed in brief compass. There is, to begin with, linear perspective, as in Figure 14. Note also here a size-distance rela-

Figure 14[15]

tion, which reminds us how drastically the perceptual system trans-forms what is presented to it. As the little man walks into depth he

14. Ditchburn Experiment, R.L. Gregory, *Eye and Brain*, 2nd ed. (New York: World University Library, 1973, 46.

15. Gombrich, *Art and Illusion,* 280.

appears to increase in size, although the three images in fact take up the same size on the plane surface. This has to do with what is called perceptual constancy, which I shall discuss in a moment.

Gradients of texture immediately produce in us the sense of continuous surfaces stretching backward (see Figure 15). Such surfaces generally provide the background against which we locate objects.

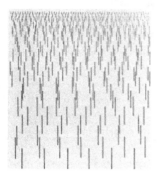

Figure 15[16]   Figure 16[17]

Then there is lighting. Light usually comes from one side, so the surfaces facing in different directions are differently illuminated. In late antiquity, four-tone mosaic floors like the one shown in Figure 16 seem to have been popular, despite the treacherous appearance of being other than flat.

There is motion perspective, the differences in apparent relative velocity of different parts of the visual field as one moves one's head. The diagram in Figure 17 is for the more drastic case of an airplane pilot approaching the landing field.

Figure 17[18]

16. Gradient of texture, in James J. Gibson, *The Perception of the Visual World* (Boston: Houghton Mifflin, 1950), 86.

17. Mosaic floor, Antioch, 2nd century A.D., in Gombrich, *Art and Illusion*, 41.

18. Motion perspective, in Gibson, *The Perception of the Visual World*, 121.

There are also *binocular* clues, impossible to illustrate in a single picture. There is the convergence of the two eyes, for example, and the accommodation of the lens in each. By far the most important is binocular disparity, a fact which only came to be recognized after Charles Wheatstone's invention of the stereoscope in the 1830s. The retinal images received by the two eyes are not simply superposable, but are notably different; and this difference by itself can produce the visual perception of depth.

(3) Scenes may be ambiguous. In particular, pictures being stationary and flat cannot provide motion perspective or the binocular clues to depth, and since the different three-dimensional shapes that can give the same projection on a plane are unlimited in number, it is evident that two-dimensional shapes can be ambiguous. The surprising thing is that we are so seldom misled or made aware of the ambiguity; the perceptual system quickly adopts the interpretation that satisfies the available clues. Shown, however, in Figure 18 is an illustration called the Necker cube, which forces its ambiguity on our attention. No clue is offered as to which is the back face and which the front; both squares

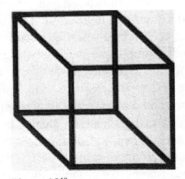

Figure 18[19]

are of the same size. By an act of concentration on one square or the other, we can cause the perceptual shift from one to the other of the two possible interpretations of the figure as cube; we can even, by putting the mind to it, cause ourselves to see the figure as flat, though the spontaneous interpretation is a three dimensional one. But suppose we gaze at the Necker cube steadily, without any attempt to have one perception or the other: then the perceptual shift occurs spontaneously. It

19. Necker cube, source unknown.

is as if perception were a matter of suggesting and testing hypotheses; of the two most satisfying ones, each is entertained in turn; but since neither is more successful than the other, neither is allowed to stay.

The ambiguities of two-dimensional representatives have been used by Escher and others to depict wonderful and impossible places and objects. Figure 19 shows an impossible object. Perception shifts

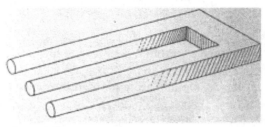

Figure 19[20]

back and forth, making sense of each part of the object, and trying but failing to make sense of the whole as an actual object.

(4) Perceptual processes exhibit a characteristic called Perceptual Constancy, and this is another indication that perception is ceaselessly and actively oriented toward the interpretation of impressions in terms of objects. A piece of coal in bright sunlight sends to our eyes maybe one hundred times more light than a sheet of typing paper in the shade; yet we interpret the former as a *black* piece of coal in the sunlight, and the latter as a *white* piece of paper in the shade. The colored light with which the retina is presented is analyzed in perception into a constant *surface* color belonging to the object, on the one hand, and the illumination to which the object is exposed, on the other, the latter being drastically variable. Besides color constancy, there is shape constancy and size constancy. Size constancy, for instance, means that when an object doubles its distance from us, so that the retinal image of it is halved in size, we perceive it not as shrinking but as retaining its objective dimensions. To experience this, look at your two hands, one held at arms' length and the other at half the distance. They will probably look almost exactly the same size. But if the near hand is brought to overlap the far one, then they will look different in size, in the way the laws of perspective require.

20. Impossible object, Gregory, *Eye and Brain*, 235.

(5) Certain well-known distortion illusions turn up ever and again in psychology textbooks. These, too, appear to be explicable in terms of the perceptual system's orientation towards interpreting patterns as objects in a world. In Figure 20 the upper horizontal bar appears longer, though it is of the same length as the lower one. The explanation that now seems likeliest is that the perceptual system is set to measure

Figure 20[21]

lengths appropriately to the normal world of three-dimensional objects as seen in linear perspective. The bar that would be more distant is scaled up on size in accordance with constancy scaling. That we do not see the inclined lines as parallel, receding railroad tracks is due to the countermanding of the three-dimensional interpretation by textural features of the surface on which the figure appears.

The same kind of explanation can be made for the famous arrow illusion, in which two identical lengths appear different because of added fins (see Figure 21). If the two lines with fins here pictured are

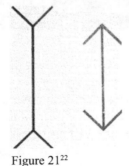

Figure 21[22]

21. Scaling illusion, Gregory, *Eye and Brain*, 137.
22. Arrow illusion, Gregory, *Eye and Brain,* 136.

constructed of wire, painted with luminous paint, and looked at in the dark, the left-hand one appears as an inside, receding corner of a rectangular room, the right-hand one as the outside projecting corner of a rectangular block or building, with sharp foreshortening. Even when, because of surface texture, three-dimensional depth is not perceived, it seems that constancy scaling goes to work on the basis of the clues that would normally indicate depth, to produce the illusion.

(6) Finally, consider the famous Distorted Room constructed by Adelbert Ames some twenty-five years ago [in 1946] (Figure 22). The far wall slopes back to the left at an angle of 45°, and the floor also slopes downward to the left, but linear perspective is used to make this oddly shaped room give, from a viewing point in the center of the front wall, the same retinal image as a normal rectangular room. The person in the far left-hand corner looks too small because the image is smaller

Figure 22[23]

than would be expected for the apparent distance of that part of the room. Evidently our perceptual system has so accommodated itself to rectangular rooms that we accept it as obvious that it is the objects— here twin sisters—that are of odd sizes, rather than that the room is of an odd shape. The perceptual system has, as it were, made a bet, the wrong one, but then, the experimenter has rigged the odds by choosing such an extremely odd shape for the room. Familiarity with the room

23. Ames's Distorted Room, Gregory, *Eye and Brain,* 178.

gained by touching its walls with a long stick, or a strong emotional relation to the persons seen in the room, will reduce the distorting effect of the room on other objects until it is finally seen for what it is—a distorted room in fact.

From all the foregoing, I conclude: Our perceptual system seems to be—behaves as if it were—an instrument acting purposively with a view to identifying, placing, classifying, and judging objects in a world. What the senses initially receive are but the slenderest and most fleeting of clues, varied and varying patterns of energy. Objects, on the other hand, have indefinitely many features beyond the immediately sensed ones. They have pasts and futures; they have hidden aspects that manifest themselves under special conditions; they change and interact with one another. How is it possible, on the basis of the fleeting clues, to perceive the objects? Our perceptual system, faced with multiple, fleeting clues, in effect makes a guess, launches into belief. Taking into account its previous beliefs, it *hypothesizes* that an object of such-and-such a kind is the invariant something of which the fleeting clues are perspectives. In some moments, some few of our perceptual processes may become conscious processes; for the most part, it is only the results that we are aware of.

And where are we, and what are we, in relation to this world that our perceptual processes lead us to posit? Is it not our primary experience that we find ourselves *in* the world, turned *toward* the world, in direct encounter with an Other that is over against us? Our observing is not neutral, from behind centimeters of bullet-proof polaroid. We are beset by what we see; we are affected, caught, seized by what confronts us. Wind, heat waves, rain, and sleet obtrude themselves upon us. Things appear attractive and repulsive.

What is presented has inherent distance, depth. The object is apprehended as *over there* in its suchness and thusness. The perceived object reveals itself insofar as it presents a surface. But while surfaces reveal, they also hide. Beyond what is directly revealed, in any experience, there remains that which is hidden, the substance of things.

Depth and distance are not merely visual. We are mobile—indeed, we believe ourselves to be self-moved. We can be purposeful, adopting goals, moving up or down or along paths, going from a Here to a There and from a Now to a Then, with changing perspectives. This is our primary mode of being. But our being is then a *becoming* in relation to an Other, the Other that is the world stretching out in depth before us, in its manifold familiarity and strangeness. We seek the unity or unities underlying the varied, perspectival perceivings, the invariant *somethings* of which we would conceive the world to be made. We are ever potentially learners; each of us, in Kepler's metaphor, a point seeking to expand into a circle—some of us no doubt more ardently than others.

Can the circle become a sphere? Kepler, you will recall, denies it. Descartes would push on, geometrizing as he goes. In the geometrized and mechanized world that he imagines, everything that happens, or almost everything, is to be accounted for in terms of displacements, *pushes.* Later, with Newton and his successors, we get *pulls* as well as pushes. But whether we take the Cartesian or the Newtonian or even a more recent quantum mechanical version of the geometrized world, we become puzzled if we try to place ourselves in it; how do we connect with it? Descartes supposed that the soul—the "thinking thing"— could act upon the extended world, namely by influencing the direction of motion without changing its quantity, which he believed to be conserved. That supposition was shown to be impossible when Newton and others demonstrated that the conserved quantity is *vectorial* or *directed.* So natural science in its advance always appears to aim at an account from which we would be absent, a silent, nonhuman world of deterministic connections. Conscious life could only be explained away, in such an account. There is something odd about the totalization of the Cartesian geometrical view of the world.

Long before Descartes, already in the thirteenth century, Henry of Ghent characterizes those in whom the geometrical imagination dominates over the cognitive faculty as suffering from an ailment:

> Whatever they think, is a quantity, or is located in quantity as is the case with the point. Therefore such men are melancholy, and become excellent mathematicians but very bad metaphysicians, for they cannot extend their thought beyond location and space which are the foundations of mathematics.[24]

This very melancholy is pictured in Dürer's famous engraving of 1513 or 1514, the *Melencolia I* (Figure 23). Here we see a winged, pre-

---

24. Henry of Ghent, *Quodlibet II, Quaestio 9* (Paris, 1518), fol. 36r.

Figure 23[25]   Figure 24[26]

sumably celestial being, staring fixedly, evidently in despair. The scene is lit with an eerie light from the moon, a comet, and a lunar rainbow. Strewn about in bewildering disorder are the tools of geometrical and architectural construction. She, the celestial being, is afflicted, we surmise, with a sense of spiritual confinement, of insurmountable barriers separating her from a higher realm of thought.

Dürer's conception of that highest realm is presented in another engraving which he made at the same time as the *Melencolia,* and distributed with it as its appropriate counterpart: *St. Jerome in his Cell* (Figure 24). Here Jerome, comfortably seated in his warm, sunlit cell, which he shares with his contented animals, is absorbed in his theological work. Even the skull looks friendly. Jerome's incorporation into the strictly mathematical projection of our perspectival system, though it reveals his contentment, in no way reveals the secret of it, which may remain for some of us inaccessible.

I would acknowledge the *perspectivity* of human knowing—using the term *perspective* here in an extended and metaphorical sense. Accordingly, I would not disregard the fact that natural science is formed by human beings. Natural science does not simply describe and explain

25. Dürer, *Melencolia I,* in Panofsky, *Albrecht Dürer,* vol. 2, 209.

26. Dürer, *St. Jerome in his Cell,* in Panofsky, *Albrecht Dürer,* vol. 2, 208.

nature, it is part of the interplay between nature and ourselves; it describes nature as exposed to our method of questioning.

Is this a skeptical conclusion? I think not. For all we know, the perspectivity revealed by the fact that we opine, and that opinion is not knowledge, may be the correct perspective. In any case, our perspectival viewing reveals a world into which we, along with others, are launched as essentially equal citizens. We recognize others, equally with ourselves, as potential measures of the truth. The claims of others call us out of our particularity into discourse, into the search for Right Opinion. This is a category unknown to skeptics, a human category revealing both our poverty and our power.

For the account of Kepler's *Paralipomena* I have depended very heavily on S. M. Straker's analysis in his unpublished work, *Kepler's Optics: A Study in the Development of Seventeenth-Century Natural Philosophy* (Indiana University, Ph.D. Dissertation, 1971). Other works to which I am much indebted are: E. Cassirer, "The Concept of Group and the Theory of Perception," *Philosophy and Phenomenological Research* 5 (1944): 1–35; Hiram Caton, *The Origin of Subjectivity: An Essay on Descartes* (New Haven: Yale University Press, 1973); James J. Gibson, *The Perception of the Visual World* (Boston: Houghton Mifflin, 1950); E. H. Gombrich, *Art and Illusion* Princeton: Princeton University Press, 1950); R. L. Gregory, *Eye and Brain* (London: Weidenfeld and Nicolson, 1966); William H. Ittelson, *The Ames Demonstrations in Perception* (Princeton: Princeton University Press, 1952); Erwin Panofsky, *Albrecht Dürer,* 2nd ed., 2 vols. (Princeton: Princeton University Press, 1945); Erwin Straus, *The Primary World of the Senses* (New York: Free Press of Glencoe, 1963) and *Phenomenological Psychology* (New York: Basic Books, 1966); John White, *The Birth and Rebirth of Pictorial Space* (New York: Thomas Yoseloff, 1958); and J. S. Wilentz, *The Senses of Man* (New York: Crowell, 1968).

Homo Loquens from a Biological Standpoint
(1975)

The words *homo loquens*, in the title I announced for this lecture, mean *speaking man, man the speaking one.* As a designation for the human species, *homo loquens* perhaps has an advantage over the official zoological designation, *homo sapiens,* man the sapient, wise, discerning one, the one who savours the essences of things. The human capacity for loquaciousness is somewhat more obviously verifiable. But what has that capacity to do with things biological? This is a complicated and problematic topic. Forgive me if I first approach it by slow stages, then attempt a gingerly step when the going becomes treacherous. I wish to begin with a small technical matter, an aspect of the physiology of speech-production.

Respiratory patterns in different species of air-breathing vertebrates differ in many details. Different species have special regulatory systems, adapted to special behavior patterns. There is the panting of dogs, specially adapted for cooling; birds, during flight, have the unique ability to increase their intake of oxygen a hundredfold; the sperm whale can go without breathing on a dive for ninety minutes, the beaver for fifteen, man for about two and a half; and so on. All these differences are species-specific.

In a human being, the respiratory patterns during quiet breathing and during speech are remarkably different (see Table I, on the following page). The volume of air inhaled, as shown in the first item of the table, increases by a factor of three or four during speech. The time of inspiration, as compared with the time for a complete cycle of inspiration plus expiration, decreases by a factor of three. The number of breaths per minute tends to decrease drastically. Expiration, which is smooth during speechless breathing, is periodically interrupted during speech, with a build-up of pressure under the glottis; it is during expiration that all normal human vocalization occurs. The patterns of electrical activity in expiratory and inspiratory muscles differ radically during quiet breathing and during speech. Both chest and abdominal musculature are utilized in breathing, but during speech the abdominal musculature is less involved, and its contractions are no longer fully synchronized with those of the chest musculature. In quiet breathing, one breathes primarily through the nose; during speech, primarily through the mouth.

TABLE I
Respiratory Adaptation in Speech

	Breathing	
	Quietly	*During Speech*
Tidal volume	500–2400cm³	1500-2400cm³
Time of inspiration / Time of inspiration + expiration	about 0.4	about 0.13
Breaths per minute	18-20	4-20
Expiration	Continuous and unimpeded	Periodically interrupted, with increase in subglottal pressure
Electrical activity in expiratory muscles	Nil or very low	Nil or very low at start of phonation; then increases rapidly and continues active to end of expiration
Electrical activity in inspiratory muscles	Active in inpiration & nil during expiration	Active in inspiration & in expiration till expiratory muscles become active
Musculatures involved	Chest & abdominal, closely synchorized	Mainly chest; slight desynchronization between chest and abdominal muscles
Airways	Primarily nasal	Primarily oral

More than you wanted to know, I'm sure. My point was to show that breathing undergoes marked changes during speech. And remarkably, humans can tolerate these modifications for almost unlimited periods of time without experiencing respiratory distress; witness filibusters in the U. S. Senate. Think now of other voluntary departures from normal breathing patterns. If we deliberately decide to breathe at some arbitrary rate, say, faster than ordinary—please do not try it here—we quickly experience the symptoms of hyperventilation: lightheadedness, giddiness, and so on. Similar phenomena may occur when one is learning to play a wind instrument or during singing instruction; training in proper breathing is requisite for these undertakings. By contrast, talking a blue streak for hours on end comes naturally to many a three-year-old. The conclusion must be that there are sensitive controlling mechanisms that regulate ventilation in an autonomous way during speech. More generally, it is evident that we are endowed with special anatomical and physiological adaptations that enable us to sustain speech for hours, on exhaled air.

Do we speak the way we do because we *happen* to possess these special adaptations, or did these adaptations develop during evolution in response to the pressures of natural selection or the charms of sexual selection? I think there is no way of answering these questions; it is difficult enough when one can refer to skeletons, which fossilize; behavioral traits do not. But whatever the answer, there is still this further question, whether the genetic programming for speech extends beyond the mere provision of vocal apparatus? Might it not, in addition, determine the make-up and structure of language in a more detailed and intimate fashion?

Such a question runs counter to views that are widely held. Is not language, after you have the voice to pronounce it with, fundamentally a *psychological* and *cultural* fact, to which biological explanations would be largely irrelevant? Do not languages consist of *arbitrary* conventions, made up in the way we make up the rules of games? Wittgenstein speaks of language as a word-game, thereby likening it to tennis or poker. Is it not apparent that the conventions of any particular language, like the rules of tennis or poker, are transmitted from generation to generation by means of imitation, training, teaching, and learning? Are not these the important facts about language, the facts that reveal to us its nature?

Until recently, students of linguistics and psychology have tended uniformly to answer these questions in the affirmative. To many, the extraordinary diversity of human tongues has seemed argument enough against any assumption of linguistic universals, that is, characteristics of language imagined to be rooted in human nature. The *reductio ad absurdum* often mentioned is the attempt of the Egyptian king Psammetichos to determine the original human language. As reported by Herodotus, Psammetichos caused two children to be raised in such a way that they would neither hear nor overhear human speech, the at-

143

tendants being instructed meanwhile to listen out for their first word. The report was, that it was Persian. The experiment is said to have been repeated in the thirteenth century by Frederick II, Holy Roman Emperor, and again around 1500 by James IV of Scotland, who was hoping that the children would speak Hebrew, and thereby establish a biblical lineage for Scotland. No result was reported.

Stress on the arbitrariness of language has been enhanced by a coalition between linguistics and behaviorist psychology. Behaviorist psychology is led, by its premisses, to the view that language is merely an arbitrary use to which the human constitution, anatomical and physiological, can be put, just as a tool can be put to many arbitrary uses by its manipulator. A recent account that views language in this way is the book *Verbal Behavior* by B.F. Skinner. Along with other behaviorist scientists, Skinner holds that all learning can be explained by a few principles which operate in all vertebrates and many invertebrates. The process is called operant conditioning. Learning the meaning of a word, Skinner holds, is like a rat's learning to press a bar which will cause a buzzer to sound, announcing "food pellets soon to come". Learning grammar, likewise, is supposed to be like learning that event A is followed by event B, which is in turn followed by event C. Many an animal can be trained to acquire associations of this kind. Skinner would hold that there is nothing involved in the acquisition of language that is not involved in learning of this kind.

Unquestionably, we would be mistaken to deny the importance or the power of the conditioned reflex, either in language acquisition or in other learning. The experimental psychologists have recently announced that even the visceral organs can be taught to do various things, on given signals, with rewards provided immediately afterward to reinforce the action. We are told that rats, with the reward held out of another shot of electrical juice in a certain center of the brain, have been taught to alter their blood pressures or brain waves, or dilate the blood vessels in one ear more than those in the other. Similar achievements in operant conditioning are held out as a bright future hope for humans. What rich experiences in self-operation are not in store for us?

On the other hand, the successes of this technology do not necessarily tell us much about the character of what it is that is being conditioned. The behaviorist treats the organism as a black box; he controls the inputs and records the outputs; what goes on in the box is not, as he claims, an appropriate concern of his. He cites the similar situation

144

in quantum physics. In the case of quantum phenomena, the physicist cannot successfully describe what is there when he is not looking, not using probes that interact with whatever it is. But, between the situation in quantum physics and the situation in the study of animal behavior, there is this difference. Animal behavior goes on, observably so, even when the animals are not being experimented on. May it not be important to try to observe this behavior, before we set out to change it, as we can, so frighteningly, do?

Those who study the behavior of animals in their natural habitats nowadays have a special name for their study, Ethology. Long hours of patient observation, much of it during the last fifty years [from about 1925 onward], have demonstrated how intricate, how unexpectedly adaptive, how downright peculiar, are the patterns of behavior specific to particular species of animals. Many of the patterns function as communication: the elaborate courtship rituals of birds, the less elaborate ones of butterflies and certain fish; the way in which two dabbling ducks, on meeting, lower their bills into the water and pretend to drink, as an indication of nonaggressiveness; and so on. Among these behaviors, there is one that has been called truly symbolic. That is the dance of the honeybee, the symbolism of which was first recognized and deciphered by Karl von Frisch in the 1940s. Let me describe it briefly (see Figure 1).

ROUND DANCE   TAIL-WAGGING DANCE

Figure 1

The dance that a forager bee performs in the dark hive gives, by a special symbolism, the distance and direction of the food source she has found. If, for the Austrian variety of bee, the food source is less than eighty meters away, she performs a *round dance,* running rapidly around in a circle, first to the left, then to the right. This in effect says to the hive bees: "Fly out from the hive; close by in the neighborhood is food to be fetched."

145

If, on the other hand, the food source is more than eighty meters away, the forager will use the *tail-wagging dance.* The rhythm of the dance tells the distance: the closer the source, the more figure-of-eight cycles of the dance per minute. The tail-wagging part of the dance, shown by the middle wavy line in the diagram, tells the direction, in accordance with a curious rule. On the vertical honeycomb in the hive, the direction *up* means *towards the sun,* and the direction *down* means *away from the sun.* If the tailwagging run points 60° left of straight up, the food source is 60° to the left of the sun, and so on. Directions with respect to the sun have been transposed into directions with respect to gravity, the directions are reported with errors of less than 3°.

This same dance is used in the springtime when half the bees move out of the hive and form a swarm, seeking a new nesting place. Scout bees fly out in all directions, then return and dance to announce the location they have hit on. It is important, of course, that the selected spot be protected from winter, winds, and rough weather, and that there be abundant feeding nearby. The surprising thing is that not just one nesting place is announced, but several at the same time. The dancing and the coming and going can continue for days. By their dances the bees engage in mutual persuasion, inciting one another to inspect this site or that site. The better the site, the longer and more vigorously the returning bee dances. The process continues until all the scout bees are dancing in the same direction and at the same rate. Then the swarm arises and departs for the homesite it has thus decided upon. Mistaken decisions are few.

The dance of the honeybee is symbolic in a genetically determined way. That human language is not genetically determined in the same way is easy to show: the language a child learns, whether Swahili, Cantonese, Urdu, or any other, depends solely on the language of those by whom he is brought up.

The vocabulary of a human language is not genetically fixed. I do not believe, however, that the discussion of the biological foundations of language can properly end at this point.

My reasons for saying this are two. In the first place, there are certain features of human speech which are not found in the natural communication systems of animals, but which are found universally in all known human languages, present or past. The existence of these features is, at the very least, consonant with the possibility that there is a genetic foundation underlying human speech. The facts appear to be most easily accounted for by assuming that there is such a foundation, forcing human speech to be of a certain basic type.

Secondly, this same assumption receives support from the study of primary language acquisition in children. It is not that Psammetichos was right, or that children if left to themselves would commence to speak proto-IndoEuropean or any language resembling an adult human language. All genetically determined traits depend for their appearance, to a greater or lesser degree, on features of the environment. The genes or genetic factors do not of themselves determine body parts or physiological or behavioral traits. Rather, they determine developmental processes, which normally succeed one another in a determinate way, but can be profoundly affected by environmental influence. These facts point to the possibility that genetically determined traits might appear only in the course of maturation, and then only in response to specific influences from outside the organism. Ethologists inform us of many instances of species-specific, genetically based behavior that emerge only in this way. An example is imprinting. Thomas More described it in his *Utopia*. Chicks or ducklings or goslings, a few hours or days after hatching, enter a critical period. Whatever object they first encounter during this period, within certain limits of size, and moving within appropriate limits of speed, they begin to follow, and continue to follow through childhood. The object followed can be, and usually is, the mother; but it can also be an ethologist like Konrad Korenz on his hands and knees, or something stuffed at the end of a stick. Failure to develop imprinted responses during infancy may cause behavioral abnormalities in the adult bird—abnormalities that cannot be corrected by later training. Imprinting is only one of many known species-specific characteristics or behaviors that appear in the course of development, in response to what are sometimes called "releasers", environmental stimuli of specified kinds. It will be my contention that important features of human linguistic capacity are of this kind.

After discussing these two points, I shall conclude with certain reflections on what they might mean.

I begin, then, with three features of human speech that do not appear to be found in the natural communication systems of animals (see Table II):
1. Phonematization
2. Concatenation
3. Grammar

TABLE II
Species-specific Features of Human Speech

1. Phonematization:

"Morphemes":	the smallest meaningful units into which an utterance can be divided. Examples: water spick and span "er" in "whiter", "taller", etc.
"Phoneme":	the smallest distinctive unit of sound functioning within the sound system of a language to make a difference. Examples: /p/ vs. /b/ /t/ vs. /d/
Phonematization:	all morphemes in all natural human languages are divisible into phonemes.
2. Concatenation:	single morphemes are strung together into sequences, rather than being used in isolation.
3. Grammar or Syntactical Structure:	in no human language are morphemes strung together in purely random order. Examples (Chomsky): Grammatical: "colorless green ideas sleep furiously" Ungrammatical: "furiously sleep ideas green colorless"

What is meant by phonematization? The vocalizations heard in the human languages of the world are always within fairly narrow limits of the total range of sounds that humans can produce. We are able to imitate, for instance, the vocalizations of mammals and birds with considerable accuracy, given a little training, but such direct imitations never seem to be incorporated in the vocabularies of human languages. In all human languages, the meaningful units, words, or more strictly speaking, *morphemes,* are divisible into successive, shorter, meaningless sounds called *phonemes. Morphemes* are the smallest meaningful units into which an utterance can be divided. A morpheme can be a single word such as "water"; it can be more than one word as in "spick and span"; and it can be less than a single word, as in the "er" in "whiter," which turns the adjective "white" into a comparative. *Phonemes* are the meaningless sounds into which morphemes can be divided. A phoneme is not, strictly speaking, a single sound, but rather a small class of sounds; it can be defined as the smallest distinctive unit functioning within the sound system of a language to make a difference. Refinements aside, the central fact I wish to convey is this: in all languages, morphemes are constituted by sequences of phonemes. This is a fact that the inventors of the alphabet were probably about the first to come to understand.

The fact could have been different. One can imagine a language in which the symbol for a cat was a sound resembling a miaow; in which size was represented by loudness, color by vowel quality, and hunger by a strident roar. Morphemes in such a language would not be analyzable into phonemes.

All human languages are phonematized, but each language uses a somewhat different set of phonemes, in each case a small set.

Parrots and mynah birds excel other animals in the imitation of human speech, but it is doubtful that they speak in phonemes. The matter could be put to a test. A parrot that had heard only Portuguese, and had acquired a good repertory of Portuguese words and phrases, could be transferred into an environment where he would hear only English, and have the opportunity of repeating English exclamatory remarks. If these remarks emerged with a Portuguese accent, then it would be clear that the parrot had learned Portuguese phonemes, which he proceeded to use in the vocalization of English words. In the opposite case, we would conclude that the parrot had the capacity to imitate sounds accurately, but had not acquired the habit of using phonemes for the production of speech.

In the human child, speech by the same test would turn out to be phonematized.

The second general characteristic of human speech I have listed is concatenation. Human utterances seldom consist of single morphemes in isolation; in no human speech-community are utterances restricted to single morphemes; in all languages, morphemes are ordinarily strung together into sequences. To be sure, the peoples of many, perhaps most cultures, are less garrulous than we; they use language only in certain circumstances and only somewhat sparingly, while we talk a good deal of the time. It is nevertheless true that humans in all speech-communities concatenate morphemes.

The third property presupposes concatenation; it is the property of grammatical or syntactical structure. By "structure" I am going to mean a set of relations that can be diagrammed. In no language are morphemes strung together in purely random order. Native speakers of a language normally agree in rejecting certain utterances as ungrammatical, and in recognizing certain other utterances as grammatical. According to Noam Chomsky, for instance, the sentence "colorless green ideas sleep furiously" is grammatical, though meaningless or nearly so; the concatenation "furiously sleep ideas green colorless," the same words in reverse order, is ungrammatical. The one concatenation admits of a syntactical diagram, the other does not.

It is generally assumed in linguistics that the grammar of a language is completely describable by means of a finite and in fact small set of formal rules. For no natural language has such a description been achieved as yet, otherwise one could program a computer to utter the grammatical sentences in the language. Apparently the mechanism involved in the grammar of a natural language is complex. I shall return to this topic again; the point now is just the *universality* of grammar—a relatively complex kind of system—as a feature of human languages.

All three properties I have described are, so far as the available evidence indicates, without cultural histories. Phonematization, concatenation, grammatical structure, are features of all known human language, past or present. And although languages are always in process of change, it is not the case that these changes follow a general pattern from a stage that can be called primitive to one that can be called advanced. No known classification or analysis of human languages provides any basis for a theory of the development of language from aphonemic, non-grammatical, or simple imitative beginnings.

These facts are consonant with the hypothesis that there is a genetic foundation underlying human speech, forcing it to be of a certain basic type, and in particular, to have the features I have just described. In support of this hypothesis, I take up now the development of language in the child.

The first sound a child makes is to cry. Immanuel Kant says the birth cry has not the tone of lamentation, but of indignation and of aroused wrath; presumably because [the child] wants to move, and feels his inability to do so as a fetter that deprives him of his freedom.

More recently a psychoanalyst has written of the birthcry:

It is an expression of the infant's overwhelming sense of inferiority on thus suddenly being confronted by reality, without ever having had to deal with its problems.

In view of the anatomical immaturity of the human brain at birth, these adult interpretations are rather surprising. No doubt the infant in being born undergoes a rude shock. But crying is a mechanism with a number of important functions; one of the earliest is clearing fluid out of the middle ear, so that the child can begin to hear. The mechanism is ready to operate at birth, and the infant puts it to work. The sound made in crying changes slightly during childhood, but otherwise does not mature or change during one's life. Crying is not a first step in the development that leads to articulate speech; it involves no articulation; the infant simply blows his horn without operating the keys.

A quite distinct sort of vocalization begins at about the sixth or eighth week after birth: little cooing sounds that appear to be elicited by a specific stimulus, a nodding object resembling a face in the baby's visual field. A clown's face painted on cardboard, laughing or crying, will do for a while. The response is first smiling, then cooing. After about thirteen weeks it is necessary that the face be a familiar one to elicit the smiling and cooing. During cooing, some articulatory organs are moving, in particular the tongue. The cooing sounds, although tending to be vowel-like, are not identical with any actual speech sounds. Gradually they become differentiated. At six months they include vocalic and consonantal components, like /p/ and /b/. Cooing develops into babbling resembling one-syllable utterances, for instance /ma/, /mu/, /da/, /di/. However, the babbling sounds are still not those of adult speech.

The first strictly linguistic feature to emerge in a child's vocalizations is *contour of intonation*. Before the sound sequences have determinable meaning or definite phonemic structure, they come out with the recognizable intonation of questions, exclamations, or affirmations. Linguistic development begins not with the putting together of individual components, but rather with a whole tonal pattern. Later, this

151

whole becomes differentiated into component parts. Differentiation of phonemes is only approximate at first and has to be progressively refined. The child is gradually gaining control of the dozen or so adjustments in the vocal organs that are required for adult speech. By twelve months he is replicating syllables, as in "mamma" and "dada". By eighteen months he will normally have a repertory of three to fifty recognizable words.

I have described this development as though mothers were not trying to teach, but of course they normally are. It is nevertheless a striking fact that these stages emerge in different cultures in the same sequence and at very nearly the same ages, and in fairly strict correlation with other motor achievements. Detailed studies have been made of speech acquisition among the Zuni of New Mexico, the Dani of Dutch New Guinea, the Bororo in central Brazil, and children in urban U.S.A.; in all cases, intonation patterns become distinct at about the time that grasping between thumb and fingers develops; the first words appear at about the time that walking is accomplished; and by the time the child is able to jump, tiptoe, and walk backward, he is talking a blue streak. Among children born deaf, the development from cooing through spontaneous babbling to well-articulated speech-sounds occurs as with normal children, but of course the development cannot continue onward into the stage at which adult words are learned through hearing. Among the mentally retarded, these developments are chronologically delayed, but take place with the same correlation between various motor achievements. Given the variety of environmental conditions in these several cases, it seems plausible to attribute the emergence of linguistic habits largely to maturational changes within the growing child, rather than to particular training procedures.

The specific neurophysiological correlates of speech are little known, but that there are such correlates and that they mature as speech develops is supported by much evidence. The human brain at birth has only 24% of its adult weight; by contrast, the chimpanzee starts life with a brain that already weighs 60% of its adult value. The human brain takes longer to mature, and more happens as it matures, including principally a large increase in the number of neuronal connections. A large part of the discernible anatomical maturation takes place in the first two years; the process appears to be complete by about fourteen years of age. By this time the neurophysical basis of linguistic capacity has become localized in one of the two cerebral hemispheres, usually the left. If by this time a first language has not been learned, no language will ever be learned. Speech defects due to injuries to the brain that occur *before* the final lateralization of the speech-function are usually overcome; but if the injury comes *after* lateralization, the speech defect will be permanent.

Capacity for speech does not correlate uniformly with size of brain. There is a condition known as nanocephalic dwarfism, in which humans appear reduced to fairy-tale size; adult individuals attain a maximum height of between two and three feet (see Table III). Nanocephalic

TABLE III

*Comparative Weights of Brain and Body in Humans, Inducing
Nanocephalic Dwarf, Chimpanzees, and Monkeys*

	Age	Body Wt. (kg)	Brain Wt. (kg)	Ratio (body:brain)	Speech Acquisition
Human (male)	2 1/2	13 1/2	1.100	12.3	yes
Human (male)	13 1/2	45	1.350	34	yes
Human (male)	18	64	1.350	47	yes
Nanocephalic dwarf	12	13 1/2	0.400	34	yes
Chimp (male)	3	12 1/2	0.400	34	no
Chimp (female)	adult	47	0.450	104	no
Rhesus monkey	adult	3 1/2	0.090	40	no

dwarfs differ from other dwarfs in preserving the skeletal and other bodily proportions of normal adults. Brain weight in these dwarfs barely exceeds that of a normal newborn infant. The brain weight of the nanocephalic dwarf, given in the middle row, is only a little over a third of that of a two-and-a-half-year-old boy, but the ratio of body weight to brain weight is equal to that of a thirteen-and-a-half-year-old boy. These dwarfs show some retardation in intellectual growth, and often do not surpass a mental age-level of five or six years. But all of them acquire the rudiments of language, including speaking and understanding; they speak grammatically, and can manufacture sentences which are not mere repetitions of sentences they have heard. The appropriate conclusion appears to be that the ability to acquire language depends, not on any purely quantitative factor, but on specific modes of organization of human neurophysiology.

One further point concerning the neurophysiological basis of language. The main evidence here is provided by aphasias (the Greek *aphasis* derives from *a,* "not" and *phanai,* "to speak"). These are failures in production or comprehension of language, resulting from injuries to the brain. And this evidence argues, for one thing, against regarding language ability as being encoded simply in a spatial layout of some kind, say a network of associations in the cerebral cortex. Subcortical areas are involved, as well as cortex. The aphasias most frequently involve, not disruption of associations, but rather disruption of temporal order, affecting either phonemes in the production of words, as in spoonerisms, or words and phrases in the production of sentences. The patient is unable to control properly the temporal ordering of these units, and as a consequence they tumble into the production line uninhibited by higher syntactic principles. In general, the symptom is lack of availability of the right thing at the right time.

Language is through-and-through an affair of temporal patterns and sequences. The neurophysiological organization required for this cannot be simply that of associations. In the making of speech-sounds, for instance, certain muscles have to contract, the efferent nerve fibers innervating these muscles are of different lengths and diameters, and as a consequence the times required for a nerve impulse to go from brain to muscle differ for different muscles. Hence the nerve impulses for the production of a single phoneme must be fired off from the brain at different times, and the sequences of impulses for successive phonemes must overlap in complex ways. In the simplest sequential order of events, it thus appears that events are selected, not in response to immediately prior events, but in accordance with a hierarchic plan that integrates the requirements for periods of time of several seconds' duration. All this patterning in time is thought to depend on a physiological rhythm of about six cycles per second, in relation to which other events are timed. Arrangements of this complexity do not come about by learning. The evidence here, as well as the observations I have already described as to the way voice-sounds develop in children, points to the existence of an innate mechanism for the production of phonemes, one which is activated by a specific input, the appearance of the human face, and which matures in stages.

Could anything similar be argued for competence in syntax, the ability to understand and produce grammatical sentences? Here you will undoubtedly be more doubtful, for surely the grammars of different languages are different. Please recall that the sets of phonemes used in different languages are also somewhat different. The universality of phonematization is compatible with different languages employing different subsets of the humanly possible phonemes. The claim for universality of grammar must be of similar kind. The grammars of human languages are not of just any imaginable kind of ordered concatenation of morphemes. Rather, they derive from a certain subclass of the imaginable orders, a subclass involving phrase structure and what has been called "deep structure". The production of grammatical sentences turns out to pose requirements similar to those necessary for the temporal ordering of phonemes; a serial order in which one element determines the next is insufficient; there has to be hierarchical organization, in which elements connected with one another are separated temporally in the production line.

Let me return now to the description of stages in the primary acquisition of language by a child.

At about the end of the first year of life, the child normally utters his first unmistakable word. For a number of months, while the child is building up a repertory of about fifty words, he utters only single-word utterances. He frequently hears sentences like "Here is your milk," "Shall daddy take you by-by?," and so on, but he will neither join together any two words he knows nor can he be induced to do so on request. Does he lack the memory or the vocalizing power to produce a two-word utterance? The evidence is against these suppositions. Then, roughly between eighteen and twenty-four months, he suddenly and spontaneously begins to join words into two-element phrases: "up baby," "baby highchair," "push car," and so on. What explains the shift?

An important observation at the one-word stage is that these single words are given the intonations or pitch-contours of declarative, interrogative, or hortatory sentences. The single-word utterances seem to function in meaning in the same way as sentences will function later on: "Doggie" might mean, for instance, "There is a dog." When the two-word construction "push car" appears, it is not just two single-word utterances spoken in a certain order. As single-word utterances, both "push" and "car" would have primary stresses and terminal intonation contours. But when they are two words programmed as a *single utterance,* the primary stress and higher pitch come on "car"; and the unity of the whole is indicated by the *absence* of a terminal pitch contour *between* the words and the *presence* of such a contour at the end of the sequence.

What appears to be happening is that the child is by stages increasing his span, his ability to plan or program longer utterances. Grammar is already present in embryo. Further development will be a process of successive increases in span or integration, on the one hand, and progressive differentiation of the parts of utterances on the other.

Imitation plays a role in this process, but it is seldom mere parroting. In Table IV I have listed some imitations actually produced by two children, whom I shall call Adam and Eve; both were about two years old.

TABLE IV

Imitations by Adam and Eve, Two Years of Age

Model Utterance (parent)	*Child's Imiation*
Tank car	Tank car
Wait a minute	Wait a minute
Daddy's brief case	Daddy brief case
Fraser will be unhappy	Fraser unhappy
He's going out	He go out
That's an old-time train	Old-time train
It's not the same dog as Pepper	Dog Pepper
No, you can't write on Mr. Cromer's shoe	Write Cromer shoe

Contentives
Nouns: Daddy, Fraser, Pepper, Cromer;
       tank car, minute, brief case, train, dog, shoe
Verbs: wait, go, write
Adjectives: unhappy, old-time

Functions:
  the possessive inflection *'s*
  the modal auxiliary *will*
  the progressive inflection *-ing*
  the contraction of the auxiliary very *is*
  the preposition *on*
  the articles *the* and *an*
  the modal auxiliary *can*

First note that the imitations preserve the word order of the model, even when not preserving all the words. This is not a logical necessity; it is conceivable that the child might reverse or scramble the order; that he does not suggests that he is processing the utterance as a whole. A second fact to notice is that, when the models increase in length, the child's imitation is a reduction, and that the selection of words is not random. The words retained are generally nouns, verbs, and less often adjectives: words sometimes called "contentives", because they have semantic content; their main grammatical function lies in their capacity to *refer* to things. The forms omitted are what linguists call "functors", their *grammatical functions* being more obvious than their semantic content. The omission of the functors leads to a kind of telegraphic language, such as one uses in wiring home: "Car broken down; wallet stolen; send money American Express Baghdad". In the child's telegraphic utterances, how will the appropriate functors come to be introduced?

While the child engages in imitating, with reductions, the utterances of the mother, the mother frequently imitates, with expansion, the utterances of the child (see Table V). The mother's expansions, you

TABLE V

Adult Expansions of Child Pronouncements

Utterances of Child	*Mother's Expansions* (Additions boxed)
Baby highchair	Baby is in the highchair
Mommy eggnog	Mommy had her eggnog
Eve lunch	Eve is having lunch
Mommy sandwich	Mommy'll have a sandwich
Sat wall	He sat on the wall
Throw Daddy	Throw it to Daddy
Pick glove	Pick the glove up

will note, preserve the word order of the child's sentences, she acts as if the child meant everything he said, and more, and it is the "more" that her additions articulate. She adds functors. The functors have meaning, but it is meaning that accrues to them in context rather than in isolation. The functors tell the time of the action, whether it is ongoing or completed; they inform us of possession, and of relations such as are indicated by prepositions like *in, on, up, down;* they distinguish between a particular instance of a class as in "*the* highchair," and an arbitrary instance of a class, as in "*a* sandwich"; and so on.

How or to what extent these adult expansions of the child's utterances help the child to learn grammatical usage is uncertain. It has been found that immediate imitations by the child of just uttered adult sentences are less frequently well-formed than spontaneously produced utterances. The view that progress toward adult norms arises *merely* from practice in overt imitation of adult sentences is clearly wrong. The child rather appears to be elaborating his own grammar, making use of adult models, but constantly analogizing to produce new and often mistaken words or forms.

TABLE VI
Plural Inflection

Regularization of irregular forms:

Singular	vs.	Plural
mouse		mouses
foot		foots
	or:	
feet		feets
man		mans

Words ending in sibilants

First stage:	*box* (as well as *horse, match, judge,* etc.) treated as both singular and plural
Possible Second Stage:	*bok* vs. *boks*, in analogy with normal "s" pluralization, replaces *box* vs. *boxes*
Third Stage:	after *box* vs. *boxes* is produced, then we also get *foot* vs. *footses, hand* vs. *handses*

Past Tense Inflection

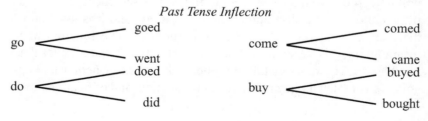

Take pluralization (see Table VI). In English there are a few irregular plurals, as of *mouse, foot, man.* The child normally regularizes these plurals: *mouses, foots, mans.* Instead of *foot* vs. *foots,* some children give *feet* for the singular, *feets* for the plural. One does not get an initial fluctuation between *foot* and *feet,* such as one would expect if only imitation of adult forms were at work.

Most English plurals are regular and follow certain formal rules. Thus we have *mat* vs. *mats,* but *match* vs. *matches.* Words ending in sibilants, such as *match, horse, box,* add a vowel before the *s* of the plural. Children have difficulty with pluralizing these words, and tend at first to use the singular form for both singular and plural. Sometimes a child will analogize in such a way as to remove the sibilant, substituting for instance, for *box* vs. *boxes,* the singular-plural pair *bok* vs. *boks.* Then at some point the child produces the regular plural of a sibilant word, say, *boxes.* Frequently when this happens he may abandon temporarily the regular plural for non-sibilant words, so that one gets *foot* vs. *footses.* What is happening? Overlaid on the child's systematic analogic forms, there is a gradual accumulation of successful imitations which do not fit the child's system. Eventually these result in a change in the system, often with errors due to over-generalizing.

Consider also the past tense inflection, which in English bears considerable similarity to the plural inflection (see Table VI again). There are regular forms like *walk-walked,* and irregular ones like *go-went.* Among the regular verbs, the form of the past depends on the final phoneme of the simple verb: so we have *pack-packed* and *pat-patted.* In the case of past-tense inflection in contrast with pluralization, however, the most frequently used forms are irregular, and the curious fact is that the child often starts regularizing these forms before having been heard to produce any other past-tense forms. Thus *goed, doed, comed* appear among the first past-tense forms produced. The analogizing tendency is evidently very strong.

The occurrence of certain kinds of errors on the level of word construction thus reveals the child's effort to induce regularities from the speech he is exposed to. When a child says, "I buyed a fire car for a grillion dollars," he is not imitating in any strict sense of the term; he is constructing in accordance with rules, rules which, in adult English, are in part mistaken. At every stage, the child's linguistic competence extends beyond the sum total of the sentences he *has* heard. He is able to understand and construct sentences which he cannot have heard before, but which are well-formed in terms of general rules that are implicit in the sentences he has heard. Somehow, genius that he is, he induces from the speech to which he is exposed a latent structure of rules. For the rest of his life, he will be spinning out the implications of this latent structure.

By way of illustration of this inductive process, and of a further stage in the achievement of grammatical competence, let me indicate some aspects of the development of the noun phrase in children's speech (see Table VII). A noun phrase consists of a noun plus modifiers of some kind, which together can be used in all the syntactic positions in which a single noun can be used: alone to name or request something, or in a sentence as subject, object, or predicate nominative. The table at the top gives a number of noun phrases uttered by Adam or Eve at about two years of age. Each noun phrase consists of one word from a small class of modifiers, M, followed by one word from the large class of nouns, N. The rule for generating these noun phrases is given below in symbols: NP is generated by M plus N.

The class M does not correspond to any single syntactic class in adult English; it includes indefinite and definite articles, a possessive pronoun, a demonstrative adjective, a quantifier, a cardinal number, and some descriptive adjectives. In adult English these words are of different syntactic classes because they have very different privileges of occurrence in sentences. For the children, the words appear to belong to a single class because of their common privilege of occurrence before nouns; the lack of distinction leads to ungrammatical combinations, which are marked in the table by an asterisk. Thus the indefinite article should be used only with a common count noun in the singular, as in "a coat"; we do not say "a celery," "a Becky," "a hands." The numeral two we use only with count nouns in the plural; hence we do not say "two sock." The word "more" we use before mass nouns in the singular, as in "more coffee", and before count nouns in the plural, as in "more nuts"; we would not say "more nut." To avoid the errors, it is necessary not only that the privileges of occurrence of words of the class M be differentiated, but also that nouns be subdivided into singular and plural, common and proper, count nouns and mass nouns.

Sixteen weeks after Time I, at Time II, Adam and Eve were beginning to make some of these differentiations; articles and demonstrative pronouns were now distinguished from other members of the class M.

Articles now always appeared before descriptive or possessive adjectives, and demonstrative pronouns before articles or other modifiers.

Twenty-six weeks after Time I, the privileges of occurrence had become much more finely differentiated. Adam was distinguishing

TABLE VII

TIME I: Noun Phases with Generative Rule

A coat	That Adam	Big boot
*A celery	That knee	Poor man
*A Becky	More coffee	Little top
*A hands	*More nut	Dirty knee
The top	*Two sock	
My Mommy	Two shoes	
My stool	*Two tinker toy	

$$NP \rightarrow M + N$$

M   a, big, dirty, little, more, my, poor, that, the, two
N   Adam, Becky, boot, coat, coffee, knee, man, Mommy, nut, sock, stool, tinker, toy, top, etc.

TIME II: Subdivision of Modifier class with Generative Rules

A. Privileges peculiar to articles

Obtained	*Not Obtained*
A blue flower	*Blue a flower
A nice cap	*Nice a nap
*A your car	*Your a car
*A my pencil	*My a pencil

Rule: $NP \rightarrow Art + M + N$ (*Not*: $NP \rightarrow M + art + N$)

B. Privileges peculiar to demonstrative pronouns

Obtained	*Not Obtained*
*That a horse	*A that horse
*That a blue flower	*A that blue flower
	*Blue a that flower

Rule: $NP \rightarrow Dem + Art + M + N$

*Ungrammatical in adult English

descriptive adjectives and possessive pronouns, as well as articles and demonstrative pronouns, from the residual class M; Eve's classification was even more complicated, though she was a bit younger. Also, nouns were being differentiated by both children: proper nouns were clearly distinct from common nouns; for Eve, count nouns were distinct from mass nouns.

Simultaneously with these differentiations, further integrations were occurring: the noun phrases were beginning to occur as constituents in longer sentences; the permissible combinations of modifiers and nouns were assuming the combination privileges enjoyed by nouns in isolation. Thus the noun phrase, for Adam and Eve, was coming to have a psychological unity such as it has for adults. This was indicated by instances in which a noun phrase was fitted between parts of a separable verb, as in "put the red hat on." It was also indicated by substitution of pronouns for noun phrases in sentences, often at first with the pronoun being followed by the noun phrase for which it was to substitute, as in "mommy get it my ladder," or "I miss it cowboy boot."

Whether any theory of learning at present known can account for this sequence of differentiations and integrations is doubtful. The process is more reminiscent of the development of an embryo than it is of the simple acquisition of conditioned reflexes or associations. What is achieved is an open-ended competence to comprehend sentences never before heard, in terms of a hierarchical structure, that embeds structures within structures.

To illustrate, let me use, not a child's sentence, but an example that Chomsky excerpts from the *Port Royal Grammar* of 1660 (see Figure 2). The sentence is: "Invisible God created the visible world." The sentence may be diagrammed as shown in the figure; Chomsky calls these diagrams *phrase markers.* There is a phrase marker for what he calls *surface* structure; this has the function of determining the phonetic shape and intonational contour of the sentence. And there is a phrase marker for what he calls *deep structure;* this shows how prior predications are embedded in the sentence, and determine its meaning.

Are formal structures like the one indicated by this diagram really operative when linguistic competence is being exercised? There are a number of indications that this is so. One indication is the extent to which the understanding of language involves resolution of ambiguities, or disambiguation as it is sometimes massively put. Consider the sentence "They are boring students" (see Figure 3). This has two different interpretations, which are represented by the diagrams of Figure 3. In interpretation A, the word "boring" is linked with the word "students"; the students are thus characterized as boring. In interpretation B, the word "boring" is linked with the word "are," which thus becomes the auxiliary verb in the present progressive tense of the verb "to bore," it is the students who are being bored, by certain other persons designated by the pronoun "they", but otherwise mercifully unidentified. In an actual conversation, the context of meaning would

FIGURE 2
Chomskian Phrase Markers

SURFACE STRUCTURE

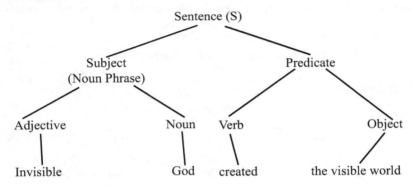

DEEP STRUCTURE

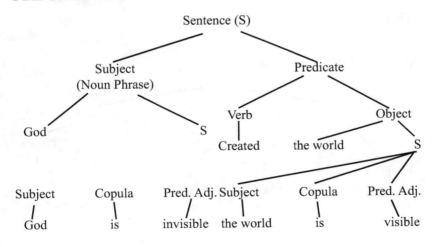

163

have led us to apply, as quick as a thought or perhaps more quickly, the correct phrase marker to the interpretation of the sequence of uttered sounds.

Other examples show how *deep structures* are essential to understanding (see Table VIII). Consider the two sentences:

> John is eager to please.
> John is easy to please.

These sentences have the same surface structure. But a moment's thought shows that the word "John" has two very different roles to play in the two sentences. John in the first sentence is the person who is doing the pleasing; in the second sentence he is the person who is being pleased. John is the underlying *subject* in the first case, and the under-

FIGURE 3
"They are boring Students": Two Interpretations

INTERPRETATION A

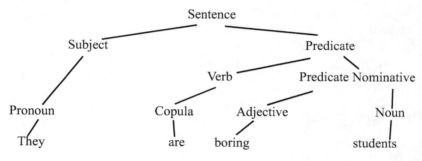

INTERPRETATION B

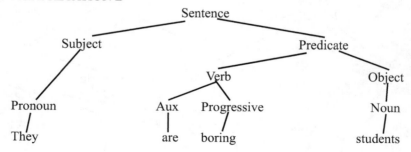

lying *object* in the second case. Deep structure or grammar is involved in understanding the difference in meaning of the two sentences.

An opposite sort of case occurs when the surface grammars of two sentences are different, although the meaning is essentially the same. Consider this sentence in the active mode: "Recently seventeen elephants trampled on my summer home." Now consider the following sentence in the passive mode: "My summer home was trampled on recently by seventeen elephants." A native speaker of English feels that these sentences are related, that they have the same or very similar meanings. Yet their surface structures are very different. Recognition that both sentences are describing the same event presupposes that speaker and hearer refer them both to a single deep structure embodying the single meaning. Something similar happens in recognition of similarity between visual patterns, where there is no point-to-point correspondence between them.

Now all of this is unlikely to seem astonishing, for it is very familiar. You and I, like the *bourgeois gentilhomme,* have been speaking and listening to more or less grammatical prose for a long time now. People

TABLE VIII
Evidence for "Deep" Structure

Surface structure the same, { John is eager to please.
deep structures different { John is easy to please.

Surface structures different, ⌠ Recently seventeen elephants
⎰ *trampled on my summer house.*
deep structures the same ⎱ My summer home was recently trampled on
⌡ by seventeen elephants.

Visual patterns recognized as similar,
although no point-to-point correspondence exists between them.

165

living at the seashore are said to grow so accustomed to the murmur of the waves that they never hear it. Aspects of things that could be important to us may be hidden by their familiarity. The point I have been seeking to make is one that is due to Noam Chomsky, a linguist I have been depending on more than once this evening. The grammaticality of human languages involves properties that are in no sense *necessary* properties of a system that would fulfill the functions of human communication. A grammar, for instance, in which statements would be generated word-by-word, from left to right, so to speak, so that any given morpheme would determine the possible classes of morphemes that might follow it, is a kind of grammar that might have been used, but was not. Instead, human speech involves dependencies between non-adjacent elements, as in the sentence "Anyone who says that is lying," where there is a dependency between the subject noun "anyone" and the predicate phrase "is lying." All operations in human languages, transforming, for instance, an active into a passive sentence, or a declarative into an interrogative sentence, operate on and take account of phrase structure. Example: we form the interrogative of the English sentence, "Little Mary lived in Princeton," by introducing an auxiliary to the verb ("Little Mary *did* live in Princeton"), then inverting the order of the auxiliary and the noun-phrase which is the subject, to get "Did Little Mary live in Princeton?" It would be entirely possible to form interrogatives in a different way independently of phrase structure. There is no a priori reason why human languages should make use exclusively of structure-dependent operations. It is Chomsky's conclusion that such reliance on structure-dependent operations must be predetermined for the language learner by a restrictive initial schematism of some sort, given genetically, and directing the child's attempts to acquire linguistic competence. Put differently, one does not so much teach a first language, as provide a thread along which linguistic competence develops of its own accord, by processes more like maturation than learning.

The Chomskian analysis requires that we take one more step. The fact that deep structures figure in the understanding and use of language shows that grammar and meaning necessarily interpenetrate. The child's grammatical competence matures only along with semantic competence, the organization of what can be talked about in nameable categories and hierarchies of categories. This process, like the development of grammatical competence, involves successive differentiations. Sensory data are first grouped into as yet global classes of gross patterns, and then subsequently differentiated into more specific patterns. The infant who is given a word such as "daddy," and has the task of finding the category labeled by this word, does not start out with the working hypothesis that a specific, concrete object, say his father, uniquely bears this name. Rather, the word initially appears to be used as the label of a general and open category, corresponding to the adult category of *people* or *men*. Infra-human animals are taught with difficulty, if at all, to make the generalizations involved in naming, whereas children fall in with the ways of names automatically. Names, other than proper names, refer to open and flexible classes, which are subject to extension and differentiation in the course of language usage. Categorization and naming involve relations between categories; nothing ever resides in a single term; a means nothing without b and probably c and d; b means nothing without a and c and d. Children go about assimilating the relations that are embodied in language, not merely imitatively, but in an active, inventive, and critical way. They are full of impossible questions:

> "How did the sky happen? How did the sun happen? Why is
> the moon so much like a lamp? Who makes bugs?"

At first, they are ultra-literal in their reactions to idioms and metaphors. When grandmother said that winter was coming soon, the grandchildren laughed and wanted to know: "Do you mean that winter has legs?" And when a lady said "I'm dying to hear that concert," the child's sarcastic response was, "Then why don't you die?" Sometimes reconciliation of adult requirements requires genius. Chukovsky reports that a four-year-old Muscovite, influenced both by an atheist father and by a grandmother of orthodox faith, was overheard to tell her playmate: "There is a god, but, of course, I do not believe in him." The active analogizing and generalizing of four- and five-year-olds is discernible in the odd questions they can put:

> "What is a knife—the fork's husband?"
>
> "Isn't it wonderful? I drink milk, water, tea, and cocoa, but
> out of me pours only tea."
>
> "What does blue look like from behind?"

For a certain period, there is a special, heightened sensitivity to the strangeness of words and their meanings; by age five or six this talent begins to fade, and by seven or eight all traces of it have disappeared. The need has passed; the basic principles of the child's native language have been mastered.

What is it that has in fact been gained? We say, knowledge of a language. But what is a language, *my* language? Thoughtfully considered, this is a well-nigh impossible question, because a language is not a simple object, existing by itself and capable of being grasped in its totality. It exists in the linguistic competence of its users; it is what Aristotle would call an actuality of the second kind, like the soul, or like knowing how to swim when you are not swimming. Through it I constitute myself a first-person singular subject, by using this short word "I," which everyone uses, and which in each seems to refer to something different, yet the same. And through it I am brought into relation with others—the ubiquitous "you"—and with the public thing that is there for both you and me, a treasury of knowledge and value transmitted through and embedded in language.

We hear language spoken of as "living language", and there is evidence enough to make it more than a metaphor. Language reproduces itself from generation to generation, remaining relatively constant, yet with small mutations, enough in fact to account for its growing and evolving, leaving vestiges and fossils behind, and undergoing speciation as a result of migrations, like Darwin's finches on the Galapagos Islands. A change here provokes an adjustment there, for the whole is a complex of relations, mediating between a world and human organisms that are a part of it. The way a word is used this year is, in biological lingo, its *phenotype;* the deep and more abiding sense in it is its *genotype.*

It is we, of course, who are accomplishing all this; but we do not know how we accomplish it. It is mostly a collective, autonomic kind of doing, like the building activities of ants and termites, or the decision-making of bees. It takes generation after generation, but we are part of it whenever and however we utter words or follow them in the sentences that we hear or read, whether lazily or intently, whether with habitual acceptance or active inquiry. Always the words are found for us, and fitted with meanings for us, by agents in the brain over which we exercise no direct control. We can either float with the stream, sometimes a muddy tide of slang and jargon and cliché, or struggle crosstream or upstream. Sometimes we can, sensing the possible presence of a meaning, attempt a raid on the inarticulate; we can launch ourselves into speech, discovering what it is that we mean as we proceed. We "articulate"; the word once meant division into small joints, then, by an effortless transition, the speaking of sentences. There are unexpected outcomes. We may find that our utterance is ungrammatical or illogical; or we may discover that the connection of ideas leads in directions we had not previously considered. In any case, phonetic, syntactical, and semantic structures are being actualized in time, without our quite knowing how. Yet we can strive after that lucidity and precision which, when achieved, make language seem transparent to what there is.

I have already been carried beyond the two propositions I set out to defend, and in doing so, I have moved into a region of ambiguity. The question as to what is determined by nature, independently of us, and what is man-made is an ancient and disturbing question, embedded in old etymologies and myths.

(See Table IX, overleaf.) In more than one language, the word "man" is derived from "earth". So it is in Hebrew: *Adam,* "man," comes from the word for "ground." As shown in the upper diagram, the IndoEuropean root for "earth" gives us "man" and "human" as well as "humus." The notion here is that of the autochthonous origin of humans, their origination from the earth itself; it is a notion found in early cultures all over the world. An implication would seem to be that man is like a plant in his naturalness. On the other hand, as shown in the lower diagram, the IndoEuropean root "wiros," "man" or "the strong one," leads not only to *virile* but, staggeringly, to *world,* suggesting that man makes himself and his world.

The dichotomy, the tension, emerges in the Theban cycle of myths (see Table X, overleaf). Following a suggestion of Levi-Strauss, I am listing elements of it in chronological order from left to right and from the top downward, but in columns, to show the repetition of similar elements. Cadmus is sent off to seek his sister; he kills a dragon, a chthonic monster, that will not permit men to live, and sows the teeth of the dragon in the earth; from the teeth sprout up armed men who kill one another, all except five who become the ancestors of the Thebans. In column I are listed events of the myth in which blood relations seem to be given too much importance. In column II are listed murders of brothers by brother, of a father by a son: here blood relations are brutally disregarded. Column I is thus opposed to column II. In column III, chthonic monsters that were killing off humans are themselves killed by men; we can interpret this as a *denial* of the autochthonous origin of man, an assertion that man has now become self-sufficient, himself responsible for his continued existence. In column IV are listed the meanings of the names of the Labdacidae, including Oedipus; the etymologies all indicate difficulty in walking or in standing upright. In myths throughout the world, this difficulty in walking or standing is

TABLE IX
Some Etymologies

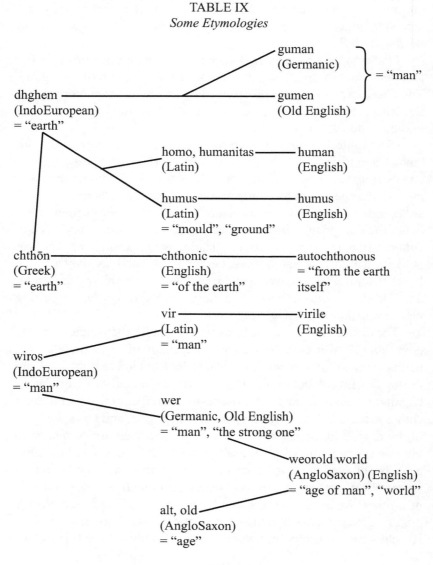

characteristic of the creature that has just emerged out of the earth; the names given in column IV thus constitute an *assertion* of the autochthonous origin of man. Column IV contradicts column III, just as column I contradicts column II. The myth deals with a difficulty of one sort, not by resolving it, but by juxtaposing it to another, parallel type of oppo-

TABLE X

I	II	III	IV
Blood relations overemphasized	Blood relations underemphasized	Chthonic monsters that would not permit men to live are slain by men	Difficulties in walking straight and standing upright
Cadmus seeks his sister Europa, ravished by Zeus		Cadmus kills the dragon	
	The Sparti (the sown dragon's teeth) kill one another		
			Labdacus (Laius' father) = "lame"
	Oedipus kills his father, Laius		Laius (Oedipus' father) = "left-sided"
		Oedipus kills the Sphinx	Oedipus = "swollen-foot"
Oedipus marries his mother, Jocasta			
	Eteocles and Polyneices, brothers, kill one another		
Antigone buries her brother, Polyneices, despite prohibition			

Column I : Column II :: Column IV : Column III

171

sition. Neither man's rootedness in nature nor his transcendence of nature is unproblematic.

The study of language and its acquisition by children indicates that our language has genetic foundations or roots. These, however, have their fruition only under appropriate conditions, only through culture. Man is *by nature* a cultural animal. He does not fabricate his linguistic culture out of whole cloth.

On the one hand, it becomes conceivable that a universal grammar and semantics might be formulated, describing the species-specific features and presuppositions that characterize human linguistic behavior universally. On the other hand, nature's gift of language brings with it an apparent freedom from deterministic necessity not previously present. Most of our sentences are quite new; it is uncommon for one sentence to come out the same as another, though the thoughts be the same. Our utterances are free of the control of detectable stimuli. The number of patterns underlying the normal use of language, according to Chomsky, is orders of magnitude greater than the seconds in a lifetime, and so cannot have been acquired simply by conditioning. While the laws of generation of sentences remain fixed and invariant, the specific manner in which they are applied remains unspecified, open to choice. The application can be appropriate. Articulate, structurally organized signals can be raised to an expression of thought.

Achievement here is subject to change and old laws, and it depends on a sensitivity to old meanings as well as new possibilities. It requires both strength and submission.

BIBLIOGRAPHY

(In the preparation of this lecture I made use of the following books: the book by E.H. Lenneberg, as well as the book edited by him, was particularly useful.)

Emile Benveniste, *Problems in General Linguistics* (Coral Gables, Florida: University of Miami Press, 1971).

Noam Chomsky, *Aspects of the Theory of Syntax* (Cambridge, Mass.: M.I.T. Press, 1965).

_____, *Cartesian Linguistics* (New York: Harper and Row, 1966).

_____, *Language and Mind* (New York: Harcourt Brace, 1968).

David Crystal, *Linguistics* (New York: Penguin, 1971).

Karl von Frisch, *The Dance Language and Orientation of Bees* (Cambridge, Mass.: Belknap Press of Harvard University, 1967).

Kurt Goldstein, *Language and Language Disturbances* (New York: Grune and Stratton, 1948.)

E. H. Lenneberg, *The Biological Foundations for Language* (New York: Wiley, 1967).

E. H. Lenneberg, ed., *New Directions in the Study of Language* (Cambridge, Mass.: M.I.T. Press, 1966).

Martin Lindauer, *Communication Among Social Bees* (New York: Atheneum, 1967).

John Lyons, *Noam Chomsky* (New York: Viking Press, 1970).

Ferdinand de Saussure, *Course in General Linguistics*, ed. Charles Bally and Albert Sechehaye, tr. Wade Baskin (New York: McGraw-Hill, 1959).

B. F. Skinner, *Verbal Behavior* (New York: Appleton-Century-Crofts, 1957).

On Knowing How and Knowing What
(1976)

My lecture this evening concerns the relation between making and knowing; between skill or craft or art–the Greeks called it *technē*–and contemplative knowledge; between knowing how and knowing what. I begin by attempting to sketch, roughly, three different, successive ways in which this relation has been lived or thought about.

First, a primitive stage, paleolithic, pre-agricultural. Certain relics of it, both living and non-living, have persisted into our time and world. Among the relics are, first of all, skeletal remains and tools, implements. Primitive humans were tool-makers. It was an amateur French geologist and antiquary, Boucher de Perthes, who in the 1840s and 50s first began to identify the chipped flint tools of the Old Stone Age, and to defend them for what they were before his disbelieving contemporaries.

Then, since the 1920s, evidence has accumulated that humans were tool makers even before they were human, that is, before they assumed the physical form of present-day *Homo sapiens.* It was not the brain that came first, but upright posture and the hand. When the hominid precursors of the human race came down out of the trees and walked upright upon the plains, they thereby freed their fingers and opposable thumbs for new uses, for the making and deployment of tools, weapons, utensils. The australopithecines of South Africa, with only 500 cubic centimeters of brain, no more than a chimpanzee or gorilla, were already walking erect and using stone implements. The more recently discovered remains of *Homo habilis,* "handy man," as his discoverers named him, show a brain of 800 cc, still not the normal size for *Homo sapiens,* which is 1200 to 1500 cc; but *Homo habilis* is already—three and three quarters million years ago according to Mary Leakey's recent find—an upright-walking tool-maker. The available evidence thus goes to show that the freeing of the hands for tool-making and tool-use preceded most human evolutionary brain enlargement. It looks as though the cerebral enlargement and the increasingly skillful use of the hands went together, the two developments reinforcing each other and yielding evolutionary advantage to the brainiest and handiest who were, so we can plausibly guess, the same.

The stone-age humans, then, came provided with the bodily and psychic equipment for seeing, grasping, and handling objects. We can guess that they were provided also with an exceptional capacity for learning. Coordination of hand and eye in handling objects, ability to learn new ways—these were what made possible the use of sticks and stones as extensions of human limbs. But probably we are thinking so far only of individual capacities; the true acquisition of a kind of tool by a group of hominids or humans implies that the making and using of it can be taught and learned, and so transmitted by tradition. There would have to be a continuing society, capable of transmitting tradition. And this is exactly what the archeological record reveals—continuity of traditions of tool production, lasting through millenia, with but minor changes.

Tool production was socially controlled. The implements of each type are practically identical in any given culture, over long periods and large areas. The hand-axes of Figure 1, for instance, were shaped

Fig. 1[1]

by a fairly elaborate process of chipping, a process that would take anyone of us a pretty long time to learn. The hand-axe is believed to have been a general purpose tool, used. mainly for cutting and scraping, as in skinning game. It seems to have been the predominant tool in the equipment of the early Stone-Age hunters. Its production and use started in southern Africa about a half million years ago, and then spread northward through Africa and Asia Minor and Europe over a period of several hundred thousand years, with only minor variations and improvements in technique. Hand-axes dug up at· sites as widely separated as the Cape of Good Hope and London are indistinguishable except for their being made of different types of rock.

The traditional character of Stone-Age craft is also represented by the paintings made by Upper Paleolithic man in a hundred caves or so

1. Photograph by Didier Descouens. En.wikipedia.org/wiki/File:Biface_Cintegabelle_MHNT_PRE_2009.0.201.1_V2.jpg. The original Figures for this lecture have not survived. The editors have substituted images that correspond to Mr. Wilson's descriptions.

of southern France and northern Spain (Figures 2 and 3). These were

Fig. 2[2]   Fig. 3[3]

started about 30,000 years ago and were kept up for 20,000 years, with increasing refinement and detail and trueness to what is seen.

The paintings were painted, of course, by lamplight and from memory. The ordering of the paintings within the caves—bison, horses, and oxen in the central chambers; deer, mammoth, and ibex in outer areas; rhinoceros, lion, and bear in the farthest recesses—seems to be fairly constant, suggesting tradition-controlled practices. Among the paintings are occasional drawings of men dressed in the skins, horns, and tails of various animals, much in the manner of the medicine men or shamans in North American Indian tribes. This suggests a connection between the paintings and ritual magic, perhaps the preparation for the hunt.

But prehistoric skeletons and artifacts reveal very little indeed as to how the primitive looked out upon his world, and viewed his own role within it. There is another kind of evidence, the whole set of observations of ethnologists on pre-literate, stone-age peoples surviving into the present or recent past. To what extent these peoples have remained untouched by civilizations, past or present, may be uncertain, but certain characteristics appear to be common. The human conmunities are small, going to several hundred at most; within them, everyone knows everyone else. They are self-contained economically. There are no full-time specialists; even the job of shaman is a part-time one; for everyone must help with the task of food getting. The women, to be sure, have different

2. This image is in the public domain. Commons.wikimedia.org/wiki/File:Lascaus,_Megaloceros.JPG.

3. This image is in the public domain. Commons.wikimedia.org/wiki/File:AltamiraBison.jpg

roles from the men, seed gathering, for instance, instead of hunting. The members intermarry and have. a strong sense of solidarity. They think their own ways better than those of others. The Bakouris in central Brazil, for instance, have one and the same word for "we," "our," and "good," and another for "not we," "bad," "unhealth." Men and women within the group are seen as persons, not as parts of mechanical operations. Groupings of people depend on status and role, not on mere practical usefulness. However pressing or demanding the business of survival, the focus is not on mere individual survival, but on the kinship unit, the personal nexus that joins human being, society, and nature in an endless round of birth, growth, decay, and rebirth. The central meaning of things lies in the linking of the deceased to the living and the yet unborn. The moral and sacred order predominates over the merely technical. The useful arts, with all the know-how they involve, are not isolated as merely useful or artful, but like an intense sport or dance or ceremony, form part of the sacred round.

As for knowledge, theory conceived as aiming at universal validity, it is absent. There is nothing for it to be of. There is no concept of a nature within which things happen according to regular, impersonal, cause-and-effect sequences. Natural events are interpreted as part of the communal life. Nature is thought of as replete with spirits, acting by social norms which can be violated only at risk of retribution. The regularity of occurrences remains in the background, does not become a theme. The primitive human is alerted only if the event is a misfortune, or other wise emotionally privileged; and then he traces it to an evil spell, or the enmity of a spirit, or a neglected ritual. But even here there is no rule-like regularity to follow in the interpretation. Everything is particular: this tree, this river, this animal, this man, this spirit. And spirits are capricious. A person may indeed attempt to exercise spiritual power over spirits; if he succeeds, he is a shaman, that is, a technician of the spirit. But woe to him who, having gained status as a shaman, fails in confrontation with another shaman to win the contest in the exercise of shamanistic power; shame, exile, possible insanity await him. The shaman's power is not founded in stable wisdom; its exercise is a risky affair.

Meanwhile, primitive consciousness is filled with knowledge, knowledge connected always in the most intimate way with know-how; knowledge in the mode of acquaintance with kinds of thing, kinds of material, kinds of process-how to coax fire into the hearth, how to use tension and twist to send the arrow hurtling through the air, and so on. Things, materials, and processes are silently recognized in their generic characters, and these recognitions form the tacit background for all the activities of everyday life. But the theoretical knower, he who would bring these things forward out of their tacitness into lucidity and artic- ulation, has not yet appeared.

Second phase. This begins with the invention of agriculture. Knowledge and utilization of the reproductive cycle of plants brings a new kind of independence of external nature, a new set of possibilities and problems. Human life ceases to be parasitic upon the animals and plants that nature happens to provide. Foresight and planning must now extend through one annual cycle to the next. New and quite different techniques replace the old: the sowing of seed, hoeing, reaping, thresh- ing, storing, grinding, baking, brewing. Permanent settlements become possible; people now live in villages. Within a relatively short span of time, considering the hundreds of thousands of years that paleolithic tribes had wandered the earth, within a very few thousand years, be- tween 8000 and 3000 B.C., the agricultural revolution passes into the urban revolution. In the river valleys of the Tigris and Euphrates, the Nile, and the Indus, that which we call civilization first emerges.

Civilization means, in the first place, a number of things added to society: a marketplace, writing, a city, public control of irrigation, pub- lic works. Different civilizations develop away from the forms of folk society in different ways, but in all, there are 'certain features distin- guish ing them from primitive, folk communities. Kinship ceases to be the basis for the organization of society, and is replaced by residence. In other words, the state has come to be. As villages turn into towns, and towns into cities, the members of society come face to face with diversity of beliefs and customs. Personal relations are replaced by im- personal, economic, utilitarian ones. Crafts became full-time occupa- tions, often engaged in under conditions of lowly servitude. The old moral orders may persist in greater or lesser degree, but they necessar- ily suffer in the midst of re cognized diversity. The common result is a state cult that draws into itself various elements of the old cults, in the effort to gain general acquiescence. Already there are those who are taking in hand the management of the moral order; these are the priests. Probably also there are scribes, those who master the calculative and notational skills required for the construction projects of the state and for the keeping of records. In brief, a literate elite has come to be, which separates itself from the world of the rural farmer and the town crafts-man. The separation has been said to be fatal to science. But the literate elite, narrow-minded and self-serving though it often may be, has the function of maintaining the lore of mathematics and astronomy and the calendar. In Babylonia, between the 6th and 3rd century B.C., the so-phistication of mathematical procedure and the accuracy of astronom-ical prediction became astounding.

Still, this was not science in our sense. Egyptian and Babylonian mathematics had nothing to do with ideas, or, in particular, with the idea of nature. At least in the west, this idea of an immanent order in the universe, independent of any arbitrary will, was first clearly artic-ulated by certain wise men, legislators and merchant princes of the 7th and 6th centuries in the commercial republics that the Greeks estab-lished along the coast of Asia Minor. It is worth noting that these wise men express themselves in the language of the administration of justice and of commercial or monetary exchange. Anaximander puts it thus:

> That from which all things are born is also the cause of their
> coming to an end., as is meet, for they pay reparations and atone-
> ment to each other for their mutual injustice in the order of time.[4]

And a century later Heraclitus is saying:

> All things may be reduced to fire, and fire to all things, just as
> all goods may be turned into gold and gold into all goods.[5]

The second part of the statement refers to coinage, which was in-vented about 610 B.C. in Lydia. So human processes and artifacts are used to express the nature of nature.

It is a man from Ionia, too, who first challenges the popular and Homeric notion that the arts were given to men by the gods from the-beginning. "The gods," says Xenophanes, "did not reveal to man all things from the beginning, but men through their own search find in the course of time that which is better." And Anaxagoras says: it is because man has hands that he became wiser than the brutes. Later on, the pre-history of the human race becomes a theme for the Sophists. Among the Greek thinkers generally, beginning with the Ionian philosophers and continuing down to Aristotle, the theories differ in detail and emphasis, but in all of them the past is viewed as

---

4. Quoted in Simplicius, *Physics,* 24.13.
5. Heraclitus, Fragment DK 90.

the history of the progressive humanization of the animal man through the invention of the arts.

Now of all the Greek discussions of the arts, the one that will have the most influence in later times, and against which the initiators of modern science will stage their revolt, is the Aristotelian account. According to Aristotle, art *imitates* or *completes* nature. This formula undoubtedly has more than one meaning and application. One way in which the arts complete nature is in giving rise to leisure, which frees humans for what Aristotle regards as their highest function, the pursuit of knowledge or science. Aristotle says:

> As more arts were invented, and some were directed to the necessities of life, others to recreation, the inventors of the latter were naturally always regarded as wiser than the inventors of the former, because their branches of knowledge did not aim at utility. Hence when all such inventions were already established, the sciences which do not aim at giving pleasure or at the necessities of life were discovered, and first in the places where men first began to have leisure.[6]

But how does art imitate nature? I think the fundamental meaning is as follows.

Things come to be, Aristotle says (and he is quoting a common view), either by nature or art or chance. Chance is an incidental cause; it means that something comes to be that could have come to be by design, but it occurred in fact by accident. Nature and art, on the other hand, are similar to one another, and unlike chance, in that each of them acts for an end.

That nature acts for an end is most obvious, Aristotle says,

> in animals other than man: they make things neither by art nor after inquiry or deliveration. Wherefore people discuss whether it is by intelligence or by same other faculty that these creatures work—spiders, ants, and the like. By gradual advance in this direction we come to see clearly that in plants too that is produced which is conducive to the end. . . . If then it is both by nature and for an end that the swallow makes its nest and the spider its web, and plants grow leaves for the sake of the fruit and send their roots down (not up) for the sake of nourishment,

6. Aristotle, *Metaphysics,* 981b16-24.

it is plain than this kind of cause is operative in .things which come to be and are by nature.[7]

Both art and nature act for the sake of an end, but they differ in that nature is an *internal* principle of motion or change, in that in which it acts, whereas art resides essentially in a subject distinct from the material on which it acts. Art is characterized by a form or idea which is present in the soul of an intelligent being and is used to direct his activity; and this form is the form of the artifact or artful result that the artist or artificer aims to realize in the material. The arts are thus principles of change belonging to an exterior agent, like the idea of health in the physician which causes outside of itself the physical health in the patient. Nature is also a form, but is internal to the natural thing; it is like the doctor doctoring himself. The tree grows by an internal principle, the house is built by the external agency of the builder.

Now art is closer and more familiar to us than nature, for it is that by which we act on the world around us. It therefore serves Aristotle as a precious intermediary for the explanation of what nature is. Nature is less easily knowable to us than art, yet it *is* knowable—so Aristotle claims. To know the nature of a thing, according to Aristotle, we must grasp the *what* of it, its *being-what-it-is*. We can do this, he says, by means of the definition. Through the verbal formula of the definition the intellect knows, has present to it, the *what* of a class of things, say swallows, spiders, or trees. This is undemonstrable but nevertheless graspable knowledge, for according to Aristotle, there are forms *in* things which are knowable. The knowing of them constitutes the starting point for all further knowledge claiming universal validity.

Third phase. The founders of modern science rejected Aristotle's claim, and along with it they rejected the notion of art as imitating nature. They begin, on the contrary, with the assertion that, as between the *products* of nature and the *products* of art, there is *no* essential difference. As Francis Bacon puts it, "men ought . . . to be firmly persuaded that the artificial does not differ from the natural in form or essence."[8] Knowledge comes to be identified with making, cognition with construction. "We know the true causes only of those things that

---

7. Aristotle, *Physics,* 199a21-31.

8. Bacon, *De augmentiis scientarum,* ch.. 2.

we can build with our own hands or intellect,"[9] says Mersenne in the [1630s] , "To men is granted knowledge only of things whose generation depends upon their own judgement,"[10] says Hobbes in the 1640s. Meanwhile, there is a reappraisal of the practices, operations, and know-how of the arts called mechanical. Says Galileo:

> I think that antiquity had very good reason to enumerate the first inventors of the noble arts among the gods, seeing that the common intellects have so little curiosity. . . . The application to great invention moved by small hints, and the thinking that under a . . . childish appearance admirable arts may be hidden is not the part of a trivial but of a super-human spirit.[11]

In the new way of looking at things, it is the *machine* which serves as the model of what can be understood and explained. What is a machine? That may not be so easy to say. The word "machine" derives from the Greek *mēchanē,* which means a contrivance for doing something, an expedient, or a remedy against ills. In antiquity the word in both Greek and Latin came to be applied particularly to devices for lifting weights, levers, pulleys, and the like. Also, in Lucretius's poem, the word *machina* is applied to the entire world in the phrase *machina mundi,* and this phrase reappears in Christian writers of the middle ages, with perhaps the connotation that the world is something made. But the machine that chiefly served as model and inspiration in the new science of the 17th century was a particular machine, invented some three centuries before: the weight-driven or mechanical clock.

Kepler,who in 1604 first introduced detailed mechanism into the theory of the heavens, wrote at the time:

> At one time I believed that the cause that moved the planets was a soul. . . . I now affirm that the machine of the universe is similar not to a divine animated being, but to a clock . . . and in it all the various movements depend upon a simple active

---

9. Marin Mersenne, *Harmonie universelle, Nouvelles observations physiques et mathématiques,* manuscript additions ca. 1638, cited in Robert Lenoble, *Mersenne ou la naissance du méchanisme* (Paris: J. Vrin, 1943), 384.

10. Thomas Hobbes, *De homine,* II, 10.

11. Galileo, *Dialogue Concerning the Two Chief World Systems,* Day Three. Translation by Mr. Wilson. See also the translation by Stillman Drake (Berkeley: University of California Press, 1953), 406-7.

material force, in the same manner that all the movements of the clock are due to the moving weight.[12]

A few years later, Descartes extends the metaphor to nature as a whole.

> There is no difference between the machines built by artisans and the diverse bodies that nature alone composes except the following: the effects of the machine depend solely upon the action of pipes or springs and other instruments which for the reason that they must have same proportion to the hands of those who build them are always so big that their figures and their motions appear visible, whereas the pipes or springs that produce natural effects are generally too small to be perceived by our senses.[13]

It was on the analogy of clocks and mills that Descartes proposed to account for the functioning of all animals as well as the functioning of the human body. Controversy over Descartes's view of animals as machines provoked one controversialist to insist that "every Cartesian, in order to be consistent, should therefore affirm, with the same seriousness with which he affirms it with respect to beasts, that the other human beings who coexist with him in the world are machines."[14] The claim that human beings are simply machines, and not in need of the immaterial rational souls with which Descartes had still seen fit to endow them, is at length asserted gleefully in the 1740s by Lamettrie, who concludes that life is solely for pleasure, becomes enormously corpulent and dies of indigestion at the court of Frederick the Great. But it is an altogether serious claim that Jacques Monod makes, in his recent book *Chance and Necessity,* when he affirms that all living things are chemical machines. In this pronouncement, I take him to be espousing the vast program of research that is molecular genetics.

I now turn to the consideration of certain machines. I shall try to make evident what makes them tick, what principles are involved. Later I shall return to the relation between knowing how and knowing what. I begin with the so-called mechanical clock, the great paradigm of the seventeenth century revolutionaries of science.

---

12. Kepler, *Mysterium Cosmographicum,* 2nd ed. (1621), .201.

13. Descartes, *Principia Philosophiae* (Amsterdam, 1644), 307.

14. P. G. Daniel, *Voyage du Monde de Descartes* (Utrecht, 1732), 343.

Instruments for keeping track of the daily passage of time have been known since very ancient times: wax candles and hemp ropes, certain lengths of which were supposed to burn in a definite time; sundials; sand-clocks; and especially water-clocks, which were in use in early Egyptian civilization and which from Alexandrian times, the third century B.C., onward, often assumed very elaborate forms, with special jackwork-actuating puppetry to mark the passage of the hours. But the clock called mechanical was invented about A.D. 1300. See Figure 4.

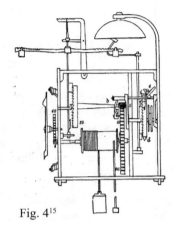

Fig. 4[15]

What you see here is an alarm clock of about 1400 from a monastery in Nuremberg. It rings the bell at settable times, and since the hand travels around the face in sixteen hours, which is exactly the length of the longest winter night in Nuremberg, the presumption is that it was used to wake the sexton, so that he might in turn call the monks to read their nightly offices.

Before going to the heart of the mechanism, let me say a few words about the gearing, which transmits measured amounts of rotational motion from one part of the apparatus to another. There is nothing novel about gear wheels in A.D. 1300. Gearing was used in the windmill, (see Figure 5), to change from a vertical plane of rotation to a horizontal plane of rotation; the windmill of this type was invented in the late twelfth century, the earliest sure date for one being 1185, in Yorkshire. Twelve centuries earlier, similar gearing was already being used in

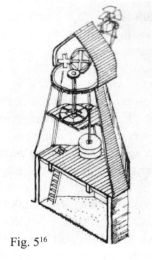

Fig. 5[16]

---

15. From *Abbott's American Watchmaker and Jeweler: An Encyclopedia for the Horologist, Jeweler, Gold and Silversmith* (Chicago: Hazlitt and Walker, 1898), 97.

16. http://www.fiddlersgreen.net/models/buildings/Windmill-Nantucket.html

waterwheels (see Figure 6), invented apparently in the first century B.C.

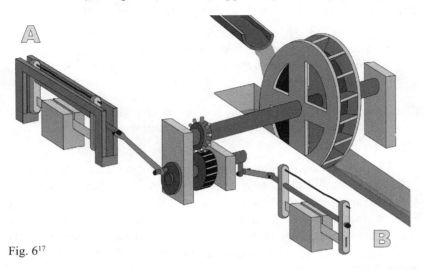

Fig. 6[17]

But even more elaborate gearing was being made as early as the third century B.C., by Archimedes and others, for calendrical computing machines and planetaria. In 1900 the sunken wreck of a Roman ship was found in the Aegean Sea by sponge divers; it has been dated to about 80 B.C. It contained, along with a lot of statuary, presumably destined for sale to the upper fluffy duff of Rome and other Italian cities, a peculiar mechanism of iron, badly rusted. (See Figure 7.) Only in 1972, with the aid of X-radiography, was sense made of it; its gear ratios, it turns out, are based on astronomical constants well-known in antiquity. It is a calendrical computer, which was turned by hand, in order to find out where the sun, moon, and

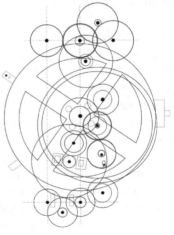

Fig. 7[18]

17. From Tullia Ritti, Klaus Grewe, Paul Kessener: "A Relief of a Water-powered Stone Saw Mill on a Sarcophagus at Hierapolis and its Implications," *Journal of Roman Archaeology* 20 (2007): 148.

18. https://en.wikipedia.org/wiki/File:Antikythera_mechanism.svg

planets would be at given times. Similar types of gearing must have been used in a mechanism described by Cicero. He writes:

> When Archimedes fastened on a globe the movements of moon, sun, and five wandering stars, he, just like Plato's God who built the world in the *Timaeus,* made one revolution of the sphere control several movements utterly unlike in slowness and swiftness. Now if in this world of ours phenomena cannot take place without the act of God, neither could Archimedes have reproduced the same movements upon a globe without divine genius.[19]

There is enough evidence to suggest a long tradition of geared calendar work and planetaria, starting with Archimedes and his contemporaries, transmitted through Islam to the West, and culminating in a number of clock driven planetaria constructed in the middle of the fourteenth century in Europe. As far as the gear-work is concerned, the clock seems to come into being fully fledged as a "fallen angel from the world of astronomy." The gear-work is one source of the clock, deriving from the heavens, but it still needs a terrestrial heart. By this I intend that which makes it, literally, tick. Look once more at Figure 4.

On the left hand, at the top, you see a bell; our word "clock" comes from the French word *cloche,* meaning "bell." Below the bell, and slightly to the left, you see a wheel shaped something like a crown, and therefore called a crown wheel. Around its axle is wound a cord, which goes to a weight that you cannot see. Suspended by a string in front of the crown wheel is a vertical rod, called the *verge,* and at the top of it is rigidly attached a bell clapper. The way this bell-ringing mechanism works will be clearer in a moment. Now this bell-ringing apparatus, it seems, was first used just by itself. You released a catch, the weight began to fall, the crown wheel to turn, and the bell to be hammered by the clapper at the top of the verge. out of one motion, the releasing of the catch, you got several motions, the bell rung several times: a helpful gadget for a sleepy or lazy bell-ringer.

But on the right you see another crown wheel, with another verge suspended in front of it, and atop the verge, a horizontal bar with weights on it, called the *foliot,* meaning "crazy dancer." The invention

---

19. Cicero, *Tusculan Disputations,* 1.25.63.

of the weight-driven clock consisted in seeing that the mechanism of the bell-ringing gadget, here on the left, could be used for a quite different purpose, to solve the problem of making a clock go by means of a weight. What is that problem? In 1271 Robert the Englishman wrote: "Clockmakers are trying to make a wheel that will accomplish a complete revolution each day, but they cannot quite perfect their work." The difficulty was that the weight as it falls tends to accelerate, and so to make the clock go faster and faster. One could of course use a brake or some form of friction to keep the weight falling at a constant rate, but very quickly the rubbing surfaces would wear smooth, so that the speed of the clock would increase. The invented solution was what is called an *escapement*: in the case of the fourteenth-century clock, it is a verge-and-foliot escapement. We shall see best how it works by turning to a simplified diagram. See Figure 8.

Here the gear-work has been eliminated for simplicity's sake, and the crown wheel has been replaced by a wheel with projecting pegs; both may be called 'scape or escape wheels. The previous picture did not show clearly the pallets or little plates that project from the verge above and below, in such a way as to mesh with the indentations in the 'scape wheel. Now suppose the 'scape wheel moving in the direction of the arrow. The peg at the top of the wheel is just striking the upper pallet. The motion of the wheel, and hence of the descending weight, is momentarily checked by the inertia of the system composed of verge, foliot, and the weights on the foliot. Then the driving weight slowly accelerates this system till the peg has pushed the top pallet out of the way, and has set the verge and foliot swinging counterclockwise as seen from above. For a brief moment the driving weight can fall freely. But now the swing of the verge and foliot brings the *bottom* pallet between the pegs of the 'scape wheel; notice that the bottom pallet projects from the verge in a different direction, something over 90° away from the direction of the top pallet. Almost immediately, the peg

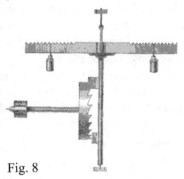

Fig. 8

20. From Pierre Dubois, *Historie de l'Horlogerie* (Paris: Administration du Moyen Age et la Renaissance, 1849), 221.

187

at the bottom of the wheel strikes the lower pallet. Now this peg at the bottom of the wheel has to be moving in the opposite direction from the peg at the top of the wheel, just because of the way wheels are. Hence the counterclockwise swing of verge and foliot is stopped, and the fall of the driving weight slowed again, until the verge and foliot are slowly accelerated into a *clockwise* rotation. Thus the fall of the driving weight is repeatedly interrupted by being compelled regularly to reverse the motion of the verge and foliot with weights. This is an instance of what is nowadays called *negative feedback:* a process produces an effect that slows down and thus regulates that very process. By means of it, the *average* overall motion of the weight, and hence of the clock, is rendered uniform.

So originated the weight-driven clock, through the invention of the escapement, which is literally what makes the clock tick. And this clock, suddenly, toward the middle of the fourteenth century, seized the imagination of the burghers and princes of Europe. Towns vied with towns to have in church or townhall the most elaborate set of planets wheeling, cocks crowing, angels trumpeting, and apostles, kings, and prophets marching and countermarching to the hourly booming of the bells. And also in the middle of the fourteenth century, Nicole Oresme, schoolman, bishop, adviser to the king of France, first enunciated the metaphor of the universe, or at least the supra-lunar part of it, as a vast mechanical clock—a metaphor that would later be extended to the whole world and become a metaphysics.

The fascination was with the mechanical marvel of the thing, with automatic, rhythmically self-acting machinery. Earlier I evaded the problem of defining the word *machine.* We need distinctions here, and with the invention of the clock, the automatic machine which is no longer a tool in the sense of a prosthetic instrument or extension of human limbs, I suggest we would do well to confine the term *machine* to devices that store *energy,* then release it in determinate ways, under various constraints and feedback mechanisms, so that particular purposes are accomplished. I should note that this term *energy* achieved its present-day sense only a little over one hundred years ago; I shall come back to the problem of its meaning.

By the constraints, the stored energy is compelled to bring about certain determinate motions, either desired in themselves, as in the clock, or for the work they can accomplish. The criterion of a good machine is completeness of constraint: the parts of the machine should so connect as to eliminate all but the desired motions.

By this criterion, the fourteenth century clock was not very good, and in fact it needed a little old lady in a black smock to reset it every day. The verge and foliot escapement, in particular, must be criticized because its swing is stopped only by an impact between the pallets and the teeth of the crown wheel, and every such impact brings with it a recoil—a source of extra friction, wear and tear, inaccuracy. Moreover, the swing of the verge and foliot has no proper period of its own; its temporal span depends simply on the successive impulses that it receives, which are unlikely to be exactly equal.

Improvements came. From the fourteenth century onwards, the craft of clockmaking begins to flourish, to develop into skilled instrument-making, a craft combining mathematical know-how with expertness at the lathe and gear-cutting machine. The clockmakers and their offspring, the instrument-makers, will have a very great deal to do with the scientific and industrial revolutions of the seventeenth, eighteenth, and nineteenth centuries. Already in the sixteenth century they were producing tiny spring-actuated watches. But the difficulty about the verge and foliot escapement is met only in the seventeenth century, with two new inventions.

The first of these inventions is Galileo's and Huygens's replacement of the foliot by the pendulum. (See Figure 9.) The verge is shown at the top of the drawing; it is now horizontal, at right angles to the pendulum to which it is rigidly attached. The advantage of the pendulum is that it has an *almost* constant, natural period of swing; the period approaches more nearly to constancy as the amplitude of swing is diminished. This clock is much more accurate than a verge and foliot clock, but unfortunately the verge with its pallets required a 40° swing to clear the teeth of the crown wheel, and with so large a swing, slight differences in amplitude make for noticeable differences in period.

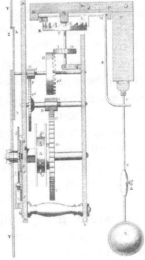

Fig. 9[21]

21. From Christian Huyghens, *Horologium Oscillatorium* (Paris: F. Muguet, 1673), 4.

The second invention reduces the angle of swing. (See figure 10.)

This is the anchor escapement, invented apparently by William Clement about 1670. Only part of the 'scape wheel is shown here. The bent lever above carries the pallets, and rotates from side to side on an axle that is rigidly attached to the pendulum's fulcrum. The arc of swing has now been reduced to 3° or 4°. With this improvement, and continued refinement of all moving metal parts to reduce recoil and friction, eighteenth-century clocks could be made that deviated from their average rate by no more than one tenth of a second per day.

Reduction of friction, elimination of impact and recoil, achievement of the smoothest working and greatest efficiency—these are machine-shop matters. But the concern with them, we shall see, leads to important conclusions: that the universe is not an eternal clock, and that change, not locomotion, is fundamental. This brings me to the second machine I shall examine, the steam engine.

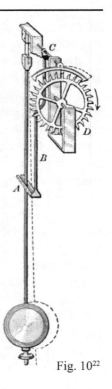

Fig. 10[22]

The first practically successful steam engine was built by a provincial iron-monger, Thomas Newcomen, between 1702 and 1712. There had been various previous efforts to use the expansive force of steam, some of them going back to Hellenistic times. The trouble with these devices was that they did not develop much power. And this they did not do because the metallurgy was not available to make boilers and pipe joints that would hold steam at high pressure. A successful steam engine built around 1700 had to use low pressure steam. The solution was to use it in conjunction with the weight of the atmosphere, which had been discovered by Torricelli and Pascal a half century before. We do not know how Newcomen came by his ideas, but in any case, his engine was an *atmospheric* engine. (See Figure 11.)

This is the 1712 version of Newcomen's engine, hooked up for pumping water from a mine, the main use to which his engine was

---

22. From Silas Ellsworth Coleman, *The Elements of Physics* (Boston: Heath, 1906), 109.

put. Below on the right is the boiler, just beneath the piston cylinder. When steam is admitted to the cylinder, the piston rises to the top, mainly because of the weight of the pump rod hanging from the other end of the rocking beam. Next, the connection between boiler and cylinder is closed, and cold water is sprayed into the cylinder, condensing the steam and so producing a partial vacuum. This allows the atmospheric pressure, acting on top of the piston, to force it back to the bottom of the cylinder, and so raise the pump rod. Then

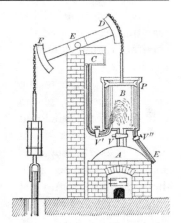

Fig. 11[23]

the next cycle is started by the admission of more steam. Notice that the force stroke is altogether due to the atmosphere; the steam pressure never rises much above one atmosphere of pressure. The working of these engines is said to have been accompanied by an extraordinary amount of wheezing, sighing, creaking, and bumping. They were compared, of course, to living things. They were dreadfully inefficient. By minor improvements, the thermal efficiency was approximately double by the 1770s, bringing it up to what we would now calculate as being about one percent.

More important improvements in efficiency were made by James Watt, during the last quarter of the century. As instrument maker to the University of Glasgow, he was asked in 1763 to repair a small model of a Newcomen engine, and was astonished by the huge quantities of steam required to make it work. Much steam was consumed just in heating up the cylinder, after it had been cooled down in the steam-condensation phase of the cycle. It would be an economy if the cylinder could be maintained always as hot as the steam entering it. "The means of accomplishing this did not inmediately present itself," Watt says; "but early in 1765 it occurred to me that, if a communication were opened between a cylinder containing steam, and another vessel which was exhausted of air and other fluids, the steam, as an elastic fluid, would immediately rush into the

---

23. From N. Henry Black and Harvey N. Davis, *Practical Physics* (New York: Macmillan, 1913), 219.

empty vessel."[24] This was the invention of the separate condenser. (See Figure 12.)

On the right is the boiler; E is the piston cylinder, which is enclosed in a steam jacket; down below it is the separate condenser F, and beside it, the vacuum pump H which is operated by a rod and chain connected to the rocking beam. When the piston is at the top of its stroke, the exhaust valve to the condenser opens, and steam begins to be drawn from the cylinder into the condenser. Steam at about

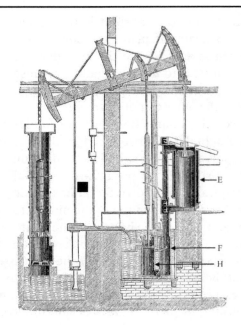

Fig. 12[25]

atmospheric pressure is simultaneously admitted to the cylinder *above* the piston, forcing the piston downward; the advantage of using steam rather than atmospheric air is that the cylinder stays hot. When the piston reaches the lower end of its stroke, the exhaust valve to the condenser is closed, the inlet valve that admits steam above the piston is also closed, and a valve is opened which allows steam to flow from the cylinder above the piston, through a pipe which is to the left of the cylinder, to the cylinder below the piston. The pressure on the two sides of the piston is thus equalized, and the piston rises, being pulled up to the top by the weight of the pump rod. The separate condenser led to about a threefold improvement in efficiency.

Watt's further improvements were aimed not at efficiency but at making the steam engine an effective replacement for the water wheel, in delivering rotary power to factory equipment. (See Figure 13.) I shall not describe this engine, except to point out that it had to deliver power

24. James Watt, cited in John Robinson, *A System of Mechanical Philosophy* Vol. 2 (Edinburgh, 1822), 116.

25. The source of this image is unknown.

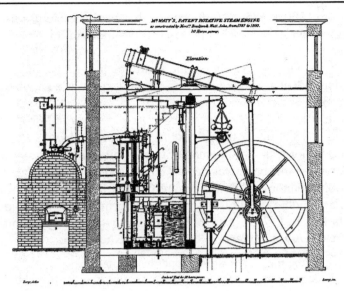

Fig. 13[26]

in both halves of its cycle, and so be double-acting, and this required that the piston rod be rigidly connected to the rocking beam, and at the same time, that it be kept moving in a straight line—no mean problem to solve, but Watt solved it by the invention of .what is called a parallel-motion linkage, and thereby initiated a whole branch of mathematical study. The large flywheel you see at the right helps by its rotational inertia to keep up a smooth delivery of power. Above the flywheel, to the left of its center, you see the centrifugal governor, whose speed of rotation is made to regulate the amount of steam entering the cylinder—another instance, like the clock escapement, of negative feedback.

What about steam engines for railway locomotives and steam boats? Engines for these purposes would have to be less massive than the Watt engine; but if they were to be a good deal smaller and still develop the required power, they would have to use high-pressure steam. Watt had always opposed the high-pressure engine, on the grounds that it was unsafe; and so high-pressure engines did not start to appear until after 1800, when Watt's various patents lapsed. The new engines *were* unsafe; life on the Mississippi and indeed the entire history of steam

26. From John Farey, *Treatise on the Steam Engine* (London: Longman, 1827), Plate XI.

power from 1800 to 1850 was punctuated by appalling explosions. The new engines were also three and more times more efficient than any earlier engines. A variety of experiments were now undertaken to discover what the most efficient engine would be like. On what did efficiency depend? Was there a limit? If so, how could it be approached or attained?

These questions receive their first *general* answer in a small book published in 1824 under the title *Reflections on the Motive Power of Fire*. The author was a young man of twenty-nine named Sadi Carnot. The thinking in this book was deeply influenced by the thinking in another book by another Carnot, Lazare Carnot, famous for his role in the military and political history of France during the 1790s, and also Sadi's father. In [1783], Lazare Carnot had written a book entitled *Essay on Machines in General*. In the preface he states:

> One of the most interesting properties of machines, which, I believe, has not yet been remarked . . . is that in order to make them produce the greatest possible effect, there must necessarily be no percussion, that is to say, that movement should always change by insensible degrees.[27]

This, of course, is an ideal condition which is impossible to attain in practice, and can only be approached. The principle is nevertheless important. It accounts, for instance, for the superior efficiency of an overshot water wheel as compared with an undershot water wheel. In the overshot wheel, the water drops into a bucket at the top of the wheel, and then acts on the wheel simply by its weight, rather than by its motion. In the undershot wheel, the water gains speed by descending along the stream bed, then impacts against the blades at the bottom of the wheel; but a good deal of the possible effect, about half, is lost in the eddies and turbulent motion of the water. Lazare Carnot had a formula for what was being lost; he called it *live force*—it was what we now call *kinetic energy*. And he shows that, in an ideal machine in which all friction, impact, and brusque motion is avoided, all the kinetic energy that is used up can appear as what we now call "work," measured by weight raised through a distance; he uses neither of the terms "work" or "energy," whose strict modern usage dates from the 1850s, but he has the ideas.

---

27. Translated in *The Philosophical Magazine* 30 (1803): 11.

Now the way in which Carnot the younger at first makes use of the elder Carnot's work is as follows. Heat, thought Sadi Carnot, is like water, in that just as water tends of itself to flow downhill, so heat tends of itself to flow from the hotter to the colder body. And just as the water wheel utilizes the live force of the descending water to do work, so, thought Sadi Carnot, the thermal machine does work by making use of the descending heat. Now the conditions for the maximum generation of power from the water wheel were that the water should enter the machine without turbulence and leave with velocity. Similarly, Carnot reasoned, the thermal engine would achieve its maximum effect if all the heat transferred from hot body to cold body had the effect of changing the volume of the gas or steam in the cylinder, and hence causing the piston to move; none of the heat should be permitted to follow its natural propensity of simply flowing from hot body to cold body without further effect. How could this condition be met?

See Figure 14, which shows what Carnot imagined. Let there be a volume V_1 of gas or steam in a cylinder, its pressure being represented in the diagram by the height of the point A. Also, let the cylinder be in thermal contact with a reservoir of heat, such as a steam jacket that can be maintained at a constant temperature θ_1; and suppose the cylinder to be at a temperature only infinitesimally less than θ_1. Heat will then flow from the heat reservoir into the cylinder; the gas will expand; and the piston will move outward. It will move very slowly, of course, because the transfer of heat will be very slow, since the temperature difference between reservoir and cylinder is only infinitesimal. Never mind; we are concerned not with speed but with thermal efficiency, with getting the most for our expenditure on fuel; and while it may take several millenia for the locomotive to progress

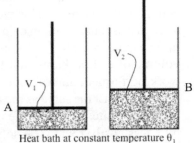

Heat bath at constant temperature θ_1

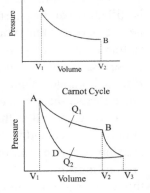

Carnot Cycle

Fig. 14

from here to Glen Burnie [about 17 miles], we are in no hurry, of course, and can do a bit of extra thinking in the interim.

What I have been describing is the isothermal expansion indicated in the diagram by the line from A to B. It is the most efficient of all ways of getting work from heat, so why not use it, letting the gas expand forever? Of course, we would need an infinitely long cylinder, which is an inconvenience. Also, we had better note that as the gas in the cylinder expands, its pressure falls, in accordance with a well-known law called Boyle's law; the falling pressure is indicated in the diagram by the falling of the curve AB from left to right. By-and-by the pressure of the gas will have fallen to the level of atmospheric pressure, and then the piston will stop. So this won't do; what we need, clearly, is a series of processes in which the system is brought back to its initial state; that is, we need a cycle, so that the isothermal expansion can be started over again.

What about simply compressing the gas isothermally, back from B to its initial state A? This won't do, either, because we should have to do just as much work in compressing it as it had originally performed in its expansion. Those of you familiar with plots of pressure against volume of a fluid know that the area under such a curve represents work performed; and of course the area under the curve AB, namely V_1ABV_2, is just the same as the area under BA, the same curve traversed in the opposite direction. We need to return the gas to its initial state by a less costly route. The solution to the problem is called a Carnot cycle. Here is the way of it.

Stop the isothermal expansion at B, while the pressure of the gas is still above atmospheric; remove the cylinder from contact with the. heat reservoir at temperature θ_1, and immediately insulate it thermally, so that no heat can pass in or out; then let the gas expand further. Because heat is not allowed to pass in or out, this further expansion, from B to C, is called an *adiabatic* process. Note that the adiabatic curve is much steeper than the isothermal curve; this means that the temperature is dropping as well as the pressure. Let there be a cold reservoir, containing, say, ice and water at temperature θ_2, and let the adiabatic expansion continue until the gas almost reaches this lower temperature, or is infinitesimally above it. Next place the cylinder in contact with this cold reservoir, and compress the gas from C to D. During the isothermal compression from C to D, we are having to do work on the gas, and heat is flowing out of the cylinder into the reservoir. However, we do less work than we would have to have done to compress the gas at the higher temperature. Finally, compress the gas adiabatically from D to A, so that its temperature rises to the original temperature θ_1, and its pressure and volume assume the original values indicated by the point A. We are now ready to begin a new cycle.

What have we gained? A certain amount of heat has been taken from the hot reservoir; call it Q_1. A certain net amount of work has been done by the engine, say, in raising a weight; call it W. W is the difference between the work the expanding gas does, represented by the area V_1ABCV_3 and the work done *on* the gas in compressing it, namely V_1ADCV_3; evidently the net work is represented by the area of the curvilinear quadrilateral ABCD. For an expenditure of coal or oil or wood yielding the heat Q_1 we have gained the work W. And Carnot asserts that no thermal engine working between the same two reservoirs at temperatures θ_1 and θ_2 could be more efficient.

Carnot proves this assertion, but before showing how he does so I wish to correct an error that his argument contains, one which follows from the analogy of the water wheel. He assumes that all the heat Q_1 that enters the cylinder during the isothermal expansion from A to B also leaves the cylinder during the isothermal compression from C to D; he assumes, in other words, that Q_1 is equal to what I have labelled Q_2. Actually, the energy to do the work W is extracted from the heat Q_1, and so W is equal to Q_1 minus Q_2. This conclusion, or its equivalent, was reached simultaneously by more than a dozen Europeans thinking and experimenting independently during the 1830s and 40s; Sadi Carnot himself reached it before his early death in 1831. Natural philosophers were pushed to it both by a conviction in the unity of nature, and by a variety of observed instances of what we would now call transformations of energy. What is this energy that is being transformed? All that can be said, I believe, is that it is something capable of doing work, capable of raising weight through a distance, and as such capable of being treated quantitatively. There is no single mathematical formula for it. But the postulate that there is this entity called *energy* which is conserved in all the transformations of nature has come to be basic in all scientific accounting, all our dealings with nature, all scientific thought about the economy of nature. It is called the first law of thermodynamics.

Now for the proof, appropriately corrected, of Carnot's theorem. (See Figure 15.) The efficiency of Carnot's engine is given by $\eta = W/Q_1$, where we can measure work and heat in the same units of energy. Let there be, if possible, a more efficient engine, working between the same heat reservoirs, and let it produce the same amount of work W

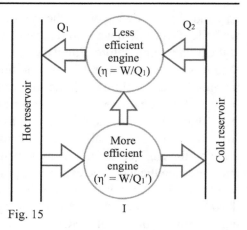

Fig. 15

while extracting from the hot reservoir a smaller amount of heat, Q_1'. Then its efficiency will be $\eta' = W/Q_1'$, where Q_1' is less than Q_1, so that η' is greater than η. Now let us use this more efficient engine to run the Carnot engine in reverse, which we can do, since all the processes that go on in the Carnot engine are reversible. Run in reverse, the Carnot engine becomes a refrigerator; a net amount of work W is put into it; it extracts heat Q_2 from the cold reservoir and rejects the larger amount of heat Q_1 to the hot reservoir. And to run the Carnot engine in reverse, we can use the work W produced by the new and supposedly more efficient engine, which I have labelled I in the diagram. Coupling these two engines together in this way, we obtain a rather peculiar device. There is no net input or output of work. Heat, in amount equal to $Q_1 - Q_1'$, is extracted from the cold reservoir and rejected to the hot reservoir. That is all. This result does not violate the first law of thermodynmanics. Yet surely, Carnot and the physicists who followed him judged, it is impossible; heat of itself does not flow up a temperature gradient. And therefore Clausius and Kelvin in the 1850s formulated a law, the second law of thermodynamics, of which this result would be the violation. As Clausius put it:

> It is impossible to construct a device that, operating in a cycle, will produce no effect other than the transfer of heat from a cooler to a hotter body.[28]

28. This statement is traditionally attributed to Rudolf Clausius, but the editors have not been able to locate it in his published writings.

Let me now try to formulate the main implications of this law and of the reasonings that accompany it.

First, as we can see from the diagram of the Carnot cycle, it is never possible to convert any quantity of heat completely into work, without further effect; some of the heat must always be ejected to a colder reservoir. This means that, even in an ideal heat engine, the efficiency is always less than 100%. I mention in passing that the efficiency depends on the temperatures of the two heat reservoirs, and is improved by raising the temperature of the hot reservoir and lowering that of the Gold reservoir. This accounts in part for the superior efficiency of the high pressure steam engine, which provides a higher difference in temperature between the boiler and the atmosphere.

Second, the Carnot engine represents an ideal limit; no actual engine can reach that limit, or come close to it. The temperature difference between reservoir and cylinder cannot be made infinitesmial. The adiabatic containers never insulate perfectly. Always some of the heat follows its natural propensity and flows from the hotter to the colder bodies without causing any motion of the piston, without doing any work. This means that if a heat engine does some work, raising a weight, say, and if we then undertake to have the weight fall and the engine run in reverse as a refrigerator or heat pump, in an effort to restore the exact initial conditions from which we started, we will not succeed except by investing extra energy .in the process. In this sense, the processes that go on in the heat engine are *irreversible.*

Third, by a series of particular arguments dealing with each kind of process encountered in the world, chemical, electrical, nuclear, and so on, it results that *all* natural processes are irreversible, in the sense just explained. In each case, some heat is dissipated, and the reversal of the process would require that we extract this heat and convert it completely into work done, say into the lifting of a weight, without any further effects ensuing. But this would violate the second law of thermo dynamics. Each process, then, has a *natural* direction, towards a more stable configuration or state. To be sure, any given process may be run in the reverse direction, by making special arrangements,but these always involve the irreversible expenditure of available energy. If we consider all the changes that occur in the surroundings as a result of any process, then the second law assures us that of the total energy with which we began, some will have become unavailable for the production of useful work. As a measure of the transformation of free into unavailable energy, the physicists use a quantity called *entropy,* which increases as the trans formation proceeds. If in any natural process we consider *all* the energy exchanges involved, we find that the net result is an increase in entropy. By computing the changes in entropy in any process, we can determine its natural direction, which is the direction of entropic increase.

From anyone moment, then, to any later moment, the world changes irreversibly. Perhaps we are in some sense aware of this, just in being aware of being alive; but it is a different matter to assert it as a fundamental fact of natural science. A number of physicists during the nineteenth century felt the second law of thermodynamics to be disturbing, and attempted to reduce it to mechanics, that is, to derive it from mechanics. But this cannot be done, for the simple reason that the equations of mechanics are indifferent as to whether time runs backwards or forwards, and therefore irreversibility is not derivable from them. What the physicists in fact did was to construct a kind of *analogue* to thermodynamics, called statistical mechanics. It turns out to be quite as irreducibly and fundamentally statistical as it is mechanical. The kind of statistics used must be chosen so as to fit the system studied, and lead to the known empirical consequences that thermodynamics predicts. In any case, the second law remains, as Eddington called it, "time's arrow," signifying not how fast change will occur, but the overall direction in which it will irrevocably go, toward configurations that we may call more stable.

Thermodynamically, then, the world changes irreversibly; but in special circumstances, it appears that it does so in ways that especially interest us. Let there be, for instance, a sun, radiating energy unremittingly into the unfillable sink of outer space. But in its flow from hot to cold, let 'some of the energy pass by way of an earth, an assemblage of certain chemicals, a temporary trap for the energy in its inevitable entropic descent. In such case, Morowitz has recently argued, with high probability the improbable happens; that is, order arises—symmetry, cyclical transformation, process that has a shape and pattern. Matter which left to itself, in the dark, without a sun to shine upon it, remains inanimate, random, chaotic, now under the surge of solar energy is trans formed into an ordered dance of living forms. Are they forms that we will recognize, feel convivial with, if it comes to be the point of our being introduced? Or is life as we know it a very unique thing, perhaps a species of some more inclusive genus, put nevertheless a quite distinct species? The question is very speculative, but if one examines the delicate balance of conditions our earth has enjoyed up to now, and if one considers the extent to which chance events, events that were not determined mechanistically to happen, have entered irreversibly into biological evolution, then the likelihood that human-like beings exist elsewhere in the universe looks small, nothing worth gambling on. Living systems could have employed right-handed proteins, instead of left-handed ones, and perhaps that would have made little difference. But the evolution of the human brain, into which thousands of irreversible events have entered, has happened only once that we know of; and the alternative possibilities seem countless. A conclusion on which I expect us therefore to agree is that life-stuff as we know it, and the biosphere within which and with which it evolves, are to be cherished as our proper heritage. And thermodynamics, wearing the human smudge and sharing the human smell, is the.economic science that must guide us in the management of this our household, warning us of the irreversible character of our transactions with nature, the finitude of the resources upon which we draw, the ineluctable price of degradation of energy that must be paid for every maintenance or achievement of order or form or value.

I have been engaged in what can have seemed a long digression from my original theme, but I think I am not too far from my starting point; something that also happens with random walkers. Modern science in its inception, I have said, set up for itself the program of science as construction, the sublimation of our age-old capacities for lifting, heaving, pushing, pulling, taking apart, rearranging. Perhaps there is something inescapable about our imagining that program as carried to completion—the completed description of the world and of ourselves as an assemblage of spatially and temporally located, deterministically interacting parts—a machine. Yet surely the result is bizarre, a bad metaphysical dream, a world of bare fact from which problems and persons, learning and knowing and valuing are absent.

If asked to argue against this image on the basis of scientific results, I should say that there are no doubt certain bridges over which the effects of molecular happenings—the deterministic ones and also (please remember!) the chanceful ones—move into our world of sweet and bitter, hot and cold, painful and pleasurable, clumsy and skillful. And I should propose that if chance has acted in the development of the biosphere, it must also be active in the normal functioning of the

201

living body and can be expected to lead to its most significant effects in the functioning of the human brain. Long ago Epicurus, knowing that otherwise knowing and willing were impossible, postulated an alternative to necessity in the swerve of the atoms. The present-day version of that alternative in physics allows us to speculate how decisions and deliberations *utilize* (but do not constitute!) the chance-like forking of the causality of elementary events, how morally and logically a chance of alternative sequence can become significant in allowing us to will yes or no, to give way to or to "stand up to temptation." Of course I do not know this; it is speculation.

Less speculatively, I would redirect your attention from the constructed, or what is assumed to be constructed, to the constructing, that is, the practice of skills that is everywhere entwined in the activity of science. Now these skills involve, to begin with, the use of our body and of tools. Our own body is the only thing in the world that we never normally experience as an object; we experience it rather in terms of the world to which we are attending *from* our body. It is by making this intelligent use of our body that we feel it to be *our* body, and not simply an object. When we adopt a tool for use, we transform it from an object into a sentient extension of our body. Suppose, for instance, we are using a probe to explore a dark cavern. If we are using it for the first time, we feel its impact against our fingers and palm; but as we become accustomed to its use, our awareness of its impact on the hand is transformed into a sense of its point touching the objects we are exploring. We become aware of the feelings in our hand in terms of their meaning located at the tip of the probe or stick to which we are attending. We attend *from* the feelings in our hand *to* their meaning at the tip of the probe. So, in the exercise of this and other skills, there is a tacit background of perception and rule-following, and a focal awareness directed to an object.

Similarly, in the vocal expression of a thought, I rely on an ability to produce syllabic sounds, on an acquaintance with vocabulary and a grammatical skill in stringing words together to form sentences, but all of this muscular and linguistic know-how is tacit and subsidiary to the meaning that I am attempting to convey. All thought contains components—rules that are being followed, perceptions, dispositions to act or respond—on which we depend but of which we are not focally aware. Thought dwells in these components as if they were parts of our body. Thinking is not only *of* something, though it is always and necessarily that; it is also fraught with the roots from which it springs. Like a muscular skill, it has a *from-to* character.

So do we keep expanding our body into the world, by assimilating to it sets of particulars which we intergrate into comprehensive entities. So do we form, intellectually and practically, an interpreted universe populated by entities, the particulars of which we have interiorized for the sake of comprehending their meaning in the shape of the wholes to which they belong.

So do we recognize a problem, Meno's paradox to the contrary notwithstanding; to recognize a problem is to recognize that something is present though hidden; it is to have an intimation of the coherence of hitherto uncomprehended particulars.

So also do we come to recognize a person, in a gesture or in the performance of a skill. Indeed, we cannot recognize a skill unless we understand that we are faced with a coordinated performance, and proceed to pick out the features that are essential to it, the action that is at work within it. So we get to know the intimate parts of a skill and the powers of the person behind it.

Finally, what about objectivity, objective science in the view I am taking? I should answer, first, that we stand on no platform, from which a strictly detached knowing is possible. The zero-point of our history is not accessible to us; even as knowers, we are subject to irreversible time. But standing within our world, the world that has come to be for us, we can once more entertain, following the example of certain Ionians, the idea of knowing as universally valid, true knowledge as a ideal, limiting notice. This will be a moment of suspension of practical activity. It will be a moment of wonder, in which there emerges the idea of the essential whatness of things, their being. To entertain the idea of such knowing is to enter consciously a tradition that is embodied but dormant within us; it is to accept, not a model, but an unfinished and unfinishable task. To seek to uncover the original meaning of this idea is an essential step toward the discovery of what we are.

In a mathematics tutorial, ca. 1948

At a commencement ceremony, ca. 1960

At a question period, sometime suring the 1970s

In a language tutorial, ca.1990

At a Road Scholar event, during the 2000s

In the Coffee Shop, ca. 2010

**Some Additional Essays and Lectures
for the St. John's Community**

On the Origins of Celestial Dynamics
(1971)

I wish to consider two moments in the emergence of celestial dynamics, a Keplerian moment and a Newtonian one, seeking to explore what the development of such a dynamics meant to its authors. Before Kepler, astronomy was a branch of applied mathematics, employing arithmetic and geometry, but having nothing to do with physics or forces (in Greek, *dynameis*). It was Kepler who introduced forces into the heavens, and thus founded celestial dynamics. David Gregory, a follower of Newton, writing in 1702, spoke of the new celestial physics that "the most sagacious Kepler had got the scent of, but the Prince of Geometers Sir Isaac Newton brought to such a pitch as surprizes all the world." Actually, the Keplerian dynamics and the Newtonian dynamics differ in important respects, but Gregory's singling out of Kepler: and Newton makes sense. Kepler introduces a dynamics into the heavens in the sense of hypothesizing a quantifiable influence of one celestial body on the motion of another, and Newton's universal gravitation does the same kind of thing. Moreover, the mathematical results Kepler arrives at by pursuing his hypothesis nearly coincide with Newton's results, derived from a different dynamics.

Meanwhile, in the period intervening between the appearance of Kepler's hypothesis in 1609, and the appearance of Newton's *Principia* in 1687, there were various attempts at proposing what may be called mechanical causes for the celestial motions, but none of them allowed of mathematical formulation, or led to an astronomical calculus, a way of predicting positions of planets. The egregious Thomas Hobbes imagined that, as the southern and northern hemispheres of the Earth differ with respect to the proportion of dry land and ocean, therefore the aethereal vortex or whirlpool that moves about the Sun, having now more solid land to press against and now more of the yielding ocean, would drive the Earth in a path differing from a circle, perhaps approaching an ellipse. Descartes figured out a reason why the suns or stars are off-center in their vortices, so that in the solar vortex the planetary paths are eccentric to the sun, but as in Hobbes's case, the hypothesis did not lend itself to mathematization; on the contrary, because Descartes believed the universe to be packed with vortices inclined at various angles to one another, vortices that fill all space and inter act with one another by transference of matter and motion, any simple mathematical rule for the planetary orbits and motions becomes implausible.

In Kepler's and Newton's cases, we can ask how the dynamical hypothesis and its quantification come about, what they presuppose, what they mean to their authors.

The first sprouts of Kepler's celestial dynamics make their appearance in his first venture into print, his *Cosmographic Mystery* of 1596, published when he was just turning twenty-five. Since April 1594, Kepler had been holding the position of district mathematician in Graz, with the task of teaching mathematics to the boys in a Protestant school, and making up an annual astrological calendar for the province. The calendar was to show when to plant crops, and what to expect of the weather and the Turks. He was, let me mention, marvelously successful with his first calendar: the cold spell he had predicted was so grievous that herdsmen in the mountains lost their lives or their noses from frostbite, and the invasions of the Turks he had predicted were also grievous; the provincial magistrates therefore added a bonus to his stipend. But Kepler was not satisfied with this kind of astrological hackwork. Beginning on the Sunday of Pentecost in 1595, we find him concerned with, and indeed thinking unceasingly about, three large cosmological questions.

At the start of the *Cosmographic Mystery*, Kepler says, "there were three things above all of which I sought the causes why they were thus and not otherwise: the number, size, and motions of the planetary orbs. That I dared this was brought about by that beautiful harmony of the quiescent things, the Sun, fixed stars, and intervening space, with God the Father, the Son, and the Holy Ghost." That is, Kepler sees the spherical layout of the cosmos, with the Sun at the center, and the stars at the periphery, as an image or signature of the triune God, the Creator, His Being, Knowledge, and Love. And with this vision in his head, he makes bold to seek the number, spacings, and motions that the Creator gave to the mobile bodies, the planets, occupying the intermediate space between Sun and stars.

Obviously, Kepler is at this point a Copernican, a heliocentrist. But he does not have a thorough knowledge of the details of Copernicus's planetary theory. As he begins his speculations, he has not read and does not even possess Copernicus's book; he does not even know Rheticus's *Narratio prima*, the book in which, in 1540, three years before the appearance of the *De revolutionibus*, Rheticus had communicated to the world the major outlines of Copernicus's theory and given an account of its superiority over the Ptolemaic theory. Kepler says that he had learned partly from his teacher Maestlin at Tubingen, and partly from his own thinking, the mathematical advantages that Copernicus has over Ptolemy. The Copernican arrangement, simply by its layout, accounts for certain phenomena that are left unaccounted for, are left as coincidences, in the Ptolemaic arrangement. Why do the Sun and the Moon not retrograde, while the other planets do? Why do Mercury and Venus always keep relatively close to the Sun, while the other planets can be at any angular distance? Why are the superior planets always lowest in their epicycles, when in opposition to the Sun? For these questions and a few more, the Copernican arrangement provides an answer; the Ptolemaic does not.

By the time he had finished his Cosmographic Mystery, Kepler had apparently read the famous tenth chapter of Book I of the *De revolutionibus*, where Copernicus says, in his brief commendation of the heliocentric arrangement, "We find in this arrangement a marvelous symmetry of the world and a harmony in the relationship of the motion and size of the orbits, such as one cannot find elsewhere." But even before, Kepler was asking not merely in what the symmetry and harmony consist, but also: On what are they founded? How does man come to recognize them? And already at the start, Kepler has answers to which he will always adhere: The world carries in itself the features of the omnipotent creator and is his copy, his signature. To man, God gave a rational soul, thereby stamping him in His own image. It is with that soul that man can recognize the symmetry and harmony of the Copernican world. Seeing that spherical Copernican world in terms of an idea of Nicholas Cusanus, as a kind of quantitative representation of the indissoluble triune essence of God, Kepler is encouraged to raise and pursue his bold, naive questions.

One of the questions was not new. If you were a Copernican, there were six circumsolar planets, not seven planets as with Ptolemy, since Copernicus leaves the Moon as a satellite of the Earth. Rheticus in his *Narratio prima* had explained this sixfold number by the sacredness and perfection of the number six: six is the first perfect number, i.e., equal to the sum of its factors, 1, 2, and 3. A little later in the sixteenth century, Zarlino will be using this same idea to explain the role of the first six numbers in musical consonances; he will be the first musical theorist to include thirds and sixths among the consonances, as they needed to be included for polyphony's sake. Kepler will be the second such musical theorist, but here as in the case of the number of the planets he will reject the notion of particular numbers as causes. He rejects number-mysticism in that sense. Numbers for him are only abstractions from the created things, and hence posterior to Creation; they could not therefore be used by God as archetypal forms for cosmopoiesis, the making of the world. Not satisfied with Rheticus's answer, Kepler has to face the question afresh: why are there just six planets, no more and no less?

The second and third of Kepler's questions were new.

They had to do with the causes for the relation of the Sun-planet distances to one another, and for the ratios of the planetary periods. In August of 1595 Kepler wrote to Maestlin, his former teacher at Tubingen, telling of his investigations, and asking whether he had ever heard or read of anyone who went into the reason of the disposition of the planets, and the proportions of their motions. In the margin, Maestlin wrote in answer: "No."

Let me remark here that no analogous questions are likely to arise in what can be called, and indeed came to be called, the Ptolemaic system, which was what Kepler had been officially taught at the university. By this term I mean not the set of planetary theories in Ptolemy's *Almagest*, but rather the world picture, current in the Middle Ages and Renaissance, according to which the planetary spheres are nested to fill exactly without remainder the space between the highest sublunary element, fire, and the fixed stars. There is no trace of this picture in the *Almagest*, but in 1967 it was discovered that it is given in Ptolemy's *Hypotheses of the Planets*, the relevant passage having been omitted from Heiberg's standard edition of Ptolemy, apparently from some confusion among the translators; most of the work, including this passage, exists only in Arabic MSS, of which Heiberg gives only a German translation.* What I now say is based on this recovered portion of the *Hypotheses*.

The Ptolemaic system, Ptolemy freely admits, involves conjecture, but he also insists on its plausibility, as did his followers through the

---

*The discoverer was Bernard R. Goldstein; see his "The Arabic Verson of Ptolemy's *Planetary Hypotheses*", *Transactions of the American Philosophical Society* vol. 57, Part 4 (1967).

Middle Ages and Renaissance. Tycho Brahe was still accepting it in the 1570s. The plausibility is as follows. Ptolemy "gives certain arguments in the *Almagest*, and again in amplified form in the *Hypotheses*, for a certain order of the planets, beginning Moon, Mercury, Venus, Sun, and going on to the superior planets; I won't repeat the arguments here. (The Sun, note, is the central one of the seven planets or wandering stars.) He had a very good value for the maximum distance of the Moon from the Earth, determined from observations, namely sixty-four Earth radii. Assume now that the maximum distance of one planet from the Earth is equal to the minimum distance of the planet next above it; take from the Almagest the ratios of nearest approach to farthest distance for each planet, and start constructing outward, using the Ptolemaic order. After the Moon comes Mercury and then Venus. The maximum distance of Venus turns out to be 1,079 Earth-radii, and the Sun is to come next. But there was an independent method for determining the relative distances of the Sun and the Moon, a way invented by Hipparchus, described in the *Almagest*, using eclipses. The method is unreliable, but it did not come to be distrusted till the seventeenth century. The result of that method, reported in the *Almagest*, was that at its closest approach to the Earth, the Sun was 1,160 Earth-radii distant, 81 Earth-radii beyond the high est point of Venus's orb. Is this a big gap? Ptolemy shows in the *Hypotheses*, that by a very slight change in the data of this determination, a change within the limits of observational error, the Sun at nearest approach will be found to use up the extra 81 Earth-radii, and everything fits. Moreover, this is the only order in which the planets can be made to fit in such a sequence of nested spheres, using the Ptolemaic numbers. In further justification Ptolemy adds that "this arrangement is most plausible, for it is not conceivable that there be in Nature a vacuum, or any meaningless and useless thing."

This Ptolemaic system was very well known during the sixteenth century, owing to the description of it in Peurbach's *Theoricae planetarum*, which went through many editions. I suspect it was widely accepted as filling out the heavens, and allowing for the strange motions of these divine beings-motions which, according to Ptolemy, follow from the essence of the planet and are like the will and understanding in man. Copernicus, and also Kepler in the *Cosmographic Mystery*, explicitly reject this system, but I do not think any really forceful argument was made till Kepler showed, some years later, that the

Hipparchic method for the Sun's distance, based on observation of lunar eclipses, and in particular of the width of the Earth's shadow, was practically useless, a small error in the observations leading to an enormous error in the final result.

The question of the reason of the spacing of the planetary orbs does not, then, arise in the Ptolemaic system, because all the available space has been used up in the placing of the orbs. In the Copernican theory,

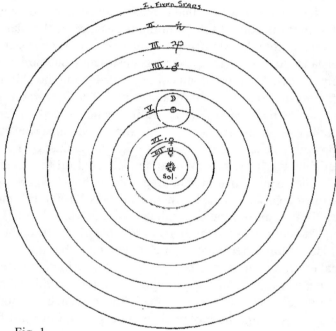

Fig. 1

on the contrary, there are unused spaces, not only a huge one beyond Saturn, separating the solar system from the stars, but also unused spaces between the hoop-shaped regions of space that the individual planets pass through in their motions (Figure 1). That is the effect of the *economy* of the Copernican system, the elimination of the large epicycles. Copernicus speaks of planetary orbs and spheres; whether he believed them to be real or imaginary, solid bodies or merely geometrical figures, remains a subject of scholarly debate. Kepler thought Copernicus believed the spheres to be real and solid, but in the Cosmographic Mystery he is already pointing out some of the difficulties with this conception. By what chains or struts is the Earth with its atmosphere held within the solid spherical shell to which it belongs? We are already in the heavens, and they aren't solid. But in either case, whether the spheres are real or not, there have come to be apparently functionless spaces, and the question can be raised as to the reason of the spacing of the planetary orbs.

Copernicus does not raise this question. He is apparently seeking to redo not the cosmography of Ptolemy's *Hypotheses of the Planets*, an exercise in geometrical arrangement or layout, but the mathematical, predictive astronomy of Ptolemy's *Almagest*, and he wishes to do this job in a manner consistent with the first principles of the astronomical art. A primary principle is that there must be only uniform circular motion; this is required if there is to be strict periodicity, if the motions are not sometimes to fail, owing to their dependence on a changing and thus changeable motor virtue. The intellect abhors such an idea, Copernicus says.

The Copernican insistence on uniformity of circular motion will be taken up by later astronomers—Tycho, Longomontanus, Bullialdus, and others—and echoed for a hundred years and more by both heliocentrists and non heliocentrists. Not only had Ptolemy failed to keep to the principle but new phenomena, discovered since Ptolemy's time, showed that there was an inequality in the precession of the equinoxes that Ptolemy had not suspected. This was called the *trepidation,* supposedly proved by observations of the Arabs collated with those of Ptolemy and Hipparchus. According to one scholar (J. E. Ravetz), it was this supposed phenomenon that pushed Copernicus into setting the Earth in motion. For, argued Ravetz, if the precession of the equinoxes is due to the motion of the stars, if this motion is non-uniform, and if the standard of time by which equality is judged is provided by the diurnal rotation of those very same stars, then the standard of time has been vitiated, and the entire system has become logically incoherent.

The Copernican revolution, Ravetz wanted to argue, was a logical necessity, forced on Copernicus if he was to avoid logical incoherence in the measurement of time. But this is wrong. The truth is that the uniformity of the diurnal rotation would be vitiated slightly whether one assigned the trepidation to the Earth or to the stars; in either case, one would have to *calculate* one's way back to a uniform measure of time—something astronomers had long been doing with respect to the apparent diurnal motions of the Sun. Fortunately, the trepidation is unreal.

Sometime after 1588 Tycho Brahe convinced himself that it is merely the effect of the large errors in the times of the equinoxes that Ptolemy reports in Book III of the *Almagest*; and this is the conclusion of modem astronomy. As for Copernicus, it was not the supposed inequality in the precession, or the problem of measuring time, that led him to cast the Earth into motion.

No, Copernicus's original motive appears to have been opposition to the Ptolemaic equant-that point, not the center of the circle, about which Ptolemy assumes the motion on the deferent circle to be uniform. This violated the first principle of the astronomical art, the assumption of only uniform circular motions. With this idea primarily in mind, Copernicus redoes the *Almagest*. Year after year, from the time he first sketched out his idea until his death, he labored over the revision of numerical constants, trying to obtain an astronomy that would be accurately predictive, fitting all the available, recorded observations. One recent biographer (Arthur Koestler) has judged him to be timorous and myopic. What is more certain is that in his efforts he met with discouragement: he could not get the numbers to come right. And in any case, he is not primarily looking at the emergent system with the eye of a cosmologist; and he is not, like the young enthusiast Kepler in 1595 and 1596, asking for the archetypal, a priori reasons in the mind of God that will account for the layout of the heliocentric world.

Between Copernicus's death in 1543 and 1596, the date of Kepler's *Cosmographic Mystery*, there were very few Copernicans who spoke out. The ill-fated Bruno; a poet or two in the entourage of Henry III of France; Benedetti, Galileo's precursor in mechanics; a mystically minded Englishman named Thomas Digges—they were few.

An overwhelming chorus of denunciation opposed them. Melanchton (1497-1560), Luther's lieutenant and a professor at Wittenberg, referring in 1541 to the Copernican doctrine, said, "Really, wise governments ought to repress impudence of mind." Maurolycus, a very competent and indeed innovative mathematician of Messina, said that Copernicus "deserves a whip or a scourge rather than a refutation" (*Opera Mathematica*, 1575). Pyrrhonist skeptics like Montaigne and his followers were fond of citing Copernicus and Paracelsus to show that there can be found people to deny even the most universally accepted principles. In these references they desired to show that we are so ignorant that it is even excessive to assert that we know that we know nothing. And Tycho Brahe wrote: "What need is there without any justification to imagine the Earth, a dark, dense and inert mass, to be a heavenly body undergoing even more numerous revolutions than the others, that is to say, subject to a triple motion, in violation not only of all physical truth but also of the authority of Holy Scripture, which ought to be paramount" (*Progymnasmata*, 1602). And the list of denunciations could be greatly extended.

Kepler turns out to be one of the early Copernicans, one of a handful, to speak out; he does so before Galileo does, and before his own teacher Maestlin. Maestlin praises Kepler for his first book, saying, "at last a learned man has been found who dared to speak out in defense of Copernicus, against the general chorus of obloquy." And Kepler's defense has a unique character, starting as it does from the notion of the spherical, Sun-centered world as symbol of God, a geometrical reflection of His triune essence, a signature of the Creator in the created world. It is this symbol, Kepler explicitly states, that encourages him to seek the reasons of the number, spacings, and ratio of motions of the planetary orbs. This symbol of God remains central in Kepler's thought; everyone of his major undertakings and achievements can be related to it.

Let me mention in passing that, just as Kepler's question about the spacings is inappropriate to the Ptolemaic system, so it is unlikely to arise for a follower of Tycho's system, which resembles the Copernican except that the Earth remains stationary, and the Sun with the remaining planets moves about the Earth (Figure 2). In letters written in the late

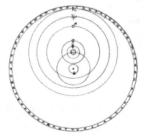

Figure 2

1580s, Tycho says that he was induced to give up the Ptolemaic system by the discovery, from measurements of the parallax of Mars when it is in opposition to the Sun, that it is closer to the Earth than the Sun is. This is possible in the Copernican system, but not in the Ptolemaic; the

Tychonic system accommodates the fact by preserving the Copernican spacings (again, see Figure 2). Actually, Kepler found later that Tycho could not have determined, from his observations, the parallax of Mars; it was too small for observational discrimination by the means at his disposal. And poring over Tycho's MSS, Kepler concluded that some assistant of Tycho had misunderstood instructions and computed the parallax, not" from observation, but from the numerical parameters of Copernicus's system. In any case, if you do accept the Tychonic system, then the path of Mars cuts across the path of the Sun—not impossible, because Tycho knows by now from his study of comets that there are no solid orbs, but still inelegant. And the entire set-up lacks the centered symmetry that provoked the Keplerian inquiry.

The answer Kepler finds to the first two of his questions, concerning the number and spacing of the planets, is well known (Figure 3);

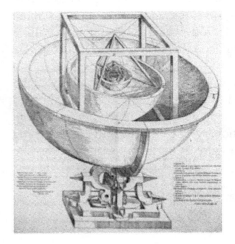

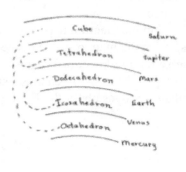

Figure 3   Figure 4

the discovery comes after he has tried many different schemes, and it comes, he tells Maestlin, accompanied by a flood of tears. It is based on the five regular solids or polyhedra. That there are just five polyhedra, with all faces consisting of equal, regular polygons and with all solid angles convex and equal, was one of the discoveries of *Theaetetus*, and the proof of it forms the culmination of Euclid's *Elements*. A beautiful paradigm, this, of completeness of understanding: we can prove that there are these five, and we can see why there are no more.

216

Kepler's answer as to why there are just six planets is a structure in which the regular polyhedra are encased in one another like Chinese boxes, but with spheres in between, and with a sphere circumscribing the largest polyhedron and another sphere inscribing the smallest polyhedron, so that there are six spheres in all (Figure 4). His arrangement is: sphere of Saturn-cube-sphere of Jupiter-tetrahedron-sphere of Mars-dodecahedron-sphere of Earth-icosahedron-sphere of Venus-octahedron-sphere of Mercury. Each solid is inscribed in a sphere which passes through all its vertices, and at the same time has inscribed within it a sphere which touches the centers of all its faces. The structure is not built outward from the Sun: it is built inward and outward from the Earth's sphere, which divides the five regular solids into two groups. The cube, tetrahedron, and dodecahedron Kepler calls primary; each has vertices formed by three edges, each has its own special kind of face-square, triangle, or pentagon. The other solids, called secondary because built out of the primary, are the octahedron and icosahedron; these have their vertices formed by four and five edges, respectively, and have triangular faces. The octahedron is formed from the cube by replacing square faces by the points at their centers; the icosahedron is similarly formed from the dodecahedron. A similar transformation performed on the tetrahedron yields only another tetrahedron.

Kepler therefore speaks of the secondary bodies, octahedron and icosahedron, as offspring of the cube and dodecahedron, respectively; and he calls the latter bodies their fathers, as the chief determiners of their forms. But he also calls the tetrahedron their mother, as the one from whom they receive their triangular faces. The tetrahedron, meanwhile, is hermaphroditic in its production of tetrahedra. Of the primary solids, the cube has to come first, because, Kepler says, it is "the thing itself," meaning, I believe, that it presents to us the very idea of corporification, the creation of body by the regular filling-out of space in the three dimensions. The transformation of cube into tetrahedron is carried out by subtraction, replacing each square face by one of its diagonals; the transformation of cube into dodecahedron is carried out by addition, roofing over the cube, turning each edge into the diagonal of a pentagon.

Out of the 120 possible orders of the five bodies, Kepler can say that he has chosen the one that singles out, as a starting point, the very notion of corporification or the creation of body, that singles out the Earth's sphere as the very special place it is, the home of the image of

God, and that, given these conditions, has the most complete symmetry. And it shows at once why the number of the planets must be just six; there are only five regular solids, as Euclid proves, hence only six circumscribing and inscribing spheres; the number has been deduced from the very idea of the creation of body, of the world, by an ever geometrizing, and let me add, echoing Kepler, a playful God. And man was meant to understand these things. Kepler says:

> As the eye was created for color, the ear for tone, so was the intellect of humans created for the understanding not of just anything whatsoever but of quantities. . . . It is the nature of our intellect to bring to the study of divine matters concepts which are built upon the category of quantity; if it is deprived of these concepts, then it can define only by pure negations.

Thus the five regular solids, the being of which depends on quantitative ratios, form the basis of the layout of the world; and man, the contemplative creature, was meant to see and appreciate this beautiful structure.

But is it true? To know that, we must know that the distances in the construction jibe with the distances determined by the astronomers, and moreover, jibe rather exactly. Kepler at different times expresses the thought that the imposed forms might not fit the world quite exactly, but in that case he hopes to find reasons even for the deviations.

The problem Kepler faces in testing his hypothesis is, first of all, to know which distances to take from the Copernican theory. The sphere of each planet must be of such a thickness as to accommodate the planet's approaches to and recessions from the Sun; but should one, for instance, allow space for Copernicus's equatorial epicycle, which sticks out beyond the planet's path at aphelion? And can one trust Copernicus's theories for Venus and Mercury, which involve some peculiar hypocyclic and epicyclic motions that keep time with the Earth's motion? Moreover, Kepler thinks it incongruous that Copernicus computes the planetary distances from the center of the Earth's orbit rather than from the Sun itself. It is with such considerations that Kepler begins his critique of the details of the Copernican theories. But in disallowing the equatorial epicycles, and in shifting to the real Sun as reference point, Kepler is able to make a preliminary comparison of distances. The ratios for the intervals between Mars and Jupiter and between Venus and the Earth come out with zero percent error; for the

Earth-Mars interval the error is five percent, for the Jupiter-Saturn interval about nine percent. For the Mercury-Venus interval, with Copernicus's numbers, the error is unfortunately twenty percent. Kepler persuades himself-on the ground of Mercury's very unusual situation and motion-that for Mercury the sphere to be used is that inscribed, not in the octahedron itself, but in the three squares formed by the twelve edges of the octahedron-the octahedron is the only regular solid that can be sliced through along its edges in such a way as to yield regular polygons (Figure 5). With this concession, the Mercury-Venus error is

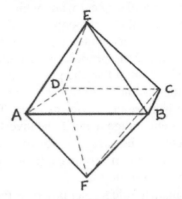

Figure 5: The squares in the octahedron are ABCD, BDEF, AECF.

reduced to two percent; the largest error remains that for Saturn, whose distance is the greatest and therefore most difficult to measure; the next largest error involves the Earth, and Kepler has reason to believe that Copernicus's theory of the Earth is in need of a major revision; and the average error for all the intervals is but 3.3 percent.

Seeing how closely the numbers derived from observation and those derived from his model agree, Kepler has his initial moment of elation; later on, as he calculates, there are doubts, and then again moments of elation. He writes to Maestlin that he suspects a tremendous miracle of God. Older, more cautious, Maestlin, widely known as a competent astronomer, comes to agree with him, comes to suppose that it will be possible to obtain the distances of the planets a priori, from Kepler's model. He assists extensively in the preparation and publication of the book, in which Kepler calls upon all astronomers to help in working out the details of the hypothesis. Among the readers were those who, like Johann Praetorius of Altdorf, said that even if the numbers came out exactly, it would not mean a thing: astronomy should go back to its practical business of predicting the planetary positions on the basis of observations. Tycho's reaction was less hostile: of course there are harmonies, he said, but one must work out the planetary theories on the basis of exact observations first, before investigating the harmonies. Tycho understands here that the theories must employ uniform circular motion, in accordance with the Copernican insistence on that principle; and in contrast to Kepler he assumes that the Earth is at rest.

This brings me to another theory that is contained in Kepler's book, one which Tycho will object to, and which even Maestlin finds, he says, too subtle. From the very beginning, Kepler had had a third question: he had wanted to account not only for the number and spacing of the planets, but for the proportion of their motions. From the very beginning, he had noted that the periods of the planets increase more rapidly than the distances, so that the period of the planet twice as far from the Sun is more than twice as great. This observation had been one of Kepler's encouragements in the investigation of the reason of the distances, because, he says, if God adapted the motions to the orbs according to some law of distances, then surely He also accommodated the distances to some rule.

The first mathematical rule Kepler proposes for the periods is given in a diagram (Figure 6). Note that the diagram is pretty. S is the Sun,

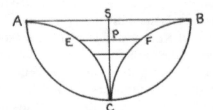

Figure 6

ACB the sphere of fixed stars. AEC and BFC are quadrants of circles with radius equal to the radius of the stellar sphere, SC. To a given distance of a planet from the Sun, SP, Kepler imagines that there would correspond a "vigor of motion" or speed, proportional to the line EF. In the Sun would be the moving soul, and an infinite force of motion; at the periphery are the motionless stars, providing by their distance the space for the planetary motions, and by the *non-uniformity* of their distribution, a background against which the contemplative creature, man, can locate the planets.

The difficulty with the scheme is that Kepler has no clue as to the radius of the cosmos, SC, and without a value for that radius, there is no possibility of calculating the consequences of the hypothesis, and so subjecting it to empirical test. This hypothesis, for Kepler at this point, has a status similar to that of the other one about the five regular solids, in the sense that it arises from the same thought, of the world as symbol of God. The five-regular-solid theory had the assumed fact of spacings to work with; this hypothesis has the assumed *fact* of some kind of inverse relation between distance and speed.

Kepler tries another hypothesis for the motions which is more testable, and in a rough way correct, although it is not the right one (the right one is the third law that he will discover only in 1618). I shall not describe it, but will only remark that here again Kepler is looking for a pure mathematical form, graspable by the mind because mathematical. He is looking for a form which will somehow make the action of the Sun on the planets a symbol of the creating and radiative activity of the Godhead. He will therefore speak of the decrease in motive vigor with increasing distance from the Sun as *suitable;* it was *fitting* that God should have arranged matters thus.

Kepler also begins to compare, the spreading-out of the motive virtue to the spreading out of light from a center; light, as he will say later, is a kind of mediating thing, intermediate between bodies and souls. Kepler is the first to quantify light's intensity, to say that it varies inversely as the square of the distance. It is by a similar quantification of the Sun's motive virtue that he will arrive at his celestial dynamics. He is already onto an important clue to it, in eliminating Copernicus's equatorial epicycle, which was totally incompatible with the five-regular-solid theory, and in thinking about the individual planet as slowing up at aphelion, in some proportion that he is not sure of.

Between the time of completing the *Cosmographic Mystery* in 1596, and going to Prague to work with Tycho Brahe in 1600, Kepler became involved in the study of musical harmony, and a word must be said about this investigation as it relates to his study of the planets. Kepler loved polyphonic music, which he regarded as one of the most important discoveries of modern times, ranking with the compass and printing. In his *Harmonic of the World*, published in 1619, he will write:

> It is no longer a marvel that at last this way of singing in several
> parts, unknown to the ancients, should have been invented by
> Man, the Ape of his Creator; that, namely, he should, by the ar-

tificial symphony of several voices, play out, in a brief portion
of an hour, the perpetuity of the whole duration of the world,
and should to some degree taste of God the Creator's satisfac-
tion in His own works, with a most intensely sweet pleasure
gained from this Music that imitates God.

Kepler refers here to the potentially infinite structure of the polyphonic music he was familiar with; the *Missa Papae Marcelli,* for instance, like a rope of many intertwined strands, might be imagined as
going on indefinitely; nothing in the internal structure requires that it
come to an end at this point or that.

Now for the production of polyphony, one needs to be aiming at
thirds and sixths as consonances; and these intervals involve the ratios
4:5, 5:6, 3:5, and 5:8. The ancient derivation of the consonances, as for
instance in Plato's *Timaeus*, does not treat these ratios as consonances.
The trouble with Plato and the rest, Kepler says, is that they didn't listen carefully enough, before setting out to make their theory. Kepler
sets out to make a new theory, without invoking the causal efficacy of
numbers, or the perfection of the number six (Zarlino, we recall, had
claimed to derive the consonances from just this perfection of the number six). Kepler's solution involves the regular polygons constructible
with straight-edge and compass, which divide the circumference of the
circle into equal parts. If one imagines the circle stretched out into a
straight line, and transformed into a monochord, one has the divisions
giving the consonances required for polyphony, including thirds and
sixths, fundamentally because of the constructibility of the pentagon.

The pentagon depends for its construction on the division of a line
in extreme and mean ratio, the golden section. If you are familiar with
that division, and know how it can be indefinitely reproduced by sub-
tracting the smaller from the greater segment, or by adding the greater
to the whole, you may understand why Kepler views this division as
imaging sexual generation, and you will thus gain an explanation of
the tender feelings that accompany thirds and sixths in polyphonic
music. Kepler did not suppose, and I do not believe that any theorist
before him supposed, that the inquiry into the physical conditions for
the production of certain intervals would account for the shades of feeling that those intervals arouse in consciousness. On the one side we
have instruments like the monochord, from which we can get numbers;
on the other, we have subtle perceptions of harmony, dissonance,
restoration of consonance. There is a strange correspondence between

the soul and the bodily, as for instance, when the interval of a sixth following on certain dissonances triggers a particular perception of sweetness; but the bodily, in Kepler's view, does not account for the psychic in the sense of constituting its intelligible cause.

Kepler's explanation for the correspondence between soul and body takes us back to the sphere, image of the triune God. By creative radiation from the center, one gets the straight line, the element of bodily form, the beginning from which all body comes to be. A straight line, rotated about one of its points, describes a plane, representing in this image the bodily. When the sphere is cut by the plane, the result is a circle, the true image of the created mind, which is assigned to govern the body. As the circle lies both on the sphere and in the plane, so is the mind at the same time in the body, which it instructs, and in God as a radiation which, so to speak, flows from God's countenance. Since now Kepler conceives the circle as the bearer of pure harmonies, and believes these harmonies to be based in the nature of the soul, he comes to speak of the soul as a circle, supplied with the marks of the constructible divisions, the divisions that can be concluded with ruler and compass. It is an infinitely small circle, a point equipped with directions, a qualitative point. This is no doubt a metaphor or symbol, but it is by such means alone that we can understand (insofar as that is possible) how body, soul, and God are related.

The harmonic divisions of the circle apply, of course, in the heavens, as in music; it is from these divisions that Kepler develops his astrological doctrine, and also his harmonic theory for the planetary eccentricities. I cannot take time to describe these here. Kepler comes to see the five-regular-solid theory as inexact, an archetypal form used to determine the number of planets, but not thereafter used in its exact quantitative relations by the Creator, but slightly modified in order to jibe with the harmonic theory of the eccentricities. A playful God, ruled by the necessities of geometry, may be forced to such expedients.

All these parts of Kepler's work are omitted, to say the least, from the corpus of scientific knowledge recognized today. Meanwhile, his great achievement in remaking planetary theory, accomplished first for Mars in the years 1600 to 1605, is praised, sometimes on the mistaken grounds that it is purely empirical. It is not. It involves assumptions that are rejected today. Alternative paths to the so-called Keplerian laws are conceivable, but neither could they have been purely empirical. The empirical evidence is too inexact; some reasoned guesses are required.

Kepler's study first of optics and then of the motions of Mars in the years 1600 to 1605 leads to the development of a possibility already present in his thought. He is the first to quantify the intensity of light, in accordance with the inverse square of the distance from the source. (This is a purely *a priori* derivation, involving no experirnentation.) He does not regard light as material or corpuscular; that would have meant Epicurean philosophy, which like most good Christians of the time he abhorred. Rather, he says, light is quantified according to surface, not according to corporeality. It is one of a group of immaterial emanations, whereby bodies, which are isolated from each other by their bounding surfaces, are enabled to be in communication with one another. The motive virtue issuing from the Sun, Kepler finds, must be another such emanation, distinct from light, for as Kepler discovers in about 1602, its intensity varies *inversely* as the distance from the Sun, not as the *square* of the distance. The empirical support consists in what is known as the bisection of the eccentricity, which he had been able to verify from Tycho's observations in the case of Mars and the Earth.

A further step is taken in 1605 when he discovers that that component of the planet's motion whereby it approaches and recedes from the Sun, can be regarded as simply a libration, or what we would today call a simple harmonic motion: this, he says, smells of the balance, not of mind. By this he means that it is a pattern not chosen for its aesthetic or *mathematical* beauty but determined by the law of the lever and the nature of matter. Here is introduced something that one can perhaps call mechanism: matter turns out to have inertia in the sense of being sluggish, and it turns out to be pushed by an immaterial something in an incomprehensible way. As Kepler clearly realizes, the mechanism or quasi-mechanism could not, in principle, account for everything. It accounts for the actions but it does not account for the initial conditions, the sizes of the orbits and their eccentricities. These must be works of mind, harmonically determined.

Kepler's *Harmonies of the World* (*Harmonices mundi*) of 1619 will remain his final testament. And indeed it is through the spherical symbol, ultimate source of the harmonies that he calls archetypal, that Kepler was first enabled to accept Copernicanism, and then, developing the emanative aspect of the symbol, to banish from the sky the celestial intelligences, the planetary movers of Aristotle and Ptolemy, ultimate relics of paganism (as he calls them), and to regard the planets as material, subject to quantifiable forces that man from his moving platform can measure.

Kepler wanted to dedicate his *Harmonices mundi* to James I of England. For years, very naively from a political point of view, he had looked to this monarch as the hope of Europe, the one who could bring a religious peace out of the strife of Reformation and Counter-Reformation. The relevance of the *Harmonices mundi* to this end was that it was a work of the liberal arts, the arts of peace as Kepler called them, setting forth the principles of the harmonies with which the world had been adorned by its Creator. Kepler thought that, could these things but be seen, men would be raised above the level of doctrinal dispute. But it is doubtful that James I read far into the book. And indeed, no one in the seventeenth century that I know of accepted either Kepler's dynamics as a whole (Leibniz undertook to revamp it), or his harmonic even in part. And as the book first appeared for sale in the market stalls, the Thirty Years War had already begun its terrible course.

Turning to Newton, we will probably not expect to find effusions about the celestial harmonies in his writings. True enough, in the second edition of the *Principia*, explaining his rules of reasoning in philosophy, Newton says that Nature is ever consonant with itself; and so we might imagine it as emitting some single, deep organ tone. But this is from the second edition, 1713, and the first edition, 1687, does not contain the rules of reasoning, at least in their final form, and such as it contains, it labels "hypotheses." We are thus led to suspect that Newton's understanding of his great discovery when he was in the midst of making it, was rather different from the understanding he later came to have of it, when he was defending it before the world.

Shortly after the publication of the first edition, Newton began a series of revisions, pertaining particularly to the early part of Book III. He wrote a series of scholia to accompany those propositions, 4-9, which lead to the establishment of universal gravitation. I wish to quote to you from the proposed scholium to Proposition 8.

> By what proportion gravity decreases in receding from the Planets the ancients have not sufficiently explained. Yet they appear to have adumbrated it by the harmony of the celestial spheres, designating the Sun and the remaining six planets . . . by means of Apollo with the Lyre of seven strings, and measuring the intervals of the spheres by the intervals of the tones. Thus they alleged that seven tones are brought into being . . . and that the Sun strikes the strings. Hence Macrobius says, "Apollo's Lyre of seven strings provides understanding of the

motions of all the celestial spheres over which nature has set the Sun as moderator." And Proclus (commenting) on Plato's *Timaeus*, "The number seven they have dedicated to Apollo as to him who embraces all symphonies whatsoever, and there fore they used to call him . . . the Prince of the number seven." Likewise in Eusebius's *Preparation of the Gospel*, the Sun is called, by the oracle of Apollo, the king of the seven-sounding harmony. But by this symbol they indicated that the Sun by his own force acts upon the planets in that harmonic ratio of distances by which the force of tension acts upon strings of different lengths. . . .

The same tension upon a string half as long acts four times as powerfully, for it generates the Octave, and the Octave is produced by a force four times as great. For if a string of given length stretched by a given weight produced a given tone, the same tension upon a string thrice as short acts nine times as much. For it produces the twelfth [i.e., an octave plus a fifth], and a string which stretched by a given weight produces a given tone needs to be stretched by nine times as much weight so as to produce the twelfth.

Let me briefly review the mathematical relation here (Figure 7). Imagine a series of six strings with length proportional to the distances from the Sun to the six planets; let equal weights be hung on the strings;

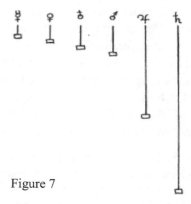

Figure 7

we thus obtain six different tones-very dissonant with one another, let me add, but Newton does not mention the fact. These tones betoken different forces, which can be measured by taking strings of equal lengths and hanging on them different weights, so as to give the same tones. Any two of the weights will be inversely as the squares of the corresponding lengths. Newton continues:

> Now this argument is subtle, yet became known to the ancients, for Pythagoras, as Macrobius avows, stretched the intestine of sheep or the sinews of oxen by attaching various weights, and from this learned the ratio of the celestial harmony. Therefore, by means of such experiments he ascertained that the weights by which all tones on equal strings [were produced] . . . were reciprocally as the squares of the lengths of the strings by which the musical instrument emits the same tones. But the proportion discovered by these experiments, on the evidence of Macrobius, he applied to the heavens and consequently by comparing those weights with the weights of the Planets and the lengths of the strings with the distances of the Planets, he understood by means of the harmony of the heavens that the weights of the Planets towards the Sun were reciprocally as the squares of their distances from the Sun. But the Philosophers loved so to mitigate their mystical discourses that in the presence of the vulgar they foolishly propounded vulgar matters for the sake of ridicule, and hid the truth beneath discourses of this kind. In this sense Pythagoras numbered his musical tones from the Earth, as though from here to the Moon were, a tone, and thence to Mercury a semitone, and from thence to the rest of the planets other musical intervals.
>
> But he taught that the sounds were emitted by the motion and attrition of the solid spheres, as though a great sphere emitted a heavier tone as happens when iron hammers are smitten. And from this, it seems, was born the Ptolemaic System of orbs, when meanwhile Pythagoras beneath parables of this sort was hiding his own system and the true harmony of the heavens.

I have to say: Newton's interpretation of the ancient texts is not a little dubious. Contrary to what all seventeenth-century Copernicans believed, the early Pythagoreans were not heliocentrists; Philolaus, a contemporary of Socrates and the first Pythagorean to write down doctrine (for which he is supposed to have been appropriately punished), did not in fact know the Earth to be round, and his Central Fire was not the Sun. Again, so far as anyone knows today, the law relating weights and string lengths for different musical intervals was first discovered not by Pythagoras but in the late 1580s by Vincenzo Galilei, the father of Galileo Calilei. Indeed, the discovery of this law, which can be verified very precisely if one has a good ear (and Vincenzo was a musician)—this discovery may have been what set Galileo on his course of experimentation, seeking exact numerical ratios in nature; he started with pendulums (again, weights hung on strings), and proceeded to motion down inclined planes, in order perhaps to analyze the motion of the pendulum.

But the incorrectness of Newton's interpretations is not my concern here. The sheer volume of the manuscripts, the many variants and revisions, in all of which Newton is seeking to show that the ancient philosophers before Aristotle understood the Newtonian system of the world, demonstrates that these views were important to Newton. Can we make that fact intelligible to ourselves or must we conclude simply that it is one of the queernesses of genius?

I want to speak briefly about the discovery of universal gravitation. I have recently changed my mind on this matter. My previous argument (which I unfortunately published) was that before 1684 Newton did not have his "proof" of universal gravitation, therefore was uncertain about the universality. I now suspect that before 1684 a good deal more was missing than just the "proof"; I suspect that the idea itself, as a clear and cogent proposal, was not yet present to his mind.

The idea of universal gravitation can seem more paradoxical than we perhaps realize. For a long time, since the 1720s, it was generally thought that Newton already in 1666 had all his principal ideas, and was held up from producing his masterpiece by the lack of a good value for the Earth's radius, or according to a nineteenth-century suggestion, by the lack of a certain mathematical theorem.

That interpretation is supported by no solid evidence whatsoever; there is no sign that Newton entertained the idea of universal gravitation before 1684. And up to 1679, all of Newton's statements about planetary motion imply either Descartes's theory of vortices, and/or an aethereal theory to keep the planets from receding from the Sun.

Newton uses Descartes's term, *conatus recedendi a centro*, the term which Huygens in 1673 replaces by the term centrifugal force. Newton's thought about planetary motion during these years, like that of Huygens, remains confined to Descartes's analogy of the stone in the sling. There is no evidence that before 1679 Newton ever conceptualizes the orbital process as the falling of the planet out of the rectilinear path it would follow if left to itself, a falling towards a central attracting body.

Now this does not mean that during these years Newton altogether rejected the possibility of attractions and repulsions as possible physical causes. He was not a Cartesian; he did not believe space to be identical with matter, and all transfer of motion to be by contact. He was familiar with Cassendi's counter-argument, according to which not everything that is, is substance or accident; thus time and space need not be the accidents of anything, but may independently subsist, and so space need not be the space of something (namely body). This argument may not have satisfied Newton, but given Torricelli's experiment with the barometer, he was willing to grant the vacuum. While this discovery does not in itself lead to the granting of real attractions and repulsions, it opens up the possibility and even the desirability of hypothesizing them. If there are spaces free of matter between the smallest parts of bodies, or the corpuscles of which ordinary bodies are composed, then in order that the parts of these ordinary bodies should cohere and various substances should have the various chemical and physical properties they exhibit, we may well be led to postulate "intermolecular" forces. No doubt, to hypothesize such forces was to depart from the accepted norm of natural philosophy established by Descartes. But Robert Hooke was doing it, and Newton began doing it, speaking of the sociability and unsociability of bodies in chemical reactions and cohesions. The forces he considered seem to have been forces acting over very small distances; his alchemical experiments were probably meant to find out about them.

In 1679 comes the famous exchange of letters between Hooke and Newton, a polite fencing between bitter enemies. Here Hooke explicitly proposes that Newton work out the path of a body under an inverse-square attraction that pulls the body away from its rectilinear trajectory. So far as the evidence goes, this is the first time that Newton faced the planetary problem in such a form. And under this provocation, he makes the great discovery that a force of attraction, directed toward a fixed center, implies the equable description of areas, Kepler's so-called second law. He applies this law, which allows him to use area to represent time, to the ellipse with center of attraction in the focus, and finds that the force follows an inverse-square law.

At one point I thought that it was Hooke who first placed in front of Newton the idea of universal gravitation, so that if Newton had not grasped it before, he did so now, and proceeded to look for a way to test it. But the fact is that Hooke himself did not believe gravitation to be universal, that is, applicable to absolutely *all* matter. He had generalized gravitation more than any previous author. Earlier authors like Kepler had regarded attraction as belonging to *cognate* bodies, that is, closely related bodies like Jupiter and its satellites, or the Earth and its moon. Thus Roberval could talk of a lunar gravity, a terrestrial gravity, a solar gravity, a jovial gravity, and so on.

Let me quote Hooke's view in 1678; he is here explaining an hypothesis about comets:

> I suppose the gravitating power of the Sun in the center of this part of the Heaven in which we are, hath an attractive power upon all the bodies of the Planets, and of the Earth that move about it, and that each of those again have a respect answerable, whereby they may be said to attract the Sun in the same manner as the Load-stone hath to Iron, and the Iron hath to the Load-stone. I conceive also that this attractive virtue may act likewise upon several bodies that come within the center of its sphere of activity, though 'tis not improbable also but that as on some bodies it may have no effect at all, no more than the Load-stone which acts on Iron, hath upon a bar of Tin, Lead, Glass, Wood, etc., so on other bodies, it may have a clean contrary effect, that is of protrusion, thrusting off, driving away . . . ; whence it is, I conceive, that the parts of the body of this Comet (being confounded or jumbled, as 'twere together, and so the gravitating principle destroyed) become of other natures than they were before, and so the body may cease to maintain its place in the Universe, where it was first placed.

Now Hooke is an inductivist of a sort, but induction is not here leading to universal gravitation. That is, Hooke is not concluding that every particle of matter attracts every other in exactly the same way. In his correspondence with Newton in the following year, Hooke suggests that Newton may be able to think of a cause of the gravitating principle: now in Hooke's understanding—and I think in Newton's, too—to say that was to imply that gravitation is not universal, for the material cause of gravitation could not itself be subject to gravitation.

In view of the passages cited and others I shall refer to later, I suspect that the idea of a truly universal gravitation became effectively present to Newton only after he had discovered the "proof." Why propose a theory which, by its very nature, precludes any mechanical explanation, which seems to preclude being tested, and which, moreover, as Newton actually suggests once he has begun to entertain it, would seem to put the calculation of a planetary orbit beyond the power of any human mind?

There is the problem, also, of explaining Newton's delay for five more years after 1679. The best explanation, I believe, is that Newton does not yet think he has discovered anything very important, and sees no direction in which to pursue his discovery. Then Halley appears, probably in August of 1684, and persuades him that his discovery of the logical relation between the inverse-square law and the Sun-focused elliptical orbit is important, and that he should publish it, to secure the invention to himself. Newton sets to work, and we have a series of MSS which can be arranged in temporal order on the basis of internal evidence.

In the first manuscript, there is no sign of the notion of universal gravitation. Newton speaks of gravity as one species of centripetal force-the term "centripetal force" making its first appearance here (it is Newton's invention). There is no hint of the problems of perturbation, the disturbance of the orbit and motion of one planet by the attraction of another planet. The inverse-square law is derived from Kepler's third law as applied to the planets and to the satellites of Jupiter and Saturn, that is, from the fact that, for both the satellites and the circumsolar planets, the squares of the periods are as the cubes of the mean distances from the central body. Newton shows that the revolving bodies must be subject to a centripetal force toward the central body which varies inversely as the square of the distance. The orbits are simply said to be elliptical. The entire development, I believe, is up in the air, in the sense that Newton does not know the cause of the attraction, does not know how exact Kepler's third law may be (he had questioned its exactitude at an earlier date), and is merely proceeding mathematically without knowing what may underlie his derivation of the inverse square law; it could be something that might lead to the results need ing to be qualified.

To mention just one possible explanation, one that Newton had thought up in the 1660s and proposed to the Royal Society in 1675: the action of the Sun on the planets might be due to the inrushing of a subtle aether, which would serve as fuel for the Sun's burning. A similar but different aether might be rushing into the Earth to produce terrestrial gravity; this aether might be transformed chemically within the Earth, then issue forth as our atmosphere. The satellite systems of Jupiter and Saturn might be sustained by similar circulations of aether.

These several centripetal forces would be explicable mechanically, that is by impacts; gravitation would not be universal, for the in-rushing aether would not itself be subject to the forces it caused in other bodies.

In the second manuscript the notion of perturbation appears.

Newton is now assuming that all the bodies of the solar system attract one another, just as Hooke had before. Can the planetary orbits still be said to be elliptical? Hardly, if the ellipses are drawn badly out of shape by the perturbations, the attractions of the different planets toward one another. What must be done is to evaluate the relative magnitude of these perturbations. How is that going to be possible?

Newton does it by considering the accelerations of the satellites of Jupiter towards Jupiter, of the Moon towards the Earth, of Venus towards the Sun. Each satellite is being accelerated towards the body round which it goes, and that acceleration depends on the *power* of the central body to attract, and so may be able to serve as a measure of that power. Of course, to be comparable measures, all three satellites ought to be at the same distance from their central body, and they aren't. But we can shift them in thought to the same distance, by using the inverse-square law. What we get, then, are the comparative attractive powers of Jupiter, the Earth, the Sun. That of the Sun is overwhelmingly larger than the others.

But do we really have attractions here or not? Thus far there has been no evidence that Newton's aethereal theory for the planets is wrong. What then happens, I think, is that Newton realizes a consequence of something he has been assuming. In his derivation of the comparative attractive powers of Jupiter, the Earth, etc., he has been assuming that the quantity of matter of the satellite or test body didn't (if you will forgive a pun) matter; it didn't matter what mass it had, it was accelerated to the same degree anyhow, the differences between the masses of the test bodies could be ignored. Is that right?

Is it so on the Earth? Did Newton know the downward acceleration of all bodies on the Earth, at a given place, to be the same? Not at this moment. Earlier we know he had assumed the rates to be slightly different for different bodies, depending on their micro-structure, and the way the down flowing aether affected them. Now, in the third MS, Newton sets out to test the constancy, and this is the most precise experiment reported in the *Principia*. He takes equal weights of nine different materials; en closes each of them—gold, salt, wool, wood, and

232

so on—in boxes of equal size and shape, to make the air resistance the same; and uses these boxes as the bobs for nine different pendulums, with very long but equal suspensions. The pendulums, he says, played exactly together for a very long time. The accelerations of these different materials, he concludes, cannot differ from one another by more than one part in a thousand. Essentially the same experiment, the Eötvös experiment, has been performed in this century with a precision of one part in one billion. Another way of stating the result, you may know, is that inertial mass is proportional to weight.

At this point in the manuscript series, there appears for the first time in history, so far as I know, a statement of Newton's third law of motion, the equality of action and reaction. Let me now put these two results together—Newton's Eötvös experiment, and his third law, as they are put together in the *Principia*. The first implies that bodies on the Earth are accelerated downward by a force that is strictly proportional to what Newton now calls their *mass,* by which he means their resistance to being accelerated. (If the proportionality had not been exact, the pendulums would not have played together, would not have had the same periods.) If the same thing holds with respect to Jupiter, with respect to Saturn, and with respect to the Sun, then one can compare the attracting powers of these different bodies in the way we have already seen: by taking a test body, it doesn't matter of what mass, placing it at a fixed distance from the attracting body, and seeing how much it is accelerated. Newton couldn't do this physically, as we've said, but assuming the inverse-square law he could find from the actual acceleration of a body at one distance what the acceleration would be if the satellite were placed at any stipulated distance.

Now comes the final step. Since the mass of the test body can be ignored, in the comparison of the attracting forces of two bodies, one can use each as a test body for the other. Then

$$\frac{\text{A's power of attraction}}{\text{B's power of attraction}} = \frac{\text{accelleration of B}}{\text{accelleration of A}}$$

By the third law of motion, these accelerations are in versely as the inertial masses:

$$\frac{\text{accelleration of B}}{\text{accelleration of A}} = \frac{\text{mass of B}}{\text{mass of A}}$$

Putting the two results together,

$$\frac{\text{A's power of attraction}}{\text{B's power of attraction}} = \frac{\text{mass of B}}{\text{mass of A}}$$

All right, that's it. The gravitational force is proportional to both the mass of the attracting and the mass of the attracted body. Inertial mass belongs to bodies merely because they are bodies. Therefore gravitational force goes with all bodies; all bodies attract gravitationally. Gravitational attraction is therefore inexplicable by any mechanical model of matter in motion. The mechanical philosophy, Newton concludes in the 1680s and 1690s, is dead; he has rediscovered the ancient mystic Pythagorean truth of the harmony of the spheres. Gravitation, he concludes, is the result simply of the immediate action of God.

There was a tradition in seventeenth-century England, pursued particularly by the so-called Cambridge Platonists Henry More and Ralph Cudworth, having to do with the *prisci theologi* or ancient theologians— Hermes Trismegistus, Orpheus, Pythagoras, Thales, Plato, and so on— whose pagan wisdom, it was claimed, was really derivative from that of the Hebrew prophets, especially Moses. More and Cudworth'developed their interpretation of these ancient doctrines into a justification for a new and revolutionary natural philosophy, that is, for modern science as it was coming to be in the works of Galileo and Descartes. Newton, influenced by these men in earlier years, now believes he has found the right interpretation of the ancient wisdom precisely because he has found the *right* natural philosophy. And so he writes:

> Since all matter duly formed is attended with signs of life and all things are framed with perfect art and wisdom and nature does nothing in vain; if there be an universal life and all space be the sensorium of a thinking being who by immediate presence perceives all things in it, as that which thinks in us, perceives their pictures in the brain; those laws of motion arising from *life* or *will* may be of universal extent. To some such laws the ancient philosophers seem to have alluded when they called God Harmony and signified his actuating matter harmonically by the God Pan's playing upon a Pipe. . . . To the mystical philosophers Pan was the supreme divinity inspiring this world with harmonic ratio like a musical instrument and handling it with modulation according to that saying of Orpheus "striking the harmony of the world in playful song." But they said that the Planets move in their circuits by force of their own souls, that is, by force of the gravity which takes its origin from the

action of the soul. From this, it seems, arose the opinion of the peripatetics concerning Intelligences moving solid globes. But the souls of the sun and of all the planets the more ancient Philosophers held for one and the same divinity exercising its powers in all bodies whatsoever. . . . All [their gods] are one thing, though there be many names.

And so Newton goes on to argue, using passages from Plato and Lucretius and many other ancient writings, that the philosophers of antiquity—Thales, Anaximander, Pythagoras, Democritus, and so on—were really agreed upon the atomicity of matter, the inverse-square law of gravitation, the universality of gravitation, and further, true mystics that they were, held the true cause of gravity to be the direct action of God. The unity of physical, moral, and theological wisdom is thus shown to have been present in the beginnings of the world, transmitted from Adam and Eve. That unity and that wisdom were gradually lost, after the corruptions of the sons of Noah; but now they have been recovered and restored by Newton, who thus takes his place among the ptisci theologi, the ancient theologians. Newton is even able to find in the biblical book of Daniel the prophecy of his, Newton's, rediscovery of the truth.

So the first beginnings of a mathematized celestial dynamics came, with Kepler, out of a trinitarian symbol, the three-foldness of the Sun, spherical shell of stars and intervening space in a Sun-centered world; Kepler had his main idea from the beginning. With Newton it was different, and the crucial justifying discovery came late, with a precise experiment to test the exactness of the constancy of the acceleration of gravity, and a new realization of the meaning of that constancy. And in a world that has now lost its geometrical center, Newton accepts this discovery as a revelation of a mysterious, omnipresent, unitarian God, to discourse of whom from the appearances, as he will tell us in the General Scholium to the second edition of the *Principia*, does certainly belong to Natural Philosophy. But the most famous statement of the General Scholium, presented there as the outcome of inductivist caution, "I do not contrive hypotheses" (hypotheses, that is, as to the cause of gravitation)—this statement disguised rather than expressed the deeper ground of Newton's original and I suspect persisting view, that gravitation was indeed universal, and the result of the direct action of God, so that no hypotheses for it could be sucessfully contrived.

Dynamical Chaos: Some Implications of a Recent Discovery
(1994)

My subject is a peculiar behavior of dynamical systems that has come to be recognized only during the last thirty years. In 1975 James Yorke christened this behavior *chaos*—perhaps a misnomer. *Chaos* is a Greek word that has no plural. Since Hesiod it has meant the nether abyss; the first state of the universe, or total disorder. The dynamical behavior that James Yorke called *chaos* is order and apparent randomness intertwined.

A dynamical system is—what? How about this? It is a set of entities that interact so as to undergo a development. The entities could be planets or billiard balls; maybe cardiac muscle fibers or neurons; maybe even bidders in the New York Stock Exchange; but let that go. In my illustrations, the components will be chunks of matter, unbesouled.

To understand why dynamical chaos was recognized only recently takes a bit of mathematical background. Mathematically, dynamical systems are represented by *differential equations*. Differential equations are distinguished by containing instantaneous rates of change, velocities, say, or accelerations. Now empirically you cannot measure an *instantaneous* rate of change, but only changes over *finite* intervals of time. A differential equation, therefore, is a hypothesis. To verify the hypothesis you must first *solve* the equation or *integrate* it. That means, you must somehow eliminate the rate or rates of change, and obtain the value of the dependent variable—which is what you are interested in—as a function of the independent variable, which is usually time. Thus you will have a relation you can check empirically.

Procedures for solving a good many differential equations were worked out in the seventeenth and eighteenth centuries. The solutions turn on what is called the fundamental theorem of the calculus, discovered by Newton and Leibniz. The procedures, like the differential equations, assume that time is continuous.

Not all differential equations are thus soluble "analytically," as we say. It may be impossible to disentangle the dependent variable from its rate of change; or, if there is more than one dependent variable, to disentangle these variables from one another. Then you can't arrive at a formula giving each variable as a function of the time.

All *linear* differential equations are soluble. In these equations, the dependent variables and their rates of change occur only to the first power, and don't multiply one another. An equation in just two variables occurring only to the first power can be graphed as a straight line; hence the name *linear*.

But there are nonlinear differential equations, in which some of the dependent variables, or their rates of change, are raised to powers, or multiply one another. Some of these equations are analytically insoluble.

In fact *most* dynamical systems in the world can be modeled accurately only by nonlinear differential equations, most of which are insoluble. It is dynamical systems modeled by such differential equations, nonlinear and insoluble, that exhibit the behavior that James Yorke called chaos. Here a small change in the independent variable can produce a sudden large change in a dependent variable. Such phenomena, we're told, flourish in nature; for instance, in the dripping of a faucet.

How can insoluble differential equations be studied? The chief way is by what is called *numerical integration*. This differs from the analytical integration I previously spoke of, which relies on the fundamental theorem of the calculus. In numerical integration the independent variable is not varied continuously; instead, it is increased by finite jumps. Starting with certain initial values of the variables, numerical integration assumes that the initial rate of change remains constant for some small, finite interval, say a second, and on that assumption computes the values of all the variables at the end of the second. Then with the new values it goes on to compute the values of all the variables at the end of the second interval. And so on. The procedure is not strictly accurate. But if the intervals are made small enough, it can give a good idea of what is going on; it can even, in many cases, be made to yield predictions as accurate as the observations.

The first large-scale numerical integration ever performed was carried out in 1758, to compute the return date of Halley's Comet. It took six months' work by three people, morning, noon, and night. Their final prediction was a month off, and even then they were lucky, because their computation contained some partially compensating errors.

Recognition of the chaos named by James Yorke came only in the decades since 1960, with the development of high-speed electronic computers that could carry out numerical integrations no one had previously thought practical.

I am going to now illustrate this kind of dynamical chaos, and to talk about some of its characteristics. My interest in this subject arose because for some years I have been pursuing the question of how planetary astronomy became a precise predictive science, and since 1980 it has become apparent that planetary astronomy involves James Yorke's chaos.

This chaos limits predictability. Philosophers, I suspect, should learn about it. New perspectives open up if we recognize how widespread it is. More on this later.

I begin with the simple pendulum. It consists of a heavy bob, suspended by a weightless, inextensible thread—mathematical physicists love to invoke such things. If we draw the bob aside and let it go, it oscillates back and forth. Let me derive its equation of motion and show it to be nonlinear (Figure 1).

We measure θ, the departure of the thread from the vertical, in radians, defined as arc-length divided by radius. So the arc-length will be given by the radius-arm, or length of the thread, here l, times θ. In the science of dynamics as founded by Galileo and Newton, we are interested in accelerations, that is, rates of change of velocity; velocity itself being a rate of change of position. Acceleration is thus a rate of change of a rate of change. In our case, we are interested in the acceleration of the bob, hence of its position as measured by the arc $l \cdot \theta$. But l is a constant; so the acceleration we are interested in is 1 times the acceleration of θ, which I write as theta with two over-dots, $\ddot{\theta}$.

The reason the bob accelerates is that it is pulled downward by gravity. The acceleration of gravity at a given spot on the Earth is a constant, which we call g. But the bob cannot go straight down, with the acceleration g, because it is suspended by the thread. To find how much of g accelerates the bob along its path, we "resolve" g into components, one in line with the thread—this component merely tenses the thread—and the other component at right angles, along the path (Figure 2). The latter component is $g \cdot \sin \theta$. Our equation is then:

$$\ddot{\theta} = g/l \sin \theta,$$

where, remember, the double over-dot means acceleration, radians per second per second.

I say that this equation is nonlinear. Sin θ is given by an infinite series—I will not prove this, please take it on faith:
$$\theta - \theta^3/31 + \theta^5/51 - \ldots$$
If only the first term were present, we would have a linear equation. But there are higher powers, going on forever.

Now it turns out that this nonlinear equation is soluble. I will not write down the solution, which is somewhat complicated. It implies that the pendulum is not isochronous; wider-angled swings take a little longer. Galileo, gazing at the suspended lamps in the Cathedral of Pisa, guessed the pendulum was isochronous, and wanted so much to believe this, that he never made the simple experiments that would have shown this assumption false.

Of course; the simple pendulum is *approximately* isochronous, for small-angled swings. Suppose that θ is 6°, about 1/10 of a radian. In the series expansion for the sine, if the first term is 1/10, the second term is 1/6000. We can choose to ignore it, along with all the higher terms. That is called *linearizing* the equation. Mathematical physicists have been doing it for neatly three hundred years, in order to obtain neat, soluble equations. The hope is always that the linearized equations give good enough approximations. And so they do, when the system is close enough to a stable equilibrium.

Suppose, then, we limit our simple pendulum to swings of 6° or less. The effects of nonlinearity will be present, but tiny. And now let us introduce a perturbation. When the word *perturbation* is used, we mean that there is some motion we can regard as fundamental, and some other disturbing motion that is superimposed. The Earth's motion is controlled primarily by the gravitational action of the Sun, but it is perturbed detectably by the Moon, Venus, Jupiter, Saturn, and so on.

Suppose the point of suspension of our pendulum is put into a small oscillation, in the very plane in which we first set the bob to oscillating. We could use a crank mechanism for this. Let the amplitude of this perturbing motion be a small fraction of the length of the pendulum. Let the period of the forcing motion be one we can vary: call it T. And suppose we set T to be somewhere near the period of our linearized simple pendulum, which I shall call T_0. T_0 is given by a formula some of you have learned, 2π times the square root of l over g (Figure 3.)

The equation of motion for the new set-up, which I will not write down, is not soluble analytically; it will not yield a formula for θ as a function of time. But a numerical integration can be carried out. John

Figure 3

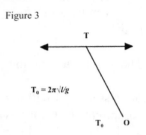

$$T_0 = 2\pi\sqrt{l/g}$$

Miles of the University of California, San Diego did this in 1984,[1] to determine the position of the bob each time the perturbing motion reaches the righthand end of its range. Starting from below T_0, he increased the period T of the perturbing motion. At a certain point, the motion of the bob, in its original direction of motion, which I will call the x-direction, became unstable. But meanwhile there were two possible motions that were stable—motions that included a sideways component, a y-component. The motion of the bob made a gradual transition to one or the other of these stable motions.

When $T = 0.9924\ T_0$, the position of the bob each time the perturbing motion comes to the righthand limit of its excursion moves in this figure (Figure 4, overleaf). You probably want to know what the whole motion of the bob is. It is in a slowly rotating ellipse with slowly varying axes.

But let me focus solely on the position of the bob each time the perturbing motion reaches the righthand end of its range. If T is 1.0150 To, our point moves in a doubled curve (Figure 5); what is called a bifurcation has occurred. A small quantitative change has produced a sharp qualitative change. Let the period be increased So that T is 1.0213 T_0 (Figure 6); another bifurcation has occurred. A cascade of further bifurcations occurs, as T is increased. When we reach $T = 1.0225\ T_0$ (Figure 7); the figure appears smudged; with many paths close together. The pattern, if accumulated over a long enough time, appears to be symmetric with respect to the x-axis, although the bob may spend substantial intervals in either the top or bottom half of this pattern, transferring from one to the other at seemingly random times.

Random is—what? The word comes from old French *randir*, to run or gallop; the French knight, having donned his armor, and drunk certain flagons of wine, was hoisted by crane onto his horse and galloped about the field, doing random mayhem. The Anglo-Saxon, by contrast, wore bearskin, drank mead, and on the battlefield went berserk, a word meaning

1. John Miles, "Resonant motion of a spherical pendulum," in *Physica D: Nonlinear Phenomena* 11.3 (1984): 309-323.

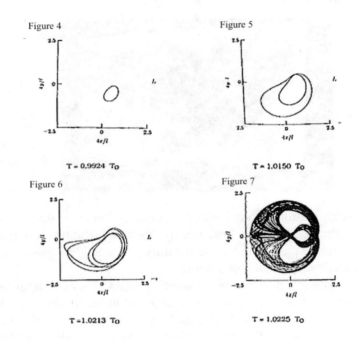

Figure 4

T = 0.9924 T₀

Figure 5

T = 1.0150 T₀

Figure 6

T = 1.0213 T₀

Figure 7

T = 1.0225 T₀

bearskin. To define *random* mathematically is something else again. Perhaps we can say what it is not. If our pattern were two or more periodic motions superimposed, it would be what is called *quasi-periodic*; if the periodic motions were incommensurable, there would be no exact repetitions, but the motion would not be random or chaotic. But our motion does not look quasi-periodic, with those sudden shifts from one part of the pattern to another.

Let's turn to an actual physical experiment. Al Toft, assisted by Otto Friedrich—machinist and carpenter for the laboratory—made this pair of double pendulums in tandem, in accordance with a description given in the *American Journal of Physics* in 1992[2] (Figure 8. Photagraphy in Figures 8-11 by John Bildahl.). Each double pendulum consists of an upper and lower part, turning on bearings, so that the friction is small. Each part of each double pendulum, both the upper and the lower, has its own natural period for small-angle oscil-

2. Troy Shinbrot et al., "Chaos in a Double Pendulum," *American Journal of Physics* 60 (1992): 491-99.

Figure 8

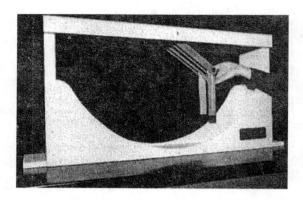

lations. The situation for the lower pendulum is similar to that of the perturbed simple pendulum I previously described. But now the perturber (the upper part) is itself significantly perturbed; we have what is called feedback, circular causation.

The two double pendulums are identical twins. They are mounted together on a sturdy support, so that neither will influence the other. If I start one of them in an oscillation, the other does not pick up the motion. If I start both together in a small oscillation, they play together nicely (Figure 9).

Figure 9

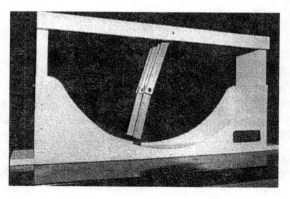

Swing it Low – Little Difference

With a large initial displacement, however, the pendulums don't stay together (Figures 10 and 11, overleaf). Nor, if we try the experiment over again, does either do exactly what it did the first time. I hope

Figure 10

Figure 11

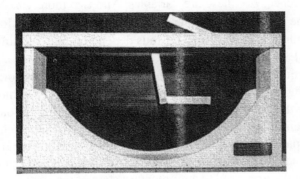

$\dot{\theta}_2$

Swing it High – Big Difference

this surprises you. Before trying to account for it, let's see how we might show quantitatively that the repetition isn't exact. We could use a rigidly mounted electromagnet to hold the pendulum in a fixed initial position, to the side. Suppose the switch releasing the pendulum started a stroboscopic flash camera, that took photos every 25th of a second. On each exposure we could measure the angular deviations from the vertical of the upper and lower parts. Then we could proceed to compare different trials. This has actually been done.

In explaining this, I shall introduce a bit of the relevant mathematics. Take a look, for just a moment, at the differential equations of the double pendulum (Figure 12). On the lefthand side, on top, you see $\ddot{\theta}_1$, the angular acceleration of θ_1, the deviation of the upper pendulum from the

Figure 12

$$\ddot{\vartheta}_1 = \frac{g(\sin\vartheta_2\,\cos(\Delta\vartheta) - \mu\sin\vartheta_1) - (l_2\dot{\vartheta}_2^2 + l_1\dot{\vartheta}_1^2\cos(\Delta\vartheta))\sin(\Delta\vartheta)}{l_1(\mu - \cos^2(\Delta\vartheta))},$$

$$\ddot{\vartheta}_2 = \frac{g\mu(\sin\vartheta_1\,\cos(\Delta\vartheta) - \sin\vartheta_2) + (\mu l_1\dot{\vartheta}_1^2 + l_2\dot{\vartheta}_2^2\cos(\Delta\vartheta))\sin(\Delta\vartheta)}{l_2(\mu - \cos^2(\Delta\vartheta))}.$$

vertical. And below, on the lefthand side of the second equation, you see $\ddot{\theta}_2$, the angular acceleration of θ_2, the deviation of the lower pendulum from the vertical.

On the righthand sides you see a number of constants: g, l_1 and l_2, the lengths of the two parts of the pendulum; μ, the ratio of their masses. Also involved are sines and cosines of the two angles, and also of their difference, $\Delta\theta$; these are variables. And two more variables are $\dot{\theta}_1$ and $\dot{\theta}_2$, the momentary angular velocities of the upper and lower parts of the pendulum. Both $\ddot{\theta}_1$ and $\ddot{\theta}_2$, thus depend on four variables, θ_1 and θ_2, as well as $\dot{\theta}_1$ and $\dot{\theta}_2$.

The equations are not soluble; we can't get separate formulas for the two angles as functions of time. But our system at any moment depends on the four variables just mentioned. In the 1830s William Rowan Hamilton proposed representing the evolution of such a system in hyperspace, with a number of dimensions equal to the number of variables on which the state of the system depends. Then a point in this space would correspond to a momentary state of the system, and a succession of points, or trajectory, would show how the system develops. The space is called phase space. In our case, the space is four-dimensional, and you can't visualize it. You can nevertheless conceive it without contradiction. In the 1890s Henri Poincaré undertook to study insoluble, nonlinear differential equations, by examining the ensemble of possible trajectories in phase space.

In our case, let's consider a few trial runs with our double pendulum, say four, and compare the points in phase space at the successive moments when the photographs are taken. For a given moment, the points in the four different trials will not be the same. There will be a "distance" between any two of them, and we can get numbers for these distances, using the four-dimensional analogue of the Pythagorean theorem; that is, we take the square root of the sum of the squares of the

244

components. In this figure (Figure 13), the experimental separations are plotted for the first half second. The solid line was obtained by numerical integration of the differential equations, using two slightly different sets of initial conditions. The separations increase on the whole. That the separations have downswings at certain places is due to the fact that, at the end of each swing, when the lower pendulum is starting down again, it pulls down on the upper pendulum, and this is a relatively stable situation.

Figure 13

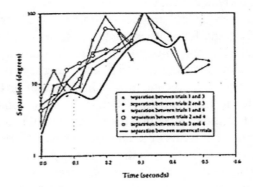

A statistical study of these numbers shows that, on the average, the separation increases geometrically. Suppose the initial separation between two trajectories in phase space is $\Delta\chi_0$. Then the separation at time t is

$$\Delta\chi_t = \Delta\chi_0 \cdot e^{\lambda t},$$

where e is a constant greater than 1, and λ is a positive constant, called Lyapunov's exponent, after the Russian who first discussed its import. The separation doesn't just increase by the same additive increment in each unit of time, but gets multiplied by the same factor, so that the increase is exponential.

Of course, our data is only for the first half-second. Strictly speaking, λ should be determined as a limit as t goes to infinity. But you can't get funding for experiments that long. The chaos is apparent, but is it real? How do we explain it?

First, however, what does a positive λ do to prediction? Recently it has been shown that the long-term orbital evolutions of the inner planets, Mercury, Venus, Earth, and Mars, are characterized by positive Lyapunov exponents. For instance, certain perturbations of the Earth are in near resonance with its annual motion, in close analogy with the case of our perturbed simple pendulum; and the same kind of chaos results. Now we never know, with infinite precision, where a planet is. By numerical integration it has been shown that initial uncertainties for the Earth increase by a factor of 3 every 5 million years. An initial error of 15 meters produces an error of 1.5 million kilometers after 100 million years.

Yes, we'll all be dead, but my concern is a theoretical one, about the nature of our knowledge. Can we understand a little better what this chaos is, and whence it comes?

Back to phase space and another of Poincaré's new techniques. Here (Figure 14) is the four-dimensional phase space of our double pendulum, somehow represented in a pseudo-diagram; q_1 and q_2 are the coordinates, in our case angles, and p_1 and p_2 are the corresponding momenta, products of velocity and mass. The presentation of the equations of dynamics in terms of the p's and q's is due, once more, to William Rowan Hamilton. H, called the Hamiltonian, is the energy of the system, expressed in terms of the p's and q's. We'll assume for the present argument that H is a constant: $H = H(p_1, p_2, q_1, q_2) = $ const. In the pseudo-diagram H is represented as a surface; but it is really a hypersurface in four-dimensional space; the pseudo-visualization is only to help you identify the terms I am using.

The constancy of H will allow us to express p_2 as a function of p_1, q_1 and q_2: $p_2 = p_2(p_1, q_1, q_2)$. We can thus consider the projection of any possible trajectory in the four-dimensional phase-space onto a three-dimensional volume. That projection will contain all the information that the four-dimensional trajectory contained. The reduction

Figure 14   Figure 15

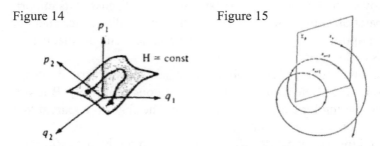

in the number of dimensions brings us back to something visualizable.

Poincaré carried the process a step further. If the trajectory is bounded—doesn't go off to infinity—then its projection in the three-dimensional space will intersect some plane in that space repeatedly, say the plane $q_2 = 0$ (Figure 15). Such a plane is called a Poincaré surface of section. What sort of pattern will the intersections make?

In the 1960s two astronomers, Hénon and Heiles, were studying a nonlinear differential equation intended to model the motion of a star around a galaxy. They used numerical integration to find successive intersections of the star's trajectory with a Poincaré surface of section. When the energy of the system was relatively small, the intersection lay on certain distinct curves (Figure 16). With an increase in energy, the pattern became this (Figure 17). There were still islands where, for certain initial conditions, the trajectory remained on nice curves; but other trajectories proved to be chaotic, giving seemingly randomly placed intersections. When the energy was increased still further, the islands disappeared (Figure 18).

The motion is deterministic; that is, given the state of the system at any moment, the equation of motion determines its state at the moments that follow. The initial conditions determine a position and a velocity, and the equation of motion then determines an acceleration. Given position, velocity, and acceleration, there is only one way to go. But the resulting pattern looks crazy. Again I ask, how should we understand that?

Consider a plane of section with the dimensions p and q; suppose the successive points of section are confined to the unit square (Figure 19). A theorem in Hamiltonian dynamics says that, if the energy of a dynamic system remains constant, the volume occupied by the allowed trajectories is also constant. But in truly chaotic dynamics, the trajectory never returns to the same point, or even to any identifiable curve. The volume of allowable paths gets dispersed, mixed up with bubbles of the unallowable.

Some Russian mathematicians have sought to describe this mixing, using a cocktail shaker, rum and cola. Sorry, we're stuck with their noxious example. Initially, the rum and cola are separate. After a few shakes, if we imagine the cola as divided up into moderately small cells, we find some rum in each cell. Later, the subdivision can be made finer and finer, with each cell containing some rum. This is called a mixing transformation. Some dynamical systems have been proved to evolve in this way, for instance a gas consisting of spherical elastic molecules. Nearby possible trajectories necessarily spread apart, defocus. But because their energy is finite, they don't go off to infinity, but

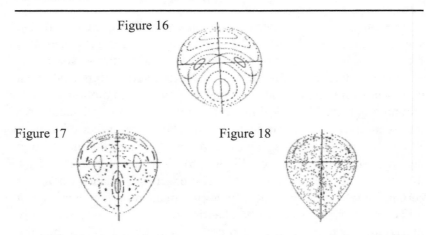

Figure 16

Figure 17

Figure 18

get folded back into the same space. The rule is Stretch and Fold. It may be characteristic of chaotic dynamics generally. For illustration, I shall present a particular mixing transformation, called the Baker's Transformation (Figure 20)

Figure 19

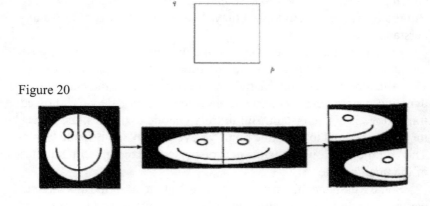

Figure 20

Let the square, as if it were dough, be first squashed down into a rectangle twice as long and half as high; then let the right half of it be set atop the left half. We get back a square. What happens to a point (p,q) within the square? Both p and q are numbers between 0 and 1. By squashing, all p's are doubled; but then, by the operation of placing the right half atop the left half of the rectangle, the p's that became greater than 1 are reduced again, by subtraction of the unit 1, to numbers between 0 and 1. We can write this: $p \rightarrow 2p$ (mod 1).

248

As for the q's, if our q was in the left half of the original square, it is simply halved: $q \rightarrow q/2$. If it was in the right half, after halving it we add $1/2$: $q \rightarrow q/2 + 1/2$.

It will be helpful to think about p and q as written in binary notion, so that each p and q will be written as, first, a zero, followed by a binary point (replacing our decimal point), followed in turn by a string of zeros and ones, infinitely long. All numbers between 0 and 1 can be written thus. 0.1 means $1/2$, 0.01 means $1/4$, 0.001 means $1/8$, and so on. We use powers of 2 instead of powers of 10. Let our initial p and q be:

$$p = 0.p_1p_2p_3\ldots, \qquad q = 0.q_1q_2q_3\ldots,$$

where the letters with subscripts are zeros and ones. Now a neat thing about binary notation is that multiplying by 2 just amounts to shifting the binary point to the right, while dividing by 2 just amounts to shifting it to the left. If our initial point was in the left half of the square, then p_1 was 0, and after the squashing, the transformed p will be

$$0.p_2p_3p_4\ldots$$

But the same result holds if our initial point was in the right half of the square, for then p_1 was 1, and after the binary point is moved to the right, 1 must be subtracted. In successive transformations p will become

$$0.p_3p_4p_5\ldots \qquad 0.p_4p_5p_6\ldots, \text{and so on.}$$

What about the q's? If our original point was in the left half of the square, then we want the, halved value of $q,$ which is

$$0.0q_1q_2q_3\ldots.$$

I have moved the binary point to the left one place. If our original point was in the right half of the square, then q first gets halved, but we must add $1/2$ when the right half of the rectangle is put atop the left half. Now, in this case p_1 was 1, which in the first binary place after the binary point, means $1/2$. So the transformed q can be written $0.p_1q_1q_2q_3\ldots.$ Actually, this works for q's in the left half of the original square as well, for there p_1 was 0. A little reflection will show you that the succession of transformed q's will be

$$0.p_1q_1q_2\ldots \qquad 0.p_2p_1q_1q_2\ldots \qquad 0.p_3p_2p_1q_1q_2\ldots, \quad \text{and so on.}$$

As the successive pairs of transformed p's and q's emerge, the important digits, the digits up front, come from ever farther to the right on the original coordinate p. Initially. they looked insignificant. Yet however far to the right they were originally, they become crucial as the returns continue. The sensitivity to initial conditions is infinite.

Suppose, though it is empirically impossible, that we knew our initial p and q exactly. The differential equation for the motion is not soluble; our only resource is numerical integration, for which we turn to a high-powered computer. The computer, however high-powered, cannot give us the chaotic trajectory precisely. That is because it is a finite-state machine. It cannot accept a number expressed by an infinite number of digits; it automatically rounds it off. With our p's and q's undergoing a mixing transformation, the rounding, after a while, will be disastrous; we will lose essential information. Whether the p or the q, at some later stage in the succession of transformations, starts with a 0 or a 1 will be as uncertain as the toss of a coin.

Let me now, by way of conclusion, state some thoughts as to the import of nonlinear dynamics.

1. According to Laplace, writing in 1814:

> An intelligence that knew, for a given instant, all the forces by which nature is animated, and the respective situation of all the beings that compose it, if it were vast enough to subject these data to analysis, would embrace in a single formula the motions of the largest bodies of the universe and of the smallest atom; nothing would be uncertain to it, and the future, like the past, would be present to its eyes. The human mind, in the perfection that it has been able to achieve in astronomy, presents a pale image of this intelligence.[3]

Laplace is expressing a universal determinism, which he equates with predictability. Such a determinism, presenting the world as a closed causal network, has often been taken as a dogma of science; but its universalism appears to make the activity of the scientist unintelligible.

Voltaire swallowed the doctrine whole.He wrote:

> Everything is governed by immutable laws . . . everything is pre-arranged . . . everything is a necessary effect. . . . There are some people who, frightened by this truth allow half of it. . . .There are, they say, events which are necessary and others which are not. It would be strange if a part of what happens had to happen

---

3. Pierre-Simon Laplace, *Essai Philosophique sur Les Probabilités* (Paris: Courcier, 1814), 2-3.

and another part did not. . . . I necessarily must have the passion to write this, and you must have the passion to condemn me; we are both equally foolish, both toys in the hand of destiny. Your nature is to do ill, mine is to love truth, and to publish it in spite of you.[4]

Post-Newtonians, impressed by the success of the new dynamics, did not consider that the new methods might prove limited in scope.

The success of this dynamics was a success in solving linearized differential equations. The world was taken to be an integrable system, each variable being finally expressible as a function of time, independent of the others. So the world would be made up of non-interacting Leibnizian monads, each experiencing its own private cinema, the harmony between them divinely preestablished.

This view, I say, was mistaken, because the differential equations required to model processes in the real world are mostly nonlinear, and most nonlinear differential equations are insoluble. It is from insoluble, nonlinear differential equations that dynamical chaos arises. Here determinism and predictability part company; Laplace's demon, to do what he required of it, would need to compute with numbers that it would take an infinity of time to write down. Successive approximations, which are the human way, wouldn't suffice.

I have spoken so far as if of a closed dynamic system, insulated from the rest of the universe. But mixing systems such as I have described are hypersensitive to initial conditions, and therefore hypersensitive to tiny perturbations. The flash of an electron in a distant star may affect our mixing system. The intelligence that Laplace imagined, however vast, being yet discursive, will suffer from overload. If there is a God that knows the future, it is by means inscrutable to human reason.

2. I want now to go beyond chaos. There is more to nonlinear science than chaos. As we have seen, the chaos we have been concerned with is not simply disorder; it is approached in an orderly way; it is describable in a coherent way. Can nonlinear dynamics lead to more interesting sorts of order? In fact, in dissipative systems far from equilibrium, new and surprising kinds of order arise. Some of these are described in the book by Prigogine and Stengers entitled *Order Out of Chaos*, cited in footnote 4.

---

4. Voltaire, from the *Dictionnaire Philosophique*, quoted in Ilya Prigogine and Isabelle Stengers, *Order Out of Chaos* (New York: Bantam Books, 1984), 257.

An example. The Bénard instability is due to a vertical temperature gradient set up in a horizontal liquid layer. The lower surface is heated to a given temperature, higher than that of the upper surface. Thus a permanent heat flux arises, from bottom to top, and for a low temperature gradient, this occurs by heat conduction alone, while the liquid remains at rest (Figure 21). But when the imposed gradient of temperature reaches a certain threshold value, a convection involving the coherent motion of ensembles of molecules is produced. Millions of molecules move coherently, forming convection cells of a characteristic size (Figure 22). At higher temperature gradients there occur periodic fluctuations in temperature and in the spatial arrangement of the cells; finally there is turbulent chaos, which, again, is not without its order.

Figure 21   Figure 22

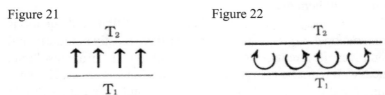

Similar kinds of order arise in dissipative chemical systems. With reactants entering and products leaving, the system may organize itself spatially, or may come to act like a chemical clock, beating rhythmically. Such coherent behaviors on the macroscopic level do not appear to be reducible to the dynamics of atoms and molecules. We have what may be called *emergence*.

3. According to a fairly broad consensus among scientists today, living things are among the entities that have so emerged. We are, in some sense of the word "are," stardust. Living things are complex, dissipative structures, to some extent self-regulating, but maintained ultimately by the flux of energy from the sun. If the geological time-scale is represented as a thirty-day month, then life appeared in the oceans by the fourth day, but became abundant only on the twenty-seventh. The first land plants and the first vertebrates appeared on the twenty-eighth day; most of human culture appeared only in the last thirty seconds. All this, at least up to the last thirty seconds, is understandable in terms of evolution by random variation and differential reproductive success.

Living things are not only embedded in the surrounding geological world, but by their activities they have altered that world; at an early stage, for instance, ancient relatives of present-day algae produced the oxygen of the atmosphere. In various degrees, living things have a circumscribed autonomy: it is wider for those with homeostasis of the blood, wider still for those who can reason before reacting. These are beings with desires, aims, purposes. The world lines of such semi-autonomous entities, with their separate agendas, may intersect. A man goes to the agora, in the case imagined by Aristotle and meets someone who owes him money. Neither planned this encounter; it is by chance. Species migrate or spread, encounter one another, interact, find new ecological niches.

A world evolving through chancelike encounters, in which new entities emerge in time, including intelligent beings, is unintelligible if, with Leibniz or Voltaire or Laplace, we take that world to be an integrable system. In an integrable system, the mere reversal of velocities sends time backwards. There is no essential distinction between future and past. The smoke can go down the chimney and reconstitute the firewood.

Why can it not? In the middle of the nineteenth century the law of entropy was discovered: in any closed system, a certain mathematical function tends to a maximum, and there is thus a forward direction to time, diametrically opposed to the backward direction. This law is of everyday use in physics and chemistry, to predict the outcome of experiments. But it contradicts dynamics, if the world of dynamics is an integrable system. Physicists like Boltzmann sought to derive the law of entropy from dynamics; irreversibility from reversibility. By the 1890s it was clear that it couldn't be done; a *statistical* or *probabilistic* assumption, distinct from the dynamics, was necessary. Chance had to be assumed to be real. This is an empirical assumption, warranted by experience. Probability theory, at its core, is an empirical science, which assumes the future to be different from the past.

Also toward the end of the nineteenth century, the American philosopher C. S. Peirce suggested that the dissipative tendencies of entropy could be balanced by the concentrative effects of chance. Chance can have an integrative role, in the emergence of new entities like us. Thus dissipative processes intertwine with integrative ones.

Nowhere is this more the case than in the activity most distinctive of humans, that of learning and communicating. It is not possible without a functioning brain, dependent on a flux of energy; the reactions involved are dissipative and therefore irreversible. When we learn a

Greek paradigm, we change the physiology of the brain. We learn not as beings detached and separate from the world, but as parts of it, by engaging in activities of exploration, hypothesis, construction, testing. Here there is an interplay between chance and reason. And this is especially true in learning about nature: such learning requires that we enter into a dialogue with nature. Thus I think that Einstein, when he sought a vision of the world from totally outside it, and denied the reality of time, was mistaken.

The perspective I am suggesting leads to a new respect for nature, of which we are not the overlords but in which we are both embedded and emergent. In this perspective, there are no guarantees; we live in a chancy world. Nevertheless, a qualified hope is rational. Human knowledge increases, not always steadily, sometimes by surprising zigzags or even reversals; but the trend is unmistakably incremental. It is not a deductive chain. It is a rope, no single strand of which is, by itself, of incorrigible strength; but different strands, by pulling against one another, constitute a fabric stronger than any of its parts.

In this lecture I have sought to signalize a mistake into which dynamicists and philosophers of the past three centuries fell, imagining the world to be an integrable system. The biological perspective I have been sketching can be a corrective. Our survival depends on recognizing and respecting our own complexity and that of the nature in which we are both embedded and emergent.

According to rabbinical commentary, the first word of Genesis, *Beréchit*, means not "In the beginning," but "In a beginning." Twenty-six attempts, say the rabbis, preceded the present Genesis; all ended in failure. *Holway shéyaanod* exclaimed God as he created the world: "Let's hope that this time it works."

Galileo Agonistes
(1999)

My topic this evening is one that fifty years ago I had aspirations of delving into, then got lured away from, and now once more seek to come to terms with. Galileo in life was a combative controversialist, and ever since he has been a subject of controversy. My talk will be an interpretation. I shall first review what I think can be said about Galileo's discovery of the law of free fall, taking care not to inject post-Galillean physics into the Galilean moment—a source of frequent errors. Then I shall speak of Galileo's struggle to keep the Church from condemning heliocentric astronomy, the failure of that struggle, and his trial before the Inquisition. My subject, I must warn you, requires attention to details. God, or the devil, is in the details.

I start with the old story in old textbooks about how the whole fabric of Aristotle's cosmology and physics came tumbling down one day in 1589, when Galileo, aged 25, newly appointed mathematics instructor at the University of Pisa, before professors and students assembled, dropped two cannonballs of different sizes from the Leaning Tower of Pisa. A simultaneous thud, we're told, heralded the birth of a new, experimental physics.

Alas, this account omits crucial details. The original story was told by Viviani, Galileo's last pupil, who likely had it from his old teacher. The experiment, says Viviani, was designed, to show

Exhibit A

that the speeds of mobile bodies *of the same material* [my emphasis] but of unequal weight, moving through the same medium, are not in the ratio of their absolute weights, as Aristotle claimed, but they move with equal speed . . . ; and neither do the speeds of a given mobile body, moving through diverse mediums, have the inverse ratio of the resistances, or densities of these mediums.[1]

The import of these details emerges from a treatise on motion Galileo was writing in 1589.[2] Consider first the second point. According

---

1. From Viviani's *Racconto istorico della vita di Galileo Galilei*, as quoted by E. A. Moody, in "Galileo and Avempace," *Joural for the History of Ideas*, 12 (1951): 167 n.8.
2. I. E. Drabkin and Stillman Drake, *Galileo Galilei On Motion and On Mechanics* (Madison: University of Wisconsin Press, 1960), 3-131.

to Aristotle's *Physics*, a given body falls in different mediums with speeds that are inversely as the resistances of those mediums.[3] If the resistance were absent, Aristotle says, the speed of the falling body would be infinite.

Exhibit B:
Aristotle's Rule of Speeds in Different Mediums

Let V_A, V_B be the body's speeds in mediums A and B; and let R_A, R_B be the resistances in those mediums. Then according to Aristotle:

$$V_A : V_B :: R_B : R_A.$$

Suppose $RA \to 0$. Then the ratio $R_B : R_A$ becomes infinite, and so must $V_A : V_B$.

Therefore $V_A \to \infty$, which is impossible.

Aristotle concludes that a medium *must* be present; the void can't exist.

Galileo in his early treatise rejects Aristotle's proportion. A piece of wood falls in air with a certain speed, call it unit speed. Galileo identifies the resistance in Aristotle's proportion with density. Let the density of air be unit density, and let the density of water be 4 (800 would be more like it, but I use Galileo's numbers). Then by Aristotle's rule, the piece of wood should fall in water with a speed of one-fourth. But it doesn't fall; it rises and floats. Galileo thinks the speed of fall or rise varies as the *difference* between the density of the body and the density of the medium.

Whence this idea? It smacks of Archimedes. In fact, Galileo has written a little book on the famous crown problem. He knows the principle of buoyancy: a body immersed in a fluid is buoyed up by a force equal to the weight of the displaced fluid. That principle is irreconcilable with Aristotle's doctrine of heaviness and lightness. According to Aristotle, heavy bodies, by their heaviness, fall toward the center of the universe; light bodies, by their lightness, recede from that center. Heaviness and lightness are the fundamental qualities of the sublunary elements in Aristotle's world. Galileo in 1589 still assumed, with Aristotle, that the center of heavy things is the center of the world. But as an Archimedean, he has had to conclude that there is no such thing as lightness; all bodies are heavy, but some are more dense than others. A body goes up or down depending on whether its density is less or greater than the density of the medium.

---

3. Aristotle, *Physics*, 215a24-216a20.

Galileo, however, is an Archimedean with an Aristotelian question. Archimedes did not deal with motion; force for him was static force, force balanced by another force. Galileo, like Aristotle, wants to know the cause of the speed of falling bodies. Aristotle had stated that the downward motion of a mass of gold or lead is quicker in proportion to its size, its weight.[4] If greater heaviness is greater downward tendency; mustn't the heavier body fall faster? Well, Galileo has learned it can't be so.

Of two pieces of the same material, suppose the heavier fell faster. Tie them together. The combination must fall more slowly than its heavier part, since it is held back by the lighter part.[5] But this combination constitutes a heavier body, so it ought to fall faster. Aristotle's idea contradicts itself.

Galileo concludes that every body of a *given* material, in a *given* medium, has a speed of fall or rise determined by the *difference* in density between the body and the medium. The greater the difference in density, the greater the body's speed. What Galileo at this time called the body's *natural speed of fall* could be determined if the medium were entirely removed, but this, he believed, was not physically possible.

Is all this right? No. For one thing, Galileo is speaking of a natural *speed,* not an *acceleration* of fall. The adjective *natural* here expresses an Aristotelian notion: a body moves naturally if the *archē* of its motion is *internal* to it, not imposed from without. Galileo at the end of his life will still be using this term "natural" as though in the Aristotelian sense, but it will have become for him a question.[6]

What about the acceleration? Galileo in 1589 considers it to be not natural but adventitious. It occurs, of course, whenever a body starts falling from rest; to reach the speed determined by its density and that of the medium, it must pass through all lesser degrees of speed. To explain this, Galileo supposes that an *impetus* was originally impressed on the body to raise it up; when it is let fall, this impetus diminishes at its own rate, the way the heat impressed on a piece of iron diminishes when the iron is separated from the fire that was the heat's source. As

---

4. Aristotle, *De Caelo*, 309b14-16.
5. "Aristotle," says Galileo, "makes this same assumption in his solution of the 24th Mechanical Problem." See Pseudo-Aristotle, *Mechanical Problems*, 855b3436.
6. Galileo Galilei, *Two New Sciences,* trans. Stillman Drake (Madison: University of Wisconsin Press, 1974), 158-59.

the impetus diminishes, the body picks up speed; and when all the impetus is gone, it moves uniformly.

This theory is scarcely testable. It looks like an intellectual trap. Where did Galileo get it?—for it is unlikely he invented the whole thing.

He had entered the University of Pisa in 1581, aiming at a medical career. In 1585 he dropped out without a degree, disgusted by his professors' standpat Artstotelianism. Mathematics had caught his fancy— Euclid and Archimedes. Natural philosophers, he said, should do as mathematicians do, deducing consequences from definitions and axioms.[7] He was out to become a university mathematician. How to gain the requisite reputation?

He published the book about Archimedes's buoyancy principle I mentioned earlier. He devised mathematical derivations, and sent them about for critique. In 1587 he visited Christopher Clavius, mathematics professor at the Collegio Romano, the Jesuit college in Rome. The Jesuits were then a new and innovative force in education. They had recently debated among themselves whether mathematics was precisely

---

7. In his early treatise on motion (see *Galileo Galilei On Motion and On Mechanics*, translated and edited by I. E. Drabkin and Stillman Drake, [Madison: University of Wisconsin Press, 1960], 50), Galileo says:

> The method that we shall follow in this treatise will be always to make what is said depend on what was said before, and, if possible, never to assume as true that which requires proof. My teachers of mathematics taught me this method. But it is not adhered to sufficiently by certain philosophers who frequently, when they expound the elements of physics, make assumptions that are the same as those handed down in [Aristotle's] books *On the Soul* or those *On the Heaven*, and even in the *Metaphysics*. And not only this, but even in expounding logic itself they continually repeat things that were set forth in the last books of Aristotle. That is, in teaching their pupils the very first subjects they assume that the pupils know everything, and they pass on to them their teaching, not on the basis of things that the pupils know, but on the basis of what is completely unknown and unheard of. The result is that those who learn in this way never know anything by its causes, but merely have opinions based on belief, that is, because this is what Aristotle said. And few of them inquire whether what Aristotle said is true.

applicable to the world, and Clavius had taken the affirmative, arguing that astronomy, music, optics, mechanics—*scientiae mediae,* "middle sciences," the schoolmen had called them—applied exactly. Many of Galileo's early manuscripts, we now know, echo lectures given at the Collegio Romano.[8] Galileo was specializing in mechanics, astronomy, and the critique of Aristotle's physics.

The exact route whereby Galileo acquired his early doctrines on motion remains unclear. A Venetian named Benedetti had held similar doctrines, but Galileo apparently didn't know his work at first hand.[9]

In 1589 Galileo obtained his first post, at Pisa; it paid pitifully little. When his father died in 1591, he became the family breadwinner. With support from Clavius and others he obtained a better-paying professorship at the University of Padua. There he remained from 1592 to 1610.

A hopeful thing in Galileo's early work on natural motion was his attempt to verify his theory on inclined planes, where the motion is slower, more easily measurable. From the principle of the lever he had derived the rule for equilibrium of weights on diversely inclined planes.

Exhibit C: Equilibration of Weights on Inclined Planes

W_1, lying on CA, and connected with the vertically hanging weight W_2, is in equilibrium with W_2 if, and only if,
$$W_1: W_2::CA:CB.^{10}$$

The larger weight W_1 is sustained on the incline by the smaller weight W_2 hanging vertically, because the downward tendency of W_1 is reduced by the constraint on its direction. To the downward tendency as reduced by the constraint in direction, Galileo gave the name *momento*.

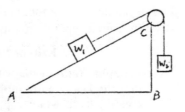

In going from statics to kinetics, Galileo makes an Aristotelian mistake: he assumes that, not the acceleration, but the speed produced is

8. William A. Wallace, *Galileo and his Sources: the Sources: the Heritage of the Collegio Romano in Gallileo's Science* (Princeton, N.J.: Princeton University Press, 1984).

9. See Stillman Drake & I.E. Drabkin, *Mechanics in Sixteenth-Century Italy* (University of Wisconsin Press, 1969), pp.204-206.

10. See *Galileo Galilei On Motion and On Mechanics*, 173-75.

proportional to the static force or *momento*. There should follow a certain ratio of the times down diversely inclined planes, but experiment disconfirms it. For a while at least, Galileo explained the disconfirmation as due to accidental causes.

To get out of this trap, Galileo needed to focus on acceleration, and then by experiment to discover the rule—famous as Galileo's discovery—that the distances traversed are as the squares of the times. Galileo had made this discovery, it appears, by October, 1604. On that date, in a letter to his friend Paolo Sarpi in Venice, he wrote as follows:

Exhibit D: Galileo's Letter to Sarpi, October 1604

Thinking over the questions about motion, in which, to demonstrate the accidents observed by me, I have been lacking a totally indubitable principle that I could take as axiom, I am reduced to a proposition which has much that is natural and evident about it; and this being supposed, I demonstrate the rest, that is, that the spaces traversed in natural motion [are as the squares of the times], and consequently the spaces traversed in equal times are as the odd numbers starting with unity. . . . And the principle is this: that the mobile body goes increasing its speed in proportion to its distance from its starting point. . . . Please consider it and tell me your opinion.[11]

For his demonstration, Galileo has only an outline of the steps, as we learn from a separate manuscript:[12]

Exhibit E: Galileo's fol. 128.

I suppose (and perhaps I shall be able to demonstrate it) that the heavy body falling naturally goes continually increasing its speed in proportion to its distance from its starting-point. . . . The speed with which the moving body has come from A to D is compounded of all the degrees of velocity it has had at all the points of the line AD, and the speed with which it has passed over AC is compounded of all the degrees of velocity it has had at all the points of the line AC. Therefore, the speed with which [the body] has passed the line AD has to the speed with which it has passed the line AD the ratio [of the square on DA to the square on CA].

---

11. See *Galileo Galilei On Motion and On Mechanics*, 173-75.
12. Ibid., 373-74.

The line AB represents distances traversed in falling. Lines at right angles to AB represent degrees of speed, increasing in proportion to the distance fallen through. A "degree of velocity" (*grado di velocità*) is a punctual speed, a speed at a point; it doesn't endure. Galileo speaks of *compounding* these punctual *gradi di velocità* to find the ratio of the speeds with which different distances are traversed. The degrees of speed thus compounded, he is saying, are measured by triangles; so the speeds with which AD and AC are traversed are as the triangles ADH and ACG. These are similar, and hence to one another as the squares on the corresponding sides.

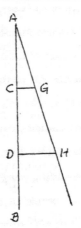

From this result, Galileo needs to get to what he knows experimentally, that the distances from the beginning of motion are as the squares of the times. The steps he proposes for this derivation are wrong - careless blunders. No mathematically legitimate steps lead from the composition of punctual degrees of speed varying as distance, to the variation of distance with the squares of the times.

Galileo himself, four or five years later, in 1608 or 1609, proved that the supposition of **Exhibit E** won't do. His argument goes as follows. Suppose AC is half AD. AD contains an infinity of points, and so does AC. At each point of AD, and at each point of AC, there is a punctual speed. The punctual speeds in the one distance can be put in one-to-one correspondence with the punctual speeds in the other, in such a way that each punctual speed in AD is twice the corresponding punctual speed in AC. Galileo infers that AD, the double distance, would be traversed in the same time as AC, its first half. But then the rest of AD would have to be traversed instantaneously, which is impossible.

The argument, I believe, is valid. A motion starting from rest, with its speed varying as distance traversed, is impossible.

Suppose, however, that the line AB in **Exhibit E** represented time. Then the two compounded sums of degrees of velocity would be given by the two triangles ACG and ADH, and these triangles are as the squares of AC and AD, that is, as the squares of the times. If the areas of the triangles were proportional to distance traversed, we would have the result that Galileo has found experimentally.

Eventually, in his *Dialogue on the Two Principal World Systems*, of 1632, Galileo will carry through just this derivation, in which an infinity of instantaneous speeds are compounded over time, and represented in a diagram as areas, and it is stated that for these areas to be as the distances traversed is *ben ragionevole e probabile*, very reasonable and probable.[13] A plausible proof but, Galileo recognizes, peculiar. It involves the very strange notion of instantaneous speed - a speed that doesn't have any duration, and so no distance is traversed by it—and it involves the adding up an actual infinity of such speeds. The serious application to the world of this questionable concept and procedure was unprecedented.

To return to 1604: Galileo at that time was unwilling to let AB represent time. Empirical evidence, he thought, showed that the speed of fall increases as distance. In the mansuscript from which I've been quoting, he says that this principle appears to him *molto naturale*, and agrees with our experience with instruments that operate by percussion: the magnitude of the effect is as the distance from which the body falls. He is thinking, for instance, of pile drivers. He has tested the principle, dropping weights on a stretched bowstring. The impact pulls the bowstring downward into a V-shape. If the weight is let fall from the double height, the V is deeper. These two V-shapes can be reproduced statically by hanging weights on the bowstring; the two weights that produce the two V's are as 1:2. Galileo assumes that the effect is proportional to speed, and so concludes that speed is proportional to distance of fall.

This is a mistake. How did Galileo correct himself? The scholars of this century have argued over whether Galileo was basically an experimentalist, or a desk mathematician.[14] He was both. In the case of

---

13. Galileo, *Dialogue Concerning the Two Chief World Systems,* tr. Stillman Drake (Berkeley, CA: University of California Press, 1962), 229.

14. Alexandre Koyré was a prominent proponent of the platonist Galileo; see his *Études Galiléennes* (Paris: Hermann, 1939) and *Metaphysics and Measurement* (Cambridge, MA: Harvard University Press, 1968), chapters I-IV. Thomas Settle showed in 1961 that Galileo's inclined plane experiment could be performed with remarkable precision, using only means that would have been available to Galileo: see Thomas B. Settle, "An Experiment in the History of Science," *Science*, 133 (1961): 19-23. Next, Stillman Drake took up the empiricist theme, and in 1973 published an article entitled "Galileo's Discovery of the Law of Free Fall" in *Scientific American* (May, 1973): 85-92. Unfortunately his enthusiasm and imagination got out of hand, and his reconstruction in this case (as in some other cases) cannot be sustained.

the motion of natural fall it was by experiment that he emerged from error. The key experiment is a highly sophisticated, indeed a masterly experiment.[15]

Galileo arrived at the correct variation of speed in free fall only in 1608 or 1609, 20 years after the Leaning Tower demonstration. This discovery presupposed the prior discovery of the parabolic trajectory of projectiles. He clinched both discoveries using the apparatus diagrammed in **Exhibit F.**

Exhibit F:

**Apparatus for testing parabolic trajectory
and law of free fall**

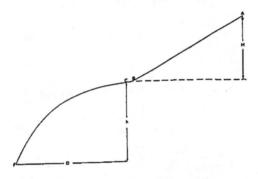

AB = grooved inclined plane of height H;
BC = grooved curve leading to horizontal projection at C;
**CF = path of projectile when falling through height h and
advancing through distance D.**

Suppose, first, that the inclination of, the inclined plane is fixed, and that the ball in repeated rolls is started each time from the same point A at height H, measured from the tabletop on which the inclined plane rests. Then on reaching C it will be projected horizontally, always with the same horizontal speed so long as the initial conditions are unchanged. It will fall in a curve. What curve? Suppose that the height h can be adjusted, by raising or lowering the horizontal board onto which the ball falls. Then it becomes possible to accumulate a series of paired values of D and h, all of them pertaining to the same curve. Galileo

---

15. I am chiefly dependent here on R. H. Naylor, "Galileo's Theory of Projectile Motion," in *Isis* 71 (1980): 550-70, and David K. Hill, "Dissecting Trajectories," *Isis* 79 (1988): 646-68.

suspected it was a semi-parabola with vertex at C. In that case h must vary as D^2. There is evidence that Galileo carried out this test. The semi-parabolic result is exactly what we would expect if h increases as the square of the time, while D increases linearly with the time.

Having confirmed the parabolic shape, Galileo turns the experiment around, and uses the parabolic shape to examine the speeds achieved in fall through different heights H along the inclined plane. The distances D turn out to vary as the *square root* of H. The relevant calculations are given in some detail in **Exhibit G.**

Exhibit G: Essential details of folio 116v

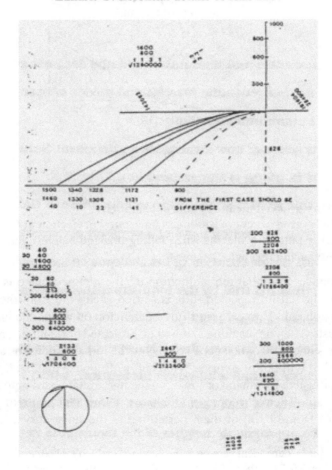

(For an explanation of these calculations, see the addendum at the end of the lecture.)

The genius of this experiment is that it avoids measuring time, so much more difficult for Galileo than for us, and it gives a direct measure of instantaneous speed in descent along an inclined plane; the final speed in the descent is turned by the curved groove BC into a uniform horizontal speed proportional to the distance D; and then D proves to be as the square root of H.

In summary, Galileo's twenty-year struggle was brought to a successful conclusion by a combination of mathematical reasoning and sophisticated experimentation. A few years later, in a letter to his former student Benedetto Castelli, Galileo wrote:

> Nature is inexorable and immutable, and she does not care at all whether or not her recondite reasons and modes of operation are revealed to human understanding.[16]

I take that to express his sense of how formidable an opponent Nature is, in the contest that consists in trying to understand her.

The remainder of this lecture is about the middle phase of Galileo's life, beginning in July 1609 with his getting news of a spyglass constructed in Flanders, and ending with the publication of his *Dialogue on the Two Chief World Systems* in 1632, then his trial by the Inquisition the following year, and his condemnation for heresy in June 1633.

In July 1609 Galileo built his own first telescope. In August he built a better one, and by December he had a twenty-power instrument, which he turned on the Moon. He saw mountains that cast shadows. From the lengths of the shadows he reckoned the approximate heights of the mountains in units of terrestrial distance. The Moon reflected sunlight, not as a polished mirror does, but as do sand, dirt, and rock; it looked rather Earth-like. Maybe plants could grow on it. Better dirt, Galileo thought, than jasper or diamond.

His report was met with incredulity. The lunar mountains, Christopher Clavius opined, were illusions of Galileo's telescope. On getting a better telescope, Clavius had to admit the appearances were as Galileo stated, but he wanted the Moon to be smooth, spherical and crystalline. Couldn't Galileo's mountains, he asked, be differences in density? To such suggestions Galileo had this reply:

---

16. Quoted from Maurice A. Finocciaro, *The Galileo Affair: a Documentary History* (Berkeley: University of California Press, 1989), 50.

[I]f we still want to let anyone imagine whatever he pleases, and if someone says that the Moon is spherically surrounded by transparent invisible crystal, then I shall willingly grant this—provided that with equal courtesy it is permitted me to say that this crystal has on its outer surface a great number of enormous mountains, thirty times as high as terrestrial ones, which, being of diaphanous substance, cannot be seen by us . . . . The only fault here is that it is neither demonstrated nor demonstrable.[17]

In January 1610 Galileo discovered four small stars accompanying Jupiter. They appeared as if situated on a straight line passing through the planet and at right angles to the line of sight. In successive hours and on successive nights they individually changed their distances from Jupiter, passing from the east to the west of it and back again. By early 1611 Galileo was able to assign fairly accurate orbital periods about Jupiter to all four, assuming uniform circular motion. Thus, contrary to Aristotle, the Earth was not the only body about which celestial bodies move in circular motion.

Galileo called these stars Medicean after the ruling family of Tuscany, and so wangled for himself a position as "mathematician and philosopher to the Grand Duke of Tuscany." He could return to his native Florence.

Late in 1610 Venus was far enough from the Sun for telescopic observation, and Galileo found that it had phases like the Moon's. It had been considered self-luminous, but the phases were evidence that it shone by reflected sunlight. Galileo found Venus's diameter to vary over time by a factor of about 6; it was largest when Venus is a crescent, and smallest when it appears as a circular disk. Hence its orbit surrounded the Sun. By similar observations of Mars, he found this planet when 90° from the Sun to be gibbous, that is, not fully round; its diameter also varied by a factor of about 5, being largest when the planet was in opposition to the Sun. The Martian orbit therefore surrounded the Sun, and the Earth as well.

Continued improvement of his tables for the Jovian satellites led Galileo to a new discovery in July 1612. Comparing an observation with his tables, he realized that one of the satellites, the outermost of

---

17. Stillman Drake, *Galileo at Work* (Chicago: University of Chicago Press, 1978), 168-69.

the four, had been eclipsed, passing into the shadow cast behind Jupiter by the Sun. He found that, if he took the mid-points of the satellite eclipses for epochs or starting-times, his tables became more accurate. The satellites were moving more uniformly with respect to the line from the Sun through Jupiter than with respect to the line from the Earth through Jupiter. The heliocentrist would expect this. It is also what would be expected under the semi-heliocentric arrangement, where the Earth remains at rest, the other planets go round the Sun, and the Sun circles the Earth: the so-called Tychonic system.[18]

Sunspots were observed from 1610 onwards. A German Jesuit, Christoph Scheiner, writing under the pseudonym Apelles, put forward the idea that they were planets, hence compatible with celestial immutability. Galileo, citing his own careful observations and measurements, destroyed this hypothesis with merciless sarcasm. The spots moved round the Sun with changes in shape and mutual distances which implied they were on or very close to the Sun's surface. The Sun must be rotating, with a period of about 25 days, about an axis through its center. The spots could be seen coming to be, coalescing, separating, ceasing to be. The Sun was a mutable body.

Following these telescopic discoveries, Galileo became for the first time a public proponent of Copernicanism. In his student days, he had opposed this doctrine, listing the standard dynamical objections against it: bodies let drop from a height would not fall vertically to the ground, and so on. His early *De Motu* shows that by 1590 he had studied Ptolemy's *Almagest* and Copernicus's *Revolutions* with care. In a late revision of the *De Motu*, he introduced the idea that a spherical body at the center of heavy things, if set rotating, would continue to rotate without the need for an internal or external mover.[19] Such a motion he called *neutral,* distinguishing it thus, both from the natural downward motion of a heavy body, where there is an internal *archē,* and from a forced motion where there is an external mover. A consequence, though Galileo does not mention it, is that the daily apparent westward rotation of the stellar sphere could be accounted for by supposing the Earth to

18. Stillman Drake, *Galileo: Pioneer Scientist* (Toronto: University of Toronto Press, 1990), 145-55, mistakenly characterizes this as compelling evidence for the Earth's annual motion about the Sun.
19. See *Galileo Galilei On Motion and On Mechanics*, 72-74; Stillman Drake, "The Evolution of *De Motu*," *Isis* 67 (1976): 245.

be rotating eastward about its polar axis. Galileo also considered as neutral the motion of a ball rolling on a polished horizontal surface concentric with the Earth's center; the smallest force would set it moving, and it would then move forever unless impeded.[20] An extension of this was that a body dropped from a tower on a rotating Earth, since it shares the tower's motion, would fall to the tower's base. The standard dynamical objections to the Earth's diurnal rotation, Galileo now realized, were without basis.

In 1597 Kepler sent Galileo a copy of his first book, *The Cosmographic Mystery*, which was outspokenly Copernican. Galileo in responding stated that he had held the Copernican view for some years, but had refrained from defending this position publicly,

> intimidated by the fortune of our teacher Copernicus, who though he will be of immortal fame to some, is yet by an infinite number (for such is the multitude of fools) laughed at and rejected.[21]

By 1613, Galileo's telescopic discoveries had shown Ptolemaic astronomy to be untenable. The semi-heliocentric arrangement, on the other hand, might still seem an option: it gave much of the economy of the Copernican system, without putting the Earth in motion. Galileo could refute arguments *against* the Earth's motion; what arguments did he have *for* its motion? He was averse to the wildly speculative theological symbolism that made Kepler a Copernican.[22] For the Earth's motion, he wanted terrestrial evidence. And already, in the 1590s, he thought he had found it: in the tides.

This is a vexed topic. Why, people ask, was Galileo so stupid as to propose a tidal theory that doesn't agree with Newton's? Galileo, of course, died 11 months before Newton was born, so couldn't learn from him. But what these people fail to realize is how much wrong guessing, by intelligent men, went on before a correct understanding of tides emerged. Galileo's theory, incidentally, was right in a respect in which

---

20. *Galileo Galilei On Motion and On Mechanics*, 171.

21. Galileo, *Opere,* Edizione Nazionale, Vol. 10, 68

22. Like Kepler, however, Galileo entertained the idea, "as not entirely unphilosophical," that the motion of the planets had its cause in the rotating Sun; see his "Letter to the Grand Duchess Christina" in *Discoveries and Opinions of Galileo,* trans. and ed. Stillman Drake (New York: Doubleday, 1957), 212-13.

Newton missed the mark.[23] First, however, what *was* Galileo's mistake? In his dialogue of 1632, he says,

> Of all great men who have philosophized on such a puzzling effect of nature [the tides], I am more surprised about Kepler than about anyone else; although he had a free and penetrating intellect and grasped the motions attributed to the Earth, he lent his ear to the dominion of the Moon over the water, to occult properties, and to similar childish ideas.[24]

In Galileo's view, attractions, sympathies, antipathies, were unscientific notions. He wanted mechanical explanations, preferably such as could be embodied in an actual model of brass, wood, water, and so on. Among seventeenth-century thinkers, he was not alone in thus restricting himself.

Among Galileo's contemporaries who attributed the tides to the Moon's attraction, none could account for there being *two* tides per day. The Moon crosses our meridian once every 25 hours, but, in Galileo's Venice as in our Chesapeake Bay, two high tides occur in that interval. So far as I know, the first to show why this should be so was Newton. What counts is differential attraction. When the Moon is on our meridian, it attracts the nearby waters more strongly than the Earth's center, and the Earth's center more strongly than the waters on the opposite side of the Earth. So when there is a high tide for us, there should be a high tide on the opposite side of the Earth as well.

Another difficulty is that high tide occurs, not when the Moon is overhead, but hours later; the delay varies from place to place. Why?

An observation that especially interested Galileo was this. In Venice, at the head of the Adriatic, the difference between high and low tide was about six feet, whereas at Dubrovnik, close to where the Adriatic opens into the Mediterranean, it was only a few inches. How explain that?

Galileo's theory had two main parts. The first of these is wrong, but not for the reasons usually given. Galileo supposes that, because

23. I am indebted here to a recent study by Paulo Palmieri, "Re-examining Galileo's Theory of Tides," *Archive for History of Exact Sciences* 53 (1998): 223-375.
24. *Galileo on the World Systems,* trans. Maurice A. Finocchiaro (Berkeley: University of California Press, 1997), 304; Galileo, *Dialogue Concerning the Two Chief World Systems* (Drake translation), 462.

the Earth has two motions, the diurnal rotation and the annual motion, the waters of the ocean are alternately accelerated and decelerated with a periodicity of one day. See **Exhibit H,** where the smaller circle is the Earth, the larger circle the Earth's annual orbit. At B, that is, at midnight, the diurnal and annual motions add together; at D, that is, at noon, the diurnal motion subtracts from the annual; at intermediate places the Earth's surface speed has intermediate values.

Exhibit H

This variation in surface speed, Galileo says, disturbs the waters; they slosh back and forth in their ocean basins, seeking to return to equilibrium. Does such a disturbance really occur? From a Newtonian point of view, Galileo's account is inadequate. Two accelerative fields are being combined. The combination would produce a disturbance, *if* the law of gravity were any other than an inverse-square law; a term from the inverse-square law cancels it out. This, to be sure, was not understood by Galileo or Newton or anybody till recently. Critics had better beware, but yes, Galileo was wrong.

But now, suppose the disturbance occurred. According to the second part of Galileo's theory, the resulting pendulum-like motion will have a characteristic period, determined by the size and shape of the basin. It will vary, Galileo says, as the length of the basin, and inversely as its depth. Actually, the period varies as the *square-root* of the length, but Galileo's second point, about the depth variation, is non-intuitive; presumably he discovered it by experiment. Apparently, by experiments with water in long basins, he observed that the water in the middle does not rise and fall, but moves back and forth, while the water at either end of the basin moves up and down. It was thus that he explained to himself the difference in the heights of the tides in Venice and Dubrovnik.

The study of the characteristic frequencies of ocean basins is a central feature of present-day tidal theory. The initial disturbance is caused, not as Galileo supposed, but by the gravitational attractions of the Moon and Sun. Newton knew that the shapes of ocean basins had something to do with the tides, but he lacked the mathematics for treating the problem. The first to give a correct mathematical formulation of the pendulum-like component of tidal motion was Laplace, a century after Newton.

Let these remarks suffice as to what is wrong and right about Galileo's theory of the tides.

Galileo's campaign for heliocentrism roused the biblical fundamentalists. In December of 1613, the Medicis invited Castelli, professor of mathematics in Pisa, to dinner. The Grand Duchess Christina pressed him to defend the compatibility of heliocentrism with the miracle reported in Chapter 10 of the Book of Joshua, where Joshua says in the sight of Israel, "Sun, stand thou still." Castelli reported the evening's discussion to Galileo, who responded with a long letter on the same theme. The letter was copied and circulated widely.

On December 21, 1614, Tommaso Caccini, a young firebrand of a Dominican, preached a sermon in Florence's Santa Marta Novella, denouncing the Galileists and all mathematicians as practitioners of diabolical arts and enemies of true religion. On the following February 7 Niccolò Lorini, a pious elderly Dominican, sent to the Inquisition in Rome a copy of Galileo's letter to Castelli, here and there altered maliciously, to be examined for heretical content. Meanwhile Galileo had expanded the letter into a longer letter to the Grand Duchess Christina.

Scripture, Galileo quotes a churchman as saying, tells how to go to heaven, not how heaven goes. It has to do with faith and morals, not with the make-up of the natural world. It is addressed to uneducated folk, and must speak their language. For the sake of theological consistency, some of its statements must be interpreted metaphorically: God does not literally stretch forth a hand, or have a backside, or get angry. As for implicit or explicit assertions in Scripture about the natural world, Galileo took his cue from Augustine's treatise, *On the literal interpretation of Genesis*. If natural philosophers have established a fact by observation or strict demonstration, their conclusion must take precedence over the literal interpretation of Scripture. And, Galileo went on to insist, the Church should remain uncommitted on matters where the fact has not yet been, but might be thus established. Such a matter, he urged, was the Copernican hypothesis.

Two comments. Galileo here assumed that faith and natural philosophy do not and cannot conflict. This seemed to him obvious. For us today, it can be a more difficult question.

Secondly, Galileo identified strict science with truths established by observation or by demonstration.[25] By observation, for instance, he had established that the Moon is mountainous. The propositions of Euclid had been established by demonstration from premises which he took to be indubitable. In the case of the Copernican hypothesis, he knew of no indubitable premises from which it could be derived. Thus in arguing from the tides, Galileo argued ex suppositione, presupposing the Earth's motion. The explanation could become an established truth only if all possible alternatives were disproved. In his *Dialogue* of 1632 Galileo has Salviati say: "We have established the impossibility of explaining the motions observed in the tides while simultaneously maintaining the immobility of the containing vessel."[26] That is a claim to have excluded the alternatives. Evidently Galileo underestimated the difficulty of doing that. Exclusion of all alternatives, in any ultimate sense, is probably impossible. But the point I want to make is that Galileo did not articulate a practicable methodology for the new science.

In December 1615, against the advice of the Tuscan ambassador, Galileo went to Rome to try to clear his name of the suspicion of heresy and to campaign against the suppression of the Copernican theory. He was an ardent campaigner. One Roman witness reported in January 1616:

> He discourses often amid fifteen or twenty guests who make hot assaults upon him, now in one house, now in another. Monday . . . he achieved wonderful feats; and what I like most was that, before answering the opposing reasons, he amplified them and fortified them himself with new grounds which appeared invincible, so that, in demolishing them subsequently, he made his opponents look all the more ridiculous.[27]

His efforts were to no avail. On 24 February 1616, theological consultants appointed by the Holy Office to assess Galileo's Copernicanism reported to the Pope as follows:

---

25. See Ernan McMullin, "The Conception of Science in Galileo's Work," in *New Perspectives on Galileo,* ed. Robert E. Butts and Joseph C. Pitt (Dordrectr: D. Reidel, 1978), 209-57.

26. *Galileo on the World Systems,* 288.

27. See Giorgio Di Santillana, *The Crime of Galileo* (Chicago: University of Chicago Press), 112-13.

Exhibit J

Propositions to be assessed:
   (1) The Sun is the center of the world and completely devoid of local motion.
   Assessment: All said that this proposition is foolish and absurd in philosophy, and formally heretical since it explicitly contradicts in many places the sense of Holy Scripture, according to the literal meaning of the words and according to the common interpretation and understanding of the Holy Fathers and the doctors of theology.
   (2) The Earth is not the center of the world, nor motionless, but moves as a whole and also with diurnal motion.
   Assessment: All said that this proposition receives the same judgment in philosophy and that in regard to theological truth it is at least erroneous in faith.[28]

On the following day, 25 February, he wrote:

Exhibit K

His Holiness [the Pope] ordered the most illustrious Lord Cardinal Bellarmine to call Galileo before himself and warn him to abandon these opinions; and if he should refuse to obey, the Father Commissary, in the presence of a notary and witnesses, is to issue him an injunction to abstain completely from teaching or defending this doctrine and opinion or from discussing it; and further, if he should not acquiesce, he is to be imprisoned.[29]

And on 26 February:

Exhibit L

   At the palace of . . . the said Most Illustrious Lord Cardinal Bellarmine, . . . and in the presence of the Reverend Father Michelangelo Segizzi, . . . Commissary of the Holy Office, having summoned the above-mentioned Galileo before himself, the same Most Illustrious Lord Cardinal warned Galileo that the above-mentioned opinion was erroneous and that he should abandon it; and thereafter, indeed immediately, . . . the aforesaid Father Commissary, in the name of His Holiness the Pope and the whole Congre-

---

28. Maurice A. Finocchiaro, *The Galileo Affair: A Documentary History* (Berkeley: University of California Press, 1989), 146.

29. Ibid., 147.

gation of the Holy Office, ordered and enjoined the said Galileo, who was himself still present, to abandon completely the above-mentioned opinion that the Sun stands still at the center of the world and the Earth moves, and henceforth not to hold, teach, or defend it in any way whatever, either orally or in writing, otherwise the Holy Office would start proceedings against him. The same Galileo acquiesced in this injunction and promised to obey.[30]

According to the Pope's command, please recall, an injunction was to be imposed *only* if Galileo refused to obey the initial order to abandon his erroneous opinions. There is no evidence that Galileo refused, so the injunction would appear to be illegal. Another suspicious circumstance is that it is not signed, as injunctions usually were. Is the document a forgery, as some scholars have supposed? Or does its remaining unsigned mean that Bellarmine and Segizzi were at loggerheads? Bellarmine was chief theological adviser to the Pope, and a Jesuit; Segizzi, head of the Inquisition, was a Dominican. The Dominicans had for centuries had charge, not only of the Inquisition, but of all questions relating to theological orthodoxy. Segizzi can have been jealous of Bellarmine, or suspected him of leniency in the Galileo matter. Bellarmine may have refused to sign the document, or told Galileo to ignore it as illegal. We do not know.

The rumor circulated that Galileo had been forced to abjure, that is, to renounce under oath his Copernicanism, and had been given salutary penances. In May Galileo asked, and received from Bellarmine, a certificate denying this rumor. It asserted that

Exhibit M

he has only been notified of the declaration made by the Holy Father and published by the Sacred Congregation of the Index, whose content is that the doctrine attributed to Copernicus . . . is contrary to Holy Scripture and therefore cannot be defended or held.

This certificate was signed by Bellarmine and given to Galileo. To say with Bellarmine that the Copernican doctrine could not be defended or held, was not to say with Segizzi that it could not be taught or discussed in any way whatever. In the schools, heretical doctrines

---

30. Ibid., 147-48.

were commonly discussed, even debated, in order that they might be understood.

In 1623 Maffeo Barbarini, an educated Florentine and a friend and admirer of Galileo, became Pope Urban VIII. The event was hailed as the dawn of a new, liberal-minded regime. In the spring of 1624 Galileo went to Rome, and obtained the pope's permission to write a dialogue on the two systems of the world, Ptolemaic and Copernican, geocentric and geokinetic. Urban did not fear that Copernicanism would be proven true. Though we might be unable to account for the tides except on the geokinetic theory, God, being omnipotent, could bring them about in a different way. Galileo, Urban ordered, should feature this argument in his dialogue. In non-theological language, it says that our explanations are always hypothetical.

An ambiguity, let me say, lurks in this word "hypothesis." Its accepted meaning, in Galileo's day, was instrumentalist. A hypothesis was a likely story, useful for prediction, but without further claim to truth. It was *mere* hypothesis. Much astronomical theory in that day cannot be viewed otherwise. But a hypothesis can have a different meaning, which I shall call fallibilist. The *fallibilist* does not know the ultimate truth of his hypothesis, but he pursues it as possibly revealing a piece of the system of the world. In support of this hope, or faith, he looks to the logical economy of the hypothesis, its aesthetic aptness, the reach of its pragmatic success. I suspect that to Galileo the initial appeal of the Copernican arrangement was a fallibilist appeal. But in Galileo's basic, Artstotelian conception of science, science consisted of empirical facts together with necessary demonstrations. Such a conception was inadequate to the needs of the new science.

Well, Galileo set about writing his dialogue. He was now 60. For 20 years arthritic attacks had kept him in bed for days at a time; progress was slow. The manuscript was at last completed toward the end of 1629. It included a new argument for heliocentrism: the changing paths of sunspots during the year showed that the Sun's axis of rotation was not perpendicular to the ecliptic, but leaned to one side by about 7°; the appearances could be saved under the Copernican hypothesis by supposing the Sun to rotate in place about a fixed axis. In a geostatic theory, by contrast, this axis had to be given both annual, and daily, conical gyrations. Galileo regarded them as implausible. In fact, they are mechanically inexplicable.

With delays and some alterations, the manuscript got through the censors, and went to press. The book was published in February 1632, when Galileo was 68.

Galileo's dialogue is a genuine dialogue, addressing issues that were current and lively in the first half of the seventeenth century. Readers open to its arguments were enthusiastic; opponents ground their teeth. Simplicio, the interlocutor whose task it is to defend the traditional geostatic cosmology, cites contemporary authors for support, Chiaramonte and Locher. But the anti-geokinetic arguments of these authors don't hold up; Salviati, the spokesman for Copernicus, annihilates them. For the major thinkers of the seventeenth century, Galileo's *Dialogue* wrote "Finito!" to the traditional dichotomy between the celestial and the sublunary worlds, and "Finito!" to the traditional arguments against the Earth's motion. Simplicio's role is to defend the indefensible. Despite befuddlement, he remains, through the 450 pages of dialogue, admirably, unbelievably, courteous.

Salviati in the *Dialogue* claims to be wearing a mask, to be playing a role in arguing for the Copernican position. He denies he is demonstrating it, but says that he does not see how sunspot paths and the tides could be suitably accounted for except on a geokinetic theory. Argument *ex suppositione*, with the alternatives dismissed.

The third interlocutor, Sagredo, is the eager listener, intent on Salviati's argument, anticipating its conclusions, objecting in order to elicit clarification. A keen observer of natural effects, he is excited by the explanatory possibilities of the heliocentric hypothesis. Mustn't Galileo have shared this openness to learning that he describes so charmingly in Sagredo?

From the beginning, unfavorable comments about the *Dialogue* circulated in Rome. And now the injunction of 1616 was brought forth from the Inquisition archives, and the Pope was informed of its content. Just at this juncture, in the spring of 1632, Urban VIII was facing an international crisis of gigantic proportions. For 8 years he had pursued balance-of-power politics in alliance with Cardinal Richilieu in France, and in accommodation of the Protestants in Germany. Now Gustavus Adolphus, the Swedish Protestant general, had invaded Bavaria, the center of German Catholicism, and had sacked the Jesuit Colleges there. Urban had to realign the papacy with the Spanish Hapsburgs. Meanwhile, the Spanish cardinals were charging Urban with leniency in the fight against heresy; he was threatened with impeachment unless he took a more forceful stand.[31]

To the pope, thus pressured, Galileo's failure to inform him of the injunction was treachery in his own backyard. He was outraged.

In the summer of 1632, sales of Galileo's *Dialogue* were stopped in the papal states, and all copies were confiscated. A specially appointed commission examined the book and concluded that Galileo had violated the injunction. The Tuscan government, of which Galileo was an employee, attempted to forestall a trial, without success. The first interrogation took place on 12 April 1633. Asked about the events in 1616, Galileo stated that he had been given an oral warning by Cardinal Bellarmine that the Earth's motion could neither be held nor defended, but only discussed hypothetically. He denied having received a special injunction prohibiting him from discussing this motion in any way whatsoever; as evidence for this, he produced Bellarmine's certificate. He denied that his *Dialogue* held or defended the Earth's motion; rather, it showed that the arguments for it were not conclusive.

Bellarmine and Segizzi were dead; Galileo was the sole surviving witness of the events of 26 February 1616. The strongest charge, of disobedience to a papally imposed injunction, was fatally weakened. But, Urban insisted, a sentence there must be. In a private conference, Maculano, now Commissary of the Holy Office and chief judge in the trial, persuaded Galileo to plead guilty to a lesser charge, promising in return a light sentence. Galileo re-read his *Dialogue*, and deposed as follows on 30 April:

Exhibit N

I freely confess that [my book] appeared to me in several places to be written in such a way that a reader, not aware of my intention, would have had reason to form the opinion that the arguments for the false side, which I intended to confute, were so stated as to be capable of convincing because of their strength, rather than being easy to answer.[32]

---

31. See Pietro Redondi, *Galileo Heretic,* tr. Raymond Rosenthal (Princeton: Princeton University Press, 1987), 228-33. Redondi's larger thesis that there was a hidden agenda in the Galileo trial, pertaining to Galileo's atomism and its relation to the Eucharist, has not been accepted by other scholars.

32. Finocchiaro, *The Galileo Affair,* 278.

Galileo's excuse for having given this impression was that he was more desirous of glory than was suitable; he wanted to appear clever:

My error then was, and I freely confess it, one of vain ambition, pure ignorance, and inadvertence.[33]

In confessing to ambition and a desire to appear clever, Galileo, I believe, was honest. He was a proud man, product of a Florentine tradition of gentility, culture, and independent thought that went back to the 15th century, before the age of despotism and excessive bowing and scraping. But had he really intended to make the arguments for Copernicanism appear weak, as he claimed? That claim was disingenuous.

What else could he have said? In his letter to the Grand Duchess Christina he had written:

Exhibit O

[T]o command that the very professors of astronomy themselves see to the refutation of their own observations and proofs as mere fallacies and sophisms is to enjoin something that lies beyond any possibility of accomplishment....Before this could be done they would have to be taught how to make one mental faculty command another, and the inferior powers the superior, so that the imagination and the will might be forced to believe the opposite of what the intellect understands.[34]

The Pope was not satisfied. He ordered that Galileo be interrogated under the formal threat of torture in order to determine his intention, a standard procedure. Whatever the outcome, he was to abjure publicly, to be held under arrest at the Inquisition's pleasure, and his book was to be banned. On 21 June the interrogation was carried out, Galileo maintaining the innocence of his intention. On the following day he was read the sentence, and he recited the formal abjuration.

So Galileo lost the battle. His writings were instrumental in winning the war. Not, to be sure, in the papal states, where natural science sputtered to a stop, but elsewhere in Europe, where Galileo's *Dialogue* appeared in Latin and in English, and his last work, The *Two*

---

33. Ibid.

34. *Stillman Drake, Discoveries and Opinions of Galileo*, 193.

New Sciences, also appeared. The problems of inertial motion that Galileo had posed within a pre inertial framework were solved within an inertial framework by Huygens, Newton, and others. Newton, who had read the *Dialogue*, says that Galileo discovered the law of free fall and the parabolic path of projectiles by applying Newton's first two laws of motion: an impossible feat of anachronism, but science was hastening on and leaving its history behind.

In articulating his *Rules of Philosophizing*, Newton quoted a line from the *Dialogue* that Galileo had given in Latin and attributed to Aristotle: *frustra fit per plura quod potest fieri per pauciora* ("in vain is that done with many that can be accomplished with fewer").[35] It was a slogan enjoining logical economy. To Galileo, logical economy was a hopeful clue to hidden system. By late 1684, Newton, applying the slogan, had reached the result that the solar system's center of gravity was not at the Sun's center but near it. He had, he claimed, "proved the Copernican system apriori." It was more than a determined sceptic would grant. But the hypothesis was showing its power, preparing for a pragmatic success that would leave its rivals in the shade. The science that Galilee initiated by showing that we can be in motion without knowing it has flourished. Generalized and formalized in successive steps by Newton, by Lagrange, by Einstein, its fruitfulness has not yet been exhausted.

---

35. Galileo, *Dialogue Concerning the Two Chief World Systems* (tr. Stillman Drake, University of California Press, 1962), 123.

Addendum

The Calculations of Folio 116v

In general: $D^2 \propto H$, or $D_1^2 : D_2^2 :: H_1 : H_2$; therefore $D_1^2 = D_2^2 \left(\frac{H_1}{H_2}\right)$.

Upper Middle: $H_1 = 600, H_2 = 300, D_2^2 = 800 \times 800$.

$D_1 = \sqrt{800 \cdot 800 \cdot 600/300} = \sqrt{1280000} = 1131$

Lower Left: $H_1 = 800, H_2 = 300, D_2^2 = 800 \times 800$.

$D_1 = \sqrt{800 \cdot 800 \cdot 800/300} = \sqrt{1706666} = 1306$

Upper Right: $H_1 = 828, H_2 = 300, D_2^2 = 800 \times 800$.

$D_1 = \sqrt{800 \cdot 800 \cdot 828/300} = \sqrt{1766400} = 1329$

Lower Middle with numbers to Right: $H_1 = 1000, H_2 = 300, D_2^2 = 800 \times 800$.

$D_1 = \sqrt{800 \cdot 800 \cdot 1000/300} = \sqrt{2133333} = 1460$

Some Reflections on Darwin and C.S. Peirce
(2012)

Curtis Wilson and Chaninah Maschler

Introduction

On a Saturday morning in the mid-1950s, I attended a St. John's faculty seminar on a selected reading from Darwin's *Origin of Species*. What chiefly remains in memory is an overall impression: the discussion was halting and desultory, failing to get airborne. In those days the available edition of the *Origin* was the sixth and last (1872); compared with the first edition of 1859, it suffers from excessive backing and filling, Darwin's attempts to answer his critics. Yet, even had our text been from the sprightlier first edition, I doubt our discussion would have got off the ground. After one spell of silence a senior tutor spoke up to ask: Isn't it [Darwin's theory] just a hypothesis? The implication, I thought, was: Can't we just ignore the whole idea?

The short answer to that second question is: we can't, because Darwin's theory is the grand working hypothesis (yes, it's a hypothesis!) of biologists everywhere, and as aspirant generalists at St. John's, we need to seek out its meaning. The search can be exhilarating *as well as* disquieting.

Major features of Darwin's theory are contained in his phrase "descent with modification through natural selection." The descent of present-day organisms from organisms of preceding generations is obvious; Darwin requires us to keep this fact in focus. The offspring inherit traits from their parents, but some variation occurs. Since far more offspring are produced than can survive and reproduce, the variants best suited to surviving and reproducing are the ones that win out. Relative to a given environment, the surviving form will be better adapted than the forms that failed. Darwin saw this process as leading to diversification of kinds, or *speciation,* as indicated by the title of his book, *On the Origin of Species.*[1]

The introduction of this article was written by Mr. Wilson, Part 1 by Ms. Maschler, and the final three parts by the two authors in collaboration.

---

1. According to Ernst Mayr in his *One Long Argument: Charles Darwin and the Genesis of Modern Evolutionary Thought* (Cambridge, Mass.: Harvard University Press, 1991), what later authors think and speak of as "Darwin's Theory" is a combination of four or five strands—evolution as such, common descent, multiplication of species, gradualism, and natural selection.

Darwin opened his first notebook on "Transmutation of Species" in July, 1837. In a sustained effort of thought from 1837 to 1844, he constructed the theory. The empirical evidence consisted chiefly of the biological specimens that he had observed and collected during his tour as naturalist aboard H.M.S. Beagle, from December 27, 1831 to October 2, 1836. (This voyage was sent out to chart the coasts of South America and determine longitudes round the globe; taking along a naturalist was an afterthought of the captain's.)

At the beginning of the Beagle voyage, Darwin was a few weeks short of his twenty-third birthday. So far in his life he had had no clear goal. Enrolled in medical school at age sixteen in Edinburgh, he dropped out, unable to endure seeing patients in pain. His father (a physician, skeptical in religion) then sent him to Cambridge with the idea that he might fit himself out to become a country parson, but young Darwin found the course of study uninteresting. He completed the A.B. degree, but later acknowledged that his time at Cambridge was mostly wasted. A chance by-product of it was a friendship with John S. Henslow, the professor of botany. Henslow it was who arranged Darwin's being offered the post of naturalist on the Beagle. Darwin's father flatly rejected the idea at first, but Josiah Wedgewood, young Darwin's maternal uncle, persuaded him to change his mind.

In hindsight, we can say that young Darwin was admirably suited to his new post. From boyhood he had been a persistent collector of a variety of objects, from stamps to beetles. As a naturalist he would prove to have an unstoppable drive toward theoretical understanding, seeking to connect the dots between his numerous observations. The voyage of the Beagle, proceeding first to the coasts of South America and the nearby islands, could not have been more aptly planned to yield observations supporting the theory that he would develop. The observations were chiefly of three types.[2] Fossils from South America were found to be closely related to living fauna of that continent, rather than to contemporaneous fossils from elsewhere. Animals of the different climatic zones of South America were related to each other rather than to animals of the same climatic zones on other continents. Faunas of nearby islands (Falkland, Galapagos) were closely related to those of the nearest mainland; and on different islands of

---

2. See Ernst Mayr's Introduction to Charles Darwin, *On the Origin of Species: A Facsimile of the First Edition* (Cambridge, Mass.: Harvard University Press, 1972), xii.

the same island group were closely related. These observations could be accounted for on Darwin's theory; on the opposing theory of fixed species they remained unintelligible.

But why the uproar over Darwin's *Origin*, and why does it still today produce uneasiness? It is not merely that it appears contrary to the creation story in Genesis. As John Dewey put it in 1910:

> That the publication of the *Origin of Species* marked an epoch in the development of the natural sciences is well known to the layman. That the combination of the very words origin and species embodied an intellectual revolt and introduced a new intellectual temper is easily overlooked by the expert. The conceptions that had reigned in the philosophy of nature and knowledge for two thousand years, the conceptions that had become the familiar furniture of the mind, rested on the assumption of the superiority of the fixed and final; they rested upon treating change and origin as signs of defect and unreality. In laying hands upon the sacred ark of absolute permanency, in treating the forms that had been regarded as types of fixity and perfection as originating and passing away, the *Origin of Species* introduced a mode of thinking that in the end was bound to transform the logic of knowledge, and hence the treatment of morals, politics, and religion.[3]

More recently Ernst Mayr has characterized Darwin's new way of thinking as "population thinking," and the mode of thinking prevalent earlier as "typological thinking":

> Typological thinking, no doubt, had its roots in the earliest efforts of primitive man to classify the bewildering diversity of nature into categories. The *eidos* of Plato is the formal philosophical codification of this form of thinking. According to it, there are a limited number of fixed, unchangeable "ideas" underlying the observed variability, with the *eidos* (idea) being the only thing that is fixed and real, while the observed variability has no more reality than the shadows of an object on a cave wall. . . .
>
> The assumptions of population thinking are diametrically opposed to those of the typologist. The populationist stresses the uniqueness of everything in the organic world. What is true for the human species—that no two individuals are alike—is equally true for all other species of animals and plants. . . . All organisms and organic

---

3. John Dewey, *The Influence of Darwin on Philosophy* (New York: Henry Holt, 1910).

phenomena are composed of unique features and can be described collectively only in statistical terms. Individuals, or any kind of organic entities, form populations, of which we can determine the arithmetic mean and the statistics of variation. Averages are merely statistical abstractions, only the individuals of which the populations are composed have reality. The ultimate conclusions of the population thinker and of the typologist are precisely the opposite. For the typologist, the type (*eidos*) is real and the variation an illusion, while for the populationist, the type (average) is an abstraction and only the variation is real. No two ways of looking at nature could be more different.[4]

Mayr's abruptly nominalist "take" on the nature of species is not required by Darwin's theory, nor do all biologists espouse it.[5] One thing the theory does require is a new attention to individual differences. Species may result from processes that are fundamentally statistical, and yet be real. For young Darwin, gentleman naturalist, noting individual differences came naturally. His curiosity about connections may also have been natural to him, but he developed it into a powerful drive toward unifying theory.

Before coming to St. John's in 1948, I had taken undergraduate courses in zoology and embryology in which Darwin's theory was referred to; I accepted the theory as established. An occasion for reading Darwin's *Origin* had not arisen. On becoming a St. John's tutor, I immersed myself chiefly in problems of the laboratory on the side of physical science, to which my interests inclined me and for which my more recent graduate studies in the history of science to some degree prepared me.

In multiple ways, during my early years at St. John's, I took my cue from Jacob Klein. My admiration for him was unbounded. I respected him for his scholarly knowledge, shrewdness, and sharp discernment. It was he who drew the College community out of its 1948-49 leadership crisis and communal slough of despond in the wake of Barr's and Buchanan's departure, and he did so single-handedly and spiritedly. During his deanship (1949-1958), he gave the College a new lease on life, a new stability, and an incentive to move forward: testing, selecting, and improving the Program. Our debt to him is incalculable.

As dean, Mr. Klein in the opening lecture each year undertook to ad-

---

4. Ernst Mayr in *Evolution and Anthropology* (Washington: Anthropological Society of Washington, 1959), 2; also given in Mayr's Introduction to Darwin's *On the Origin of Species: A Facsimile,* xix-xx.
5. See Elliott Sober, "Evolution, population thinking, and essentialism," *Conceptual Issues in Evolutionary Biology*, ed. Elliott Sober, (Boston: MIT Press, 2001).

dress the question of what we were doing here, what liberal education *was*. It was with trepidation, he told us, that he addressed this question. Typically, his lecture took a Platonic turn, as when he described the *metastrophē*, or turning round, of the prisoner in the cave of Plato's *Republic.* The former prisoner had to be brought to recognize that the shadows he had previously taken for truth were in fact only images of conventional images. Getting at the truth was a matter of penetrating beyond that scrim of images.

During the academic year 1954-55 I was co-leader with Mr. Klein ("Jasha" as we tutors called him) of a senior seminar. On one evening the assignment was from Darwin's *Origin*—this was perhaps the only place in the program where Darwin's theory was addressed in those days. I recall nothing of the discussion, but at its end Jasha asked the students: Did they consider Darwin's book important to their lives? One after another they replied with a decisive "No!"—a flood of denial.

Though failing to lodge a protest, I thought the indifference to Darwin a mistake, and I was disappointed by Jasha's standoffishness with respect to it. My opinion was reinforced in conversations I had at the time with Allen Clark, a Ford Foundation intern at the College in the years 1954-56.[6] Clark had done graduate studies at Harvard on American pragmatism, reading widely in the writings of C.S. Peirce, William James, Oliver Wendell Holmes, Jr., and the Harvard-educated Spanish émigré George Santayana. He was especially attracted to the writings of Peirce, who had been both a working scientist and a close student of philosophy, and had set himself to making philosophical sense of natural science. Peirce had embraced Darwin's theory and interpreted it.

Attempting to catch up with Clark in philosophy, I began reading such writings of Peirce as were readily available. These were two collections of essays, the earliest assembled by Morris R. Cohen under the title *Chance, Love, and Logic,* and a later one due to Justus Buchler, *The Philosophy of Peirce.* There were also the six volumes of *The Collected Papers of Charles Sanders Peirce,* published by Harvard University Press in 1931-35 under the editorship of Hartshorne and Weiss, but these were formidable, leaving the inquirer puzzled as to where to get a leg up or a handhold.

My enthusiasm for Peirce was challenged one summer evening in the later 1950s. During an informal discussion of a Peirce essay at Jasha's home, Jasha took exception to Peirce's "Monism," the doctrine that the

---

6. In the seminar described at the beginning of this essay, Clark was the sole participant to speak up in defense of Darwin's theory.

world is made of a single stuff. Jasha saw this doctrine as contradicted by the *intentionality* of human thought. What was that?

The doctrine had been put forward by the Austrian philosopher Franz Brentano in 1874.[7] According to Brentano, to think is to think *of* or *about* something. Analogously, to fear or hope entails that there are objects (Jasha sometimes called them "targets") of these modes of consciousness. Their objects need not be existents in the empirical world. I can think of a unicorn, or imagine riding like Harry Potter on a broom stick, or fear an imagined bogeyman in a closet. Brentano therefore spoke of "intentional inexistence," meaning that such an object is somehow contained in the thought (*cogitatio* à la Descartes!) of which it is the object. Brentano sought to make Intentionality definitive of the mental. He concluded that mind, because of its Intentionality, is irreducible to the physical.

Edmund Husserl, one of Jasha's teachers, had been a student of Brentano. For Husserl, Brentano's idea of intentionality became the basis of a new science which he called Phenomenology. Husserl followed Brentano in treating intentionality as coextensive with the mental, and in asserting the impossibility of a naturalistic explanation of intentional acts. Jasha's rejection of Peirce's Monism, I am guessing, stemmed from his acceptance, at least in part, of Husserlian philosophy.[8]

Jasha may have been unaware that what he regarded as Peirce's Ontological Monism was an application of the maxim *Do not block the road of inquiry.* Dualism, as Peirce saw it, drew a line in the sand; naturalistic explanations were guaranteed to be impossible beyond this line. The line in the sand inevitably becomes a dare.

But I was still far in those days from understanding how the various parts of Peirce's thinking held together—or failed to. A major difficulty with the Cohen and Buchler collections and with *The Collected Papers* was that they did not present Peirce's papers in their order of composition. The editors did not sufficiently appreciate that Peirce's ideas developed over time. Throughout his life, Peirce's thought (like science as he understood

---

7. In his book *Psychologie vom empirischen Standpunkt* (Leipzig: Duncker und Humblot, 1874).

8. Jasha spoke with admiration of Husserl's repeated efforts to start all over again from the beginning, in formulating the *archai* of philosophy. Husserl's notion of sedimentation in the sciences—our tendency to take earlier achievements for granted—was a theme that Jasha took up in his studies of the origins of algebra and of the work of Galileo. Seeking to understand Jasha's Husserlian antecedents, I read a good deal of Husserl during the years I was reading Peirce. A lecture I gave in September, 1959, was based on Husserl's *Erfahrung und Urteil.*

it) was a work in progress.[9] When he died in 1914 he had not completed any single major work. During his last active decade, however, he succeeded in resolving certain major difficulties in his earlier philosophizing. A chronological edition of his work—published papers, lectures, and unpublished notes and correspondence—has now been undertaken by Indiana University Press. Of these post-1950 developments I was made aware only recently. And their full import did not dawn on me until encountering a book by the Chairman of the Board of Advisers to the Peirce Edition Project, Thomas Short. It is *Peirce's Theory of Signs.*[10]

Parts 1 and 2 of our essay provide an account of Peirce's pragmatism and of his progress from Kantian idealism to scientific realism. In Parts 3 and 4, with the help of Short's analysis, we shall indicate how Peirce accounts naturalistically for the emergence of intentionality and conscious purposefulness in the course of evolution.

Part 1. Peirce and Pragmatism

Peirce is the man through whom the word "pragmatism" enters upon the world scene as a philosophic term. According to his own recollection,[11] confirmed by the report of his friend William James,[12] this happened in the early 1870s, in Cambridge, Massachusetts, amongst a group of young Harvard men, who used to meet for philosophical discussion. Later in the 1870s the opinions Peirce had defended viva voce were issued in print in two articles, "The Fixation of Belief" (1877) and "How to Make our Ideas Clear" (1878).[13] The first of these two essays prefigured what Peirce would in the course of a life-time come to say about science as an enterprise of ongoing inquiry rather than a collection of upshots of investigation.[14] The second was sent into the world, as the title indicates, as advice on how to go about gaining greater intellectual control over one's ideas than is furnished by

---

9. The importance of this fact was first established by Murray Murphey, in *The Development of Peirce's Philosophy*, (Cambridge, Mass.: Harvard University Press, 1961.)

10. Thomas L. Short, *Peirce's Theory of Signs* (Cambridge: Cambridge Universty Press, 2007). I was introduced to this book by Chaninah Maschler.

11. *The Essential Peirce*, edited by the Peirce Edition Project, 2 Vols. (Bloomingdale, Indiana: Indiana University Press, 1998), Vol. 2, 400. Further references to this publication will be abbreviated to *EP*.

12. Ibid., 516.

13. These articles are reprinted in *EP*, Vol. 1, 109-123, 124-141.

14. Thomas L. Short, in a forthcoming second book about Peirce, gives a detailed defense of this Peircean understanding of the sciences.

the ability correctly to apply, or even verbally to define them. The advice runs as follows: "Consider what effects, which might conceivably have practical bearings, we conceive the object of our conceptions to have. Then our conception of these effects is the whole of our conception of the object." Note that the first person plural is out front. Also, that conceiving remains irreducible!

Peirce never became a full-time professor. Not even at Johns Hopkins, where John Dewey was briefly a student in his logic class. But just about every major American author in professional philosophy—William James, Josiah Royce, John Dewey, George Herbert Mead, C.I. Lewis, Wilfrid Sellars—acknowledges being profoundly indebted to Peirce's teachings, pragmatism being one of these.

> Pragmatism is, in itself, no doctrine of metaphysics, no attempt to determine any truth of things. It is merely a method of ascertaining the meanings of hard words and of abstract concepts. All pragmatists of whatsoever stripe will cordially assent to that statement. *As to the ulterior and indirect effects of practicing the pragmatistic method, that is quite another affair.*[15]

Some of the Cambridge friends whom Peirce initially persuaded to try bringing a laboratory scientist's "let's try it and see" approach to bear on the study of "hard words," particularly those used in metaphysics, suggested that he call what he was offering "practicism" or "practicalism." No, Peirce responded, he had learned philosophy from Kant, and in Kant the terms *praktisch* and *pragmatisch* were "as far apart as the poles."[16] *Praktisch* belongs to the region of thought where no mind of the experimentalist type can make sure of solid ground under his feet. *Pragmatisch* expresses a relation to some definite human purpose. "Now quite the most striking feature of the new theory [is] its recognition of an inseparable connection between rational cognition and human purpose."[17]

Here are two more statements of what pragmatism amounts to:

> I understand pragmatism to be a method of ascertaining the meanings, not of all ideas, but only of what I call "intellectual concepts,"

---

15. *EP,* Vol. 2, 400. Italics added.
16. See Immanuel Kant, *Critique of Pure Reason,* "Of the Canon of Pure Reason," A800 = B828 ff. Kant there explains, "By the practical I mean everything that is possible through freedom."
17. *EP,* Vol. 2, 333.

that is to say, of those upon the structure of which arguments concerning objective fact may hinge. Had the light which, as things are, excites in us the sensation of blue, always excited the sense of red, and vice versa, however great a difference that might have made in our feelings, it could have made none in the force of any argument. In this respect, the qualities of hard and soft strikingly contrast with those of red and blue. . . . My pragmatism, having nothing to do with qualities of feeling, permits me to hold that the predication of such a quality is just what it seems, and has nothing to do with anything else. . . . Intellectual concepts, however, the only sign-burdens that are properly denominated "concepts"—essentially carry some implication concerning the general behavior either of some conscious being or of some inanimate object, and so convey more, not merely than any feeling, but more too than any existential fact, namely, the "*would-acts*" of habitual behavior; and no agglomeration of actual happenings can ever completely fill up the meaning of a "would be."[18]

Again,

Pragmaticism[19] consists in holding that the purport of any concept is its conceived bearing upon our conduct. How, then, does the Past bear upon conduct? The answer is self-evident: whenever we set out to do anything, we "go upon," we base our conduct on facts already known, and for these we can only draw upon our memory. It is true that we may institute a new investigation for the purpose; but its discoveries will only become applicable to conduct after they have been made and reduced to a memorial maxim. In short, the Past is the sole storehouse of all our knowledge. When we say that we know that some state of things exists, we mean that it used to exist, whether just long enough for the news to reach the brain and be retransmitted to tongue or pen or longer ago. . . . How does the Future bear upon conduct? The answer is that future facts are the only facts that we can, in a measure, control. . . . What is the bearing of the Present instant upon conduct? . . . There is no time in the Present for any inference at all, least of all for inference concerning that very instant. Consequently the present object must be an external

---

18. Ibid., 401.
19. Peirce eventually (as here) made the name of the -ism ugly, "to keep it safe from being kidnapped." Consider what Peirce writes about how his thinking does or doesn't differ from that of William James, *EP*, Vol. 2, 421.

object, if there be any objective reference in it. The attitude of the present is either conative or perceptive.[20]

Part 2. Peirce's Transition from an Initial Idealism to Scientific Realism

As Peirce has told us, he learned philosophy from Kant. Yet from the start there was one Kantian doctrine he could not stomach: the doctrine of "things-in-themselves" (*Dinge an sich*) somehow standing behind the objects we meet with in experience—inaccessible beings of which, Kant says, we must always remain ignorant. In papers of the late 1860s, Peirce insisted that all of our cognitions are signs, and that each sign refers to a previous sign:

> At any moment we are in possession of certain information, that is, of cognitions which have been logically derived by induction and hypothesis from previous cognitions which are less general, less distinct, and of which we have a less lively consciousness. These in their turn have been derived from others still less general, less distinct, and less vivid; and so on back to the ideal first, which is quite singular and quite out of consciousness. The ideal first is the particular thing-in-itself. It does not exist *as such*.[21]

According to Peirce at this stage, all thoughts are of one or another degree of generality, each referring to an earlier thought, and none immediately to its object. Only if a cognition were immediately of its object, could it be *certain,* hence an *intuition.* Our lack of intuition, as thus argued by Peirce, was his initial ground for rejecting Descartes' *Cogito, ergo sum.* The *real,* as Peirce conceived it at this time, was an ideal limit to a series of thoughts, a limit to be reached in the future:

> The real . . . is that which, sooner or later, information and reasoning would finally result in, and which is therefore independent of the vagaries of me and you. Thus, the very origin of the conception of reality shows that this conception essentially involves the notion of a COMMUNITY, without definite limits.[22]

---

20. *EP,* Vol. 2, 358f. For a lucid brief description of Peirce's later "subjunctive" version of pragmatism, one which acknowledges that "modern science . . . is practice engaged in for the sake of theory," see Short, *Peirce's Theory of Signs,* 173, second paragraph.
21. *EP,* Vol. 1, 52.
22. Ibid.

Peirce here conceived all conceiving as in an infinite sequence of thoughts, stretching backward toward the non-existent thing-in-itself (an external limit) and forward toward the real, to be achieved at some future time (a limit located *within* the thought sequence). A consequence was that any individual, considered as an "it" other than the universals true of it, is unreal.

With this consequence of his late-1860s theory of knowledge, Peirce was uncomfortable. If the aim is to get outside one's head and find a purchase on reality, it is indeed disastrous.[23]

Peirce at last found a way out in his "The Fixation of Belief" of 1877:

> To satisfy our doubts . . . it is necessary that a method should be found by which our beliefs may be caused by nothing human, but by some external permanency—by something on which our thinking has no effect. Such is the method of science. Its fundamental hypothesis . . . is this: There are real things, whose characters are entirely independent of our opinions about them; those realities affect our senses according to regular law. . . .[24]

In "How to Make Our Ideas Clear" (1878), Peirce combined the hypothesis of real things on which our thinking has no effect with his earlier notion of indefinite progress toward human knowledge of the real:

> Different minds may set out with the most antagonistic views, but the progress of investigation carries them by a force outside of themselves to one and the same conclusion. . . . The opinion which is fated to be ultimately agreed to by all who investigate, is what we mean by the truth, and the object represented in this opinion is the real.[25]

In the years 1879-1884, Peirce was a part-time lecturer in logic at the Johns Hopkins University, and he and his students O.H. Mitchell and Christine Ladd-Franklin (independently of Frege in Germany) introduced quantifiers into predicate logic and the logic of relations. Thus the familiar universal and particular propositions of Aristotelian logic, "All S is P," "Some S is P," come to be replaced by

$(x)(Sx \supset Px)$ [read: For all x, if x is S, then x is P], and

$(\exists x)(Sx \cdot Px)$ [read: There is an x such that x is S and x is P],

---

23. For other difficulties with his theory in the 1860s, see Short, *Peirce's Theory of Signs*, ch. 2.
24. *EP*, Vol. 1, 120.
25. *EP*, Vol. 1, 138-139.

where we have used a notation now standard. Note that the "x" denotes an individual in whatever universe of discourse, fictional or real, we have entered upon, without any presumption that the essence of this individual is known to us. In relational logic, which is needed for mathematics, indices are crucial for representing dyadic, triadic, n-adic relations, e.g., Rxy (read: x bears the relation R to y). All our thinking, according to Peirce in the 1880s and later, is laced with indexical elements, tying discourse to the world we're in. The index asserts nothing; it only says "There!" Like such words as "here," "now," "this," it directs the mind to the object denoted.

The discovery of the nature and indispensability of indices led to a vast extension of Peirce's understanding of signs and significance (the science of semeiotic he was seeking to build). An index is anything that compels or channels attention in a particular direction. The act of attention responding to an index does not have to be a component of a thought. For instance a driver, on seeing a stoplight go red, may brake automatically without thinking; he thus interprets the red light as a command. Therefore the effect of a sign, in triggering an interpretation, need not be a thought; it can be an action or a feeling. The extension of semeiotic to nonhuman interpreters is now in the offing, as will become apparent in Part 4 below.

At the same time, Peirce has burst out of the closed-in idealism of his earlier theory of knowledge. The result is what we may call Scientific Realism.

Part 3. Anisotropic Processes

Just twelve years after the first copies of *Origin of Species* landed in the U.S.A., Peirce wrote:

> Mr. Darwin proposed to apply the statistical method to biology. The same thing had been done in a widely different branch of science, the theory of gases. Though unable to say what the movements of any particular molecule of a gas would be on a certain hypothesis regarding the constitution of this class of bodies, Clausius and Maxwell [had been able, eight years before the publication of Darwin's immortal work], by the application of the doctrine of probabilities, to predict that in the long run such and such a proportion of the molecules would, under the given circumstances, acquire such and such velocities; that there would take place, every second, such and such a number of collisions, etc.; and from these propositions [they] were able to deduce certain properties of gases, especially in regard to their heat relations. In like manner, Darwin, while unable to say what the operation of variation and natural selection in

any individual case will be, demonstrates that in the long run they will adapt animals to their circumstances.[26]

Thus Peirce took explanation in both statistical mechanics and Darwinian natural selection to be statistical. He meant, Short argues, *irreducibly* statistical, and *not* mechanistic.[27] Analyzed logically, a mechanistic explanation starts from a *particular* disposition of certain bodies at some time, and by applying general laws of mechanics, gravitation, chemistry, electromagnetism, or other general theory, derives the *particular* disposition of these bodies at a later time. "Particular" here is opposed to "general." The explanations of Celestial Mechanics are of this kind. The celestial mechanician, starting from the positions and velocities of the bodies in the solar system at one instant, and assuming gravitational theory, computes the positions and velocities of these bodies at a later instant. If we should propose to ourselves a similar calculation for molecules of a gas confined in a container, we would find it impracticable. The number of molecules is too large (in a cubic centimeter of gas at one atmosphere of pressure and 0°C. that number is about 2.7×10^{19}, or 27 quintillion). Ascertaining the positions and velocities of all these molecules at a specified "initial" instant is humanly impossible. Moreover, the motions are not governed by a single law like gravitation, but involve collisions of the molecules with each other and the walls of the container; these introduce discontinuities that are difficult to take into account.

But the crucial conclusion is this: even if such a computation were possible, it would not yield the conclusion for which statistical mechanics argues. Statistical mechanics seeks to establish that notably non-uniform distributions of molecules in the gas will in time be replaced by a more uniform distribution, with reduction in the spread of velocities amongst the molecules. The statistical argument invokes probability.

How to understand probability in this context is by no means settled, and we shall give only a rough indication of the type of solution that is believed necessary.[28] Consider a system of n molecules of gas contained in a volume V. Let V be divided into a large number m of equal cells, m being less than n (if n is in quintillions, m could be in the millions or billions). If the molecules were distributed with perfect uniformity throughout V, then

---

26. *EP*, Vol. 1, "The Fixation of Belief," 111; for the square bracketed emendations, see ibid., 377.

27. Cf. *EP*, Vol. 1, 289f.

28. See Paul Ehrenfest and Tatyana Ehrenfest, *The Conceptual Foundations of the Statistical Approach in Mechanics* (New York: Dover, 1958.)

each cell would contain n/m molecules. This distribution is a particular microstate—an extremely special one, hence unlikely. We would expect that, in most imaginable distributions, the numbers of molecules in different cells would be different. To take this likelihood into account, consider microstates in which the number of molecules in all cells falls within the range $n/m \pm e$, where e is much less than n/m. Let the class of all microstates thus characterized be called C, and let the complementary class, or class of all microstates in which the number of molecules in some cells falls outside the range $n/m \pm e$, be called C'.

In the work of the earlier theorists, distinguishable microstates compatible with the overall energy of the gas were assigned equal probabilities, since no reason presented itself for assigning different probabilities to different microstates. Later theorists sought grounds other than "equal ignorance" for assigning probabilities to microstates. Whatever the mode of assigning probabilities, the outcome must show the gas progressing from less uniform to more uniform distributions, both spatially and with respect to the spread of velocities. For that is the empirical result: a quantity of gas under high pressure, when let into an evacuated chamber, spreads out through the chamber and is soon more homogeneously distributed, with a uniform temperature and pressure lower than the original temperature and pressure.

The Second Law of Thermodynamics extends this kind of reasoning to all natural systems. It says that *in any closed system* the processes have a direction: they progress toward greater homogeneity and reduced capacity to do mechanical work.[29] For processes that are directional in time, Short uses the term *anisotropic* (*a*-privative + *iso,* "equal" + *tropos,* "direction"). Anisotropic processes are defined by the *type* toward which they progress. We shall see that there are anisotropic processes other than those that instantiate the Second Law of Thermodynamics. All such processes, however, differ from mechanical processes, which proceed from a particular configuration to a particular configuration.

Whether the universe is a closed system we do not know, but everywhere in the observable world we see the effects of the Second Law, the "degradation of energy." Nevertheless, we also see that new forms of order, though improbable, sometimes emerge. They are produced in open systems that absorb energy from, and discard unused matter and energy to, the environment. Ilya Prigogine has described such forms of order, calling them

29. Cf. *EP,* Vol. 1, 221.

"dissipative systems."[30] Locally, in the newly created form, the second law appears to be violated, but if account is taken of the exhausted fuel and other waste materials ejected to the environment, the second law is found to hold. Higher forms of order come to be at the expense of a decrease in order elsewhere, an increase in homogeneity and a lessened capacity to produce novelty.

The first coming-to-be of living forms in the universe presumably occurred in the manner of Prigogine's "dissipative systems." Such is the hypothesis generally accepted by scientists today. Living systems differ from the cases studied by Prigogine in their greater complexity and in having the capacity to self-replicate. In 1953 the graduate student S.L. Miller under the guidance of H.C. Urey circulated a mixture of methane, ammonia, water vapor, and hydrogen through a liquid water solution, and elsewhere in the apparatus continuously passed an electrical discharge through the vaporous mixture. After several days the water solution changed color, and was found to contain a mixture of amino acids, the essential constituents of proteins. Since then, most if not all of the essential building-blocks of proteins, carbohydrates, and nucleic acids have been produced under conditions similar to those obtaining when the Earth was young (the atmosphere needs to be free of oxidizing agents such as oxygen). The sequences of conditions and chemical pathways by which these building-blocks may have been assembled into a living cell remain matters of speculation.

Darwin's evolutionary theory, taking the existence of living things as given, goes on to show how, chiefly but not solely by means of natural selection,[31] biological evolution can occur. Our little word "can" here goes to signal what Nicholas Maistrellis calls "the highly theoretical, and even speculative character" of *Origin* chapter 4, dedicated to expounding that and how Natural Selection "works."

---

30. Ilya Prigogine, *From Being to Becoming* (San Francisco: W.H. Freeman, 1980). See also *Stuart A. Kauffman,* "Antichaos and Adaptation," in *Scientific American,* August 1991, 78-84.

31. See the concluding sentence of the potent last paragraph of Darwin's Introduction to *On the Origin of Species.* Gould and Lewontin, in their famous protest against unrestrained Adaptationism ("The Spandrels of San Marco and the Panglossian Paradigm: A Critique of the Adaptionist Programme," in *Conceptual Issues in Evolutionary Biology,* ed. Elliott Sober, [Boston: MIT Press, 2001]), cite this sentence and add an approving reference to George. J. Romanes's essay "The Darwinism of Darwin, and of the Post-Darwinian Schools" (in *The Monist* 6:1 [1895], 1-27). Romanes would join Gould and Lewontin when they write: "We should cherish [Darwin's] consistent attitude of pluralism in attempting to explain Nature's complexity" (82).

> We should not expect a series of examples of natural selection de-
> signed to win us over to his theory on purely empirical grounds.
> Even if Darwin had wanted to proceed in that way, he could not
> have done so, for such examples do not exist—or at least were not
> known to Darwin. . . . Notice that all the examples of natural selec-
> tion in this chapter are, as Darwin repeatedly acknowledges, *imag-
> inary* ones.[32]

Contemporary readers of Darwin have sometimes become so blasé
about the shocking idea that order may emerge out of disorder that they
don't notice how subtle, complex, and distributed the over-all argument of
Origin is. We have found C. Kenneth Waters' "The arguments in the Origin
of Species," along with the other essays included in Part 1: *Darwin's Theorizing* of *The Cambridge Companion to Darwin,* particularly conducive
to waking us up.

Peirce wrote, in *A Guess at the Riddle* (1887):

> Whether the part played by natural selection and the survival of the
> fittest in the production of species be large or small, there remains
> little doubt that the Darwinian theory indicates a real cause, which
> tends to adapt animal and vegetable forms to their environment. A
> remarkable feature of it is that it shows how merely fortuitous vari-
> ations of individuals together with merely fortuitous mishaps to
> them would, under the action of heredity, result, not in mere irreg-
> ularity, nor even in statistical constancy, but in indefinite progress
> toward a better adaptation of means to ends.[33]

A little later in this same manuscript Peirce sums up the basic idea of Darwinian selection as follows:

> There are just three factors in the process of natural selection; to wit:
> 1st, the principle of individual variation or sporting; 2nd, the princi-
> ple of hereditary transmission . . . ; and 3rd, the principle of elimina-
> tion of unfavorable characters.[34]

Darwin and Peirce lacked the benefit of a workable theory of inheri-
tance. Nothing like our genetics was available to them. We today single

---

32. *Selections from Darwin's* The Origin of Species: *The Shape of the Argument,*
ed., Nicholas Maistrellis (Santa Fe: Green Lion Press, 2009), 43.
33. *EP,* Vol 1, 200. For a correction of this overly cheerful scenario of inevitable
progress see, e.g., Elliott Sober, "Selection-for: What Fodor and Piattelli-Palmarini
Got Wrong," 11. This essay is available on the internet at the following URL:
http://philosophy.wisc.edu/sober/Fodor%20and%20Piatelli-
Palermini%20april%209%202010.pdf
34. *EP,* Vol. 1, 272. Cf. Darwin, *On the Origin of Species: A Facsimile,* 127.

out genetic make-up as the causally significant locale of "sporting," And Peirce's phrase, "elimination of unfavorable characters," is replaced in more recent neo-Darwinian formulations by the phrase "relative reproductive success," meaning, the having of more numerous offspring. The process is statistical: If one variant of a species has more numerous offspring than do others, and if in addition these offspring survive to reproduce, the original variant, possessed of one or more genetic alleles (alternative forms of a gene), is more successful in propagating its genome to later generations.

The hypothesis of Natural Selection confers little in the way of predictive power. Its chief value is to provide a post-hoc explanation of what has occurred. For example, visual acuity is crucial to the survival of both predators and prey. Evidently predators are better off with eyes in the front of their heads as they pursue prey, and potential prey are better off with eyes on the sides of their heads to detect predators coming from any quarter. Another example: Flowers evolved as a device by which plants induce animals to transport their pollen (hence sperm) to the egg cells. The evolutionarily older plants had been pollinated by the wind. The more attractive the plants were to an insect, the more frequently they would be visited and the more seeds they would produce. Any chance variation that made the visits more frequent or made pollination more efficient offered immediate advantages.[35]

We can only guess at the detailed processes by which such adaptations have been brought about. What Darwin gives us is a heuristic for research, not a set of biological laws.[36] Partly on this account, because Darwinian explanation does not fit the model of explanation in mechanics, it has taken a long time before philosophers of science became willing to award a comparable degree of intellectual dignity to Darwinian as to Galilean and Newtonian science. The books listed in the Bibliography appended to Maistrellis's *Selections* help overcome the physics envy that stands in the way of appreciating Darwin. Particularly helpful have been Sober's persevering efforts to clarify and show the interconnections amongst the fundamental concepts of Fitness, Function, Adaptation, and Selection, while steadily reminding us of the ineliminably probabilistic character of most of the theorizing of modern evolutionary biology.

One of Sober's helps into the saddle is his distinction between *selection for* and *selection of*:

---

35. Helena Curtis, *Biology* (New York: Worth, 1979).
36. Equally important, perhaps, is the inspiration of Darwin's intellectual attitude—omni-observant, persevering, sober—to which Maistrellis calls attention.

> Selection-for is a causal concept. To say that there is selection for trait T in a population means that having T causes organisms to survive and reproduce better (so having the alternative(s) to T that are present in the population causes organisms to survive and reproduce worse). In contrast, to say that there is selection of trait T just means that individuals with T have a higher average fitness than do individuals who lack T.[37]

Here is an illustration of the contrasting terms being put to use:

> Worms improve the soil, but that does not mean that their digestive systems are adaptations for soil improvement; rather, the worm gut evolved to help individual worms survive and reproduce. The benefit that the ecosystem receives is a fortuitous benefit—a useful side-effect unrelated to what caused the trait to evolve. The gut's ability to extract nutrition for individual worms is what the gut is an adaptation *for*.[38]

To balance our earlier quotation from Maistrellis stressing the not strictly empirically encountered character of Darwin's examples in his chapter about natural selection-at-work, notice that Sober feels quite comfortable about urging against the philosopher Jerry Fodor, a critic of Darwinism, that "biologists often think they have excellent evidence for saying that agricultural pests experienced selection for DDT resistance, [or] that there has been selection for dark coloration in moths."[39]

Short adopts Sober's *selection of/selection for* contrast and, integrating it with Peircian ideas of explanation by final causes, adapts it to new uses. The context is as follows. He asks us to distinguish four kinds of physical process:

> Mechanical processes that proceed from one particular configuration to another and are reversible.
>
> The processes described by statistical dynamics, which are anisotropic and result in an increase in entropy and disorder.

---

37. Elliott Sober, *The Nature of Selection: Evolutionary Theory in Philosophical Focus* (Cambridge, Mass.: MIT Press, 1984).

38. The example stems from Williams via Elliott Sober and David Sloan Wilson, "Adaptation and Natural Selection Revisited," in the *Journal of Evolutionary Biology* 24 (February 2011), 462-8. In this article, the authors are "revisiting" George C. Williams's book on adaptation in order to make sure the world knows that the book was a landmark in the development of evolutionary theory.

39. Elliott Sober, "Fodor's *Bubbe Meise* Against Darwinism," in *Mind and Language* 23 (February 2008), 43. (*Bubbe meise* is Yiddish for "old wives' tale.") This article is also available on the internet at the following URL: bit.ly/bubbe-meise

The non-equilibrium processes studied by Prigogine, which are also anisotropic, but produce open systems that have increased order and diminished entropy. The dissipative structures can sustain themselves in the given environment for a time. Living things, we assume, are of this kind—complex open systems that metabolize and have an apparatus for replicating themselves.

With living things, a third sort of anisotropic process comes into play: Natural Selection, the selection of characteristics for types of effect that conduce to reproductive success.[40]

Given living things and their struggle for existence, given heritable variability, given phenotypic features that in a given state-of-its-world enhance a creature's relative chance of producing fertile offspring, a new kind of *directional* process comes into being, natural selection. And with it, the possibility of purpose comes on the scene.

Not that anything is a purpose or has a purpose in biological evolution before the actual occurrence of a mutation that happens to be selectively retained because of some advantage that it confers. Only at that time, that is, when a feature is selected for its effect, does the effect, say visual acuity, become a purpose. There was no purpose "visual acuity" or "adaptedness" or "survival" hanging around waiting for an opportunity. But once eyes with adjustable lenses become a feature of mammals, *then* it would only be mechanicalist prejudice that could keep us from saying that eyes exist for the purpose of seeing.[41]

Sober's polar terms *selection of/selection for* are perhaps worked harder and a little differently than they were previously:

It is *because* lenses and focusing increase visual acuity that genetic mutations resulting in lenses and focusing were retained in subsequent generations; in fact, that happened in independent lines of animal evolution. The selection in those cases was *for* the visual acuity and *of* concrete structures (or the genes that determine them) that improved visual acuity in specific ways. . . . The of/for distinction

---

40. When Herbert Spencer attempted to explain evolution on mechanical principles, Peirce countered that the endeavor was illogical. See *EP*, Vol. 1, 289. Among Peirce's arguments was this: the law of conservation of energy implies that all operations governed by mechanical laws are reversible. Whence follows the corollary that growth *is not explicable by those laws, even though they are not violated in the process of growth.*
41. Private communication from Thomas Short, March 19, 2012.

is relative to the level of analysis, but the object of 'for' is always an abstract type and the object of 'of' is always something genetic or genetically determined, hence concrete. . . . As the type selected-for is essential to explanation by natural selection, such explanation is like anisotropic explanation in statistical mechanics [in that] both explain *actual* phenomena by the *types* they exemplify. Hence it is not mechanistic. . . . It is qua adaptation—hence in that aspect—that [an adaptive feature, say S] is explained by natural selection. S could also be explained, had we knowledge enough, as a product of a complicated series of mechanical events. But, then, S's enhancing reproductive success would seem a surprising coincidence, a bit of biological luck. *S's being an adaptation would not be explained.*[42]

The "aptness" of organisms is one of the facts of life that the Darwinian program of explanation seeks to account for. Having had some success in this explanatory endeavor, we easily forget that there is no guarantee that evolution will bring about an increase in complexity or intelligence or other quality that we admire. Overstatement here, Short warns us, is common, and disastrous.[43] Notice too that natural selection was not itself selected, and therefore does not have a purpose. It just occurs.

Part 4. The Emergence of Intentionality and Conscious Purposes

Cleverly joining Peircean reasoning to the more recent formulations of neo-Darwinian theory, Short's *Theory* sketches a narrative that strives to make intelligible the eventual emergence of the possibility of deliberately produced tools and self-controlled action out of advantageous anatomy and biologically useful animal behavior. *Here one must go slow and notice that it is as the world comes to hold new kinds of entity that new kinds of explanation become applicable.*[44] Short is not reducing biological explanation to chemical explanation. Nor will he assimilate human discourse to animal signaling.[45] The last three chapters of his book are given over to exploring

---

42. Short, *Peirce's Theory of Signs,* 130. Italics in last sentence added.
43. Ibid., 145
44. Ibid., 144-145.
45. As Allen Clark wrote in a manuscript never published ("The Contributions of Charles S. Peirce to Value Theory," 4), "No philosopher . . . would be less inclined than Peirce to minimize the tremendous importance of the transformation that occurs when inquiry [or any other adaptive behavior] rises from the unconscious to the conscious level. For it is at this second stage that man transcends the animal faculty of merely responding to naturally given signs, those perceptual clues furnished by nature; he begins to *make* signs, and to respond to signs of his own making, and thus learns to provoke his own responses."

the implications of applying Peirce's ideas of sign-action (= semeiosis) to distinctively human language, thought, and life. But unless we work from the bottom up, there is no explaining of emergents.

"Working from the bottom up" means for Short that he must develop so general an account of Peirce's semiotic triad Sign-Object-Interpretant that it will be applicable both to infra-human sign-interpretation—end-directed animal responses to stimuli—and, duly amplified, to distinctively human life and thought. For Short, this behaviorist interlude is in the service of Peirce's Synechism:[46] If successful in his defense of Peirce's ways, he will have warded off both Cartesian dualism and Reductionism.[47]

Among social animals, group behavior is determined by mechanisms that cause one individual to respond to another. A forager bee, for instance, having located nectar, returns to the hive and there exhibits what look like dances. The bees in the hive react to these dances as signaling the direction and distance in which the nectar will be found. Ethologists have instructed us that there is an immense variety of animal behaviors that operate as though they were intended as communicative signs. By what criteria one determines the *intendedness* of a bird- or monkey-cry the emission of which *tends to result* in fellow-birds or fellow-monkeys reacting with behavior that *makes sense* for the creatures in question (e.g., escaping in an appropriate way from a certain kind of predator, or overcoming reluctance to approach more closely) has been a topic for ethological investigation. But every parent is familiar with the fact that infant wailing and screaming is not, in the earlier phases of its life, an expression of the infant's intention to rouse its protectors. Yet when the infant is a little older its jealous brother may justly complain: "She is not crying for a reason. She's crying for a purpose!"

We have deliberately introduced the word "intend" in its ordinary sense before returning to the topic of intentionality in Brentano's scholastic and technical sense. (Unhappiness about the lack of a non-dualist treatment of

---

46. "Synechism is Peirce's doctrine that human mentality is continuous with the rest of nature," writes Thomas Short in his exchange with the critics of his book, "Response," in *Transactions of the Charles S. Peirce Society* 43 (Fall 2007), 666.
47. Ibid. Dewey's essay of 1896, "The Reflex Arc Concept in Psychology" (*Psychological Review* 3 [July, 1896], 357-370) is offered in the same, perhaps Hegel-inspired, spirit of synechism. (This article is available on the internet at the following URL:
http://psychclassics.yorku.ca/Dewey/reflex.htm.) But a more instructive comparison would be between Thomas Short's account of Peirce and the life-long work of James J. Gibson, for instance, *The Senses Considered as Perceptual Systems* (Boston: Houghton Miffin, 1966) and *The Ecological Approach to Visual Perception* (Boston: Houghton-Mifflin, 1979.).

Intentionality was what initially motivated our exploration of Short's book on Peirce's semeiotics.) Unlike many semioticians, Short follows in Peirce's footsteps by *beginning* with *interpretive behavior,* not with the *sending* of signs.[48] This permits him to take off from *responses.* For instance:

> The deer does not flee the sudden noise that startled it, but a predator; for it is to evade a predator that the deer flees. The instinct to flee is based on an experienced correlation of sudden noises to predators; the correlation is weak, but, unless the deer is near starvation, it is better for it to risk losing a meal than to risk being one. If no predator is there, the deer's flight is a mistake, albeit justified. Mistaken or not, the flight interprets the noise as a sign of a predator.

A *response* is not merely an *effect* if it can be *mistaken.* It ranks as an *interpretation.*

In what manner and measure this idea of mistake is available to infrahuman animals is a hard question. When the dog that was, in some human observer's estimation, "barking up the wrong tree," corrects itself and, redirecting its bark to the neighboring tree, glimpses the spot where the cat in fact now is, does the dog think to itself, "Now I've got it right"? Consider two other examples of interpretive responses, both reported by the ethologist Niko Tinbergen: male sticklebacks, during the breeding season, tend to adopt a "threat posture" toward potential rivals.

> When the opponent does not flee . . . the owner of the territory . . . points its head down and, standing vertically in the water, makes some jerky movements as if it were going to bore its snout into the sand. Often it erects one or both ventral fins.[49]

Tinbergen's Plate I is a photo of a Stickleback exhibiting this posture to its own reflection in a mirror! *We* know this fish is making a mistake. Does he?

> Lorenz reports . . . an incident which demonstrates the power of [some varieties of Cichlid] to distinguish between food and their young. Many Cichlids carry the young back, at dusk, to a kind of bedroom, a pit they have dug in the bottom. Once Lorenz, together with some of his students, watched a male collecting its young for this purpose. When it had just snapped up a young one, it eyed a particularly tempting little worm. It stopped, looked at the worm for several seconds, and seemed to hesitate. Then, after these seconds of "hard thinking," it spat out the young, took up the worm and

---

48. See Short, *Peirce's Theory of Signs,* 156f.
49. Niko Tinbergen, *Social Behavior of Animals,* Methuen's Monographs on Biological Subjects, Vol. 1 (New York: Taylor and Francis, 1953), 9.

swallowed it, and then picked up its young one again and carried it home. The observers could not help applauding.[50]

The antelope that fled from a lion that wasn't there, the stickleback that threatened a rival that wasn't there, did they interpret something heard, something seen, as to-be-run-from, to-be-ousted? Their behaviors, while in error in the particular cases, were appropriate. And this holds true whether or not these individual animals "knew what they were doing." Something like this is, we take it, what Short meant when he wrote: [51]

> The purposefulness of interpretation accounts for the significance of that which is interpretable. In particular, as that which has a purpose may fail of its purpose, the purposefulness of interpretations accounts for the possibility that what is signified *is not.* Because what is signified might not be, significance exemplifies Brentano's idea of intentionality, which he defined as having an "inexistent object," i.e., an object that *is* an object independently of its existing. Brentano asserted that intentionality is unique to human mentality, but the argument of [Short's] book is that sign-interpretation occurs independently of conscious thought and, hence, that Peirce's semeiotic applies to phenomena well beyond human mentality. Thus it provides for a naturalistic explanation of the mind. But that is possible only if purposefulness can occur without consciousness. Peirce's doctrine of final causation c. 1902 provides a defense of that assumption. For it identifies causation with selection for types of possible outcome, regardless of whether that selection is conscious. And it does so consistently with modern physics and biology.[52]

But the question that arose when we considered the dog that eventually managed to bark *at the cat* is still with us: The dog, in our judgment and in fact, "corrected itself." And we know that learning, in the sense of an individual's behavior being shaped "for the better" by its experience, is a constituent of the lives of very many (all?) animals. But did the dog *know* that

---

50. Ibid., 45. The following anecdote of Darwin's in his chapter comparing the mental powers of lower animals with human mental powers seems to be to the same effect: "Mr. Colquhoun winged two wild ducks, which fell on the opposite sides of a stream; his retriever tried to bring over both at once, but could not succeed; she then, though never before known to ruffle a feather, deliberately killed one, brought over the other, and returned for the dead bird." Charles Darwin, *Descent of Man* (Princeton: Princeton University Press, 1981), 48.
51. Further clarifying remarks on Intentionality are given by Short in *Peirce's Theory of Signs*, 174-177.
52. Elliott Sober, "Fodor's *Bubbe Meise* Against Darwinism," 669.

it corrected itself? Consider Lorenz's much applauded Cichlid father, which had its worm and its baby too. Mustn't it have had some sort of "inner representation" of the alternative courses of conduct between which it chose?

We seem at last to have reached the question of when and how conscious purpose, planning, and self-control emerge. Short's entire book, not just the chapter bearing the name "Semeiosis and the Mental," is in pursuit of it. Given that Peirce regarded thought to be internalized discourse, and that an individual's power of discourse is a skill that could not have been acquired had that individual's "instinct to acquire the art" (as Darwin put it) not been activated in the course of apprenticeship to speakers, Short and Peirce are clearly right that "the capacity to think for oneself and to act in despite of society is . . . social in origin." He adds: "Individual autonomy and varied personality are further examples of the irreducibility of new realities to their preconditions."[53] Among such "new realities" are not only new means to accomplish existing purposes but also new purposes.

Because Short, under Peirce's tutelage, is wholehearted about accepting the Reality of purpose and purposiveness and is unembarrassed about following Darwin in naturalizing man, his investigation of how purpose can and has become "emancipated" from biology has real content.[54]

Conclusion

We have seen that, according to Peirce, both statistical mechanics and Darwinian natural selection entail anisotropic processes, defined by the *type* of result they lead to. The "population thinking" that Darwin and later biologists introduced into biology was aimed at accounting for the emergence of biological types or species. The new thinking differed from the typological thinking of pre-Darwinian times in that the types or species arose in time.

Among the virtues of Short's presentation of Peirce is that he gives a sufficiently detailed description of Peirce's Categories (in Ch.3) for readers to be supplied with opportunity to become persuaded that Peirce's trinitarian categorial scheme accommodates Individuals and Kinds as mutually irreducible. Here is, however, not the place to exhibit or argue the point.

Why was the reception of Darwin at St. John's so lukewarm in earlier days? The theoretical physicist's impatience with fussy descriptive details such as are dwelt on in *Origin* (and *must* be by natural historians) was probably a contributing factor; and one that would have been exacerbated if the

---

53. Short, *Peirce's Theory of Signs,* 147.
54. Ibid., 148.

assigned selection from *Origin* was pedagogically haphazard. But vague apprehensions about the moral and philosophical import of Darwin's theory may have contributed more heavily to avoiding serious intellectual engagement with it.

Darwin himself anticipated this reaction. He explains (in the Introduction to *Descent of Man*) that it was in order not to stand in the way of the reading public's making fair trial of his general views that he allowed himself just one tiny paragraph, on the final pages of *Origin,* that makes direct mention of man:

> In the distant future . . . psychology will be based on a new foundation, that of the necessary acquirement of each mental power and capacity by gradation. Light will be thrown on the origin of man and his history.

Twelve years later, in *Descent of Man,* the scope of Darwin's intellectual ambition is made manifest. In Ch.3 he takes on Kant:

> "Duty . . . whence thy original?" . . . As far as I know, no one has approached [this great question] exclusively from the side of natural history.

So "approaching it," Darwin writes:

> The following proposition seems to me in a high degree probable—namely, that any animal whatever, endowed with well-marked social instincts, would inevitably acquire a moral sense or conscience, as soon as its intellectual powers had become as well developed, or nearly as well developed, as in man.[55]

His plan is to show how, granted the rest of our mental attributes and the world's make-up, the human species does better *with* than it would *without* morality. Otherwise morality (sense of duty, conscience) and the instruments for its acquisition and maintenance could not have become "selected."

But isn't there something topsy-turvy about an explanation that subordinates, as means, something better, namely a creature competent to have a sense of duty, to an end less good, namely, mere comparative fitness for producing fertile offspring? The complaint, we urge, limps, because it fails to register that when something is fruitful and multiplies or fails to, it is *as a creature possessed of certain attributes that it does so.* Darwin freely ascribes sociability, intelligence, and emotions (sympathy, jealousy, ennui,

---

55. Darwin, *Descent of Man,* 71.

curiosity, courage, maternal affection, and so forth) to, for instance, domestic animals.[56] Nevertheless, he reserves morality for human beings:

> As we cannot distinguish between motives, we rank all actions of a certain class as moral, when they are performed by a moral being. A moral being is one who is capable of comparing his past or future actions or motives, and of approving or disapproving of them. We have no reason to suppose that any of the lower animals have this capacity; therefore when a monkey faces danger to rescue its comrade, or takes charge of an orphan monkey, we do not call its conduct moral. . . . It cannot be maintained that the social instincts are ordinarily stronger in man than, . . . for instance, the instinct of self-preservation, hunger, lust. . . . Why, then, does man regret . . . and why does he further feel he ought to regret his conduct? . . . Man, from the activity of his mental faculties, cannot avoid reflection. . . . Whilst the mother bird is feeding or brooding over her nestlings, the maternal instinct is probably stronger than the migratory; but . . . at last, at a moment when her young ones are not in sight, she takes flight and deserts them. When arrived at the end of her long journey, and the migratory instinct ceases to act, what an agony of remorse each bird would feel if, being endowed with great mental activity, she could not prevent the image continually passing before her mind of her young ones perishing in the bleak north from cold and hunger. At the moment of action, man will no doubt be apt to follow the stronger impulse. . . . But after their gratification, when past and weaker impressions are contrasted with the ever enduring social instincts, retribution will surely come. Man will then feel dissatisfied with himself, and will resolve with more or less force to act differently for the future. This is conscience; for conscience looks backwards and judges past actions, inducing that kind of dissatisfaction which, if weak, we call regret, and if severe remorse.[57]

Darwin seems to have come upon Aristotle late in life and recognized a soul-mate in him. He would, we believe, have been in delighted agreement upon reading Aristotle's observation in *History of Animals,* Book 1, 488b24, that we are the only creatures capable of deliberating *(bouleutikon):*

> Many animals have the power of memory, and can be trained, but the only one that *can recall past events at will (dunatai anamimnēskesthai)* is man.

---

56. See Charles Darwin's 1872 book *Expression of the Emotions in Man and Animals,* ed. Paul Ekman (Oxford: Oxford University Press, 2009).
57. Darwin, *Descent of Man,* Ch.3, 88-91.

Where are we then? Conscience, says Darwin in the opening sentence of *Descent of Man*, Ch.3, is the chief mark of distinction of the human race. Conscience cannot come into existence or operate without the power of recollection. The power of recollection (though no texts come to mind where anyone of our three authors says this expressly) depends upon the power to learn and employ not just a communicative medium but an articulate language.[58] Beings of this sort, Peirce the logician will come to argue ever more strenuously as he ages, are capable of acting not just in a motivated way, but in accordance with an ideal:

> Every action has a motive; but an ideal only belongs to a line of conduct which is deliberate. To say that conduct is deliberate implies that each action, or each important action, is reviewed by the actor and that his judgment is passed upon it, as to whether he wishes his future conduct to be like that or not. His ideal is the kind of conduct which attracts him upon review. His self-criticism followed by a more or less conscious resolution that in its turn excites a determination of his habit, will, with the aid of sequelae, *modify* a future action; but it will not generally be a moving cause to action.[59]

Permit us to conclude with an anecdote. A recent movie presented a small group of adults with the situation of a male high-school teacher accepting seduction by one of his beautiful girl-students. Ever intent on discussing *la difference,* one of the men in the group of movie watchers asked "Do you blame the teacher?" "Yes," was the answer, "because although it may indeed be true that it is harder for young men than for young women to resist sexual arousal, the teacher knowingly entered upon a profession that he could foresee would present him with such situations as he was now in. He should, taking advantage of the human power of imagination, have *rehearsed inwardly* how he *would* act *if* the world presented him with an opportunity that he should turn down."[60]

With Peirce's help, and instructed by Short, we hope to have shown in this essay that nothing in Darwin interferes with acknowledging the emergence of organisms competent to entertain and criticize ideals. This is the kind of organism we human beings are.

---

58. Ibid., Ch.2, 54.
59. *EP,* Vol. 2, 377. Survey the Index to *EP,* Vol. 2 under "self-control."
60. The answer is inspired by Peirce's report of his childhood memory of his younger brother's having prepped himself in imagination for preventing the spread of a small fire. See *EP,* Vol. 2, 413.

Book Reviews

A Forgotten Revolution:
Book Review of Lucio Russo,
La rivoluzione dimenticata
(Milan: Giangiacomo Feltrinelli, 1996).

This book is about the mathematics, science, and technology of the late fourth, third, and second centuries BC. In its detailed argumentation, it challenges many long and widely held views. Its thesis is extraordinary.

The "forgotten revolution" of Russo's title signifies nothing less than the first emergence of science, both exact or deductive science and experimental science. It took place, according to Russo, not in modern times but in the Hellenistic Age, with the work of Euclid, Aristarchus, Eratosthenes, Ctesibius, Herophilus, Archimedes, Apollonius, Seleucus, Hipparchus, and others. Conventionally the Hellenistic Age has been dated from the death of Alexander in 323 BC to the conquest of Egypt by the Romans in 30 BC, but in science, according to Russo, it may be said to have essentially ended by 145 BC, the year in which Ptolemy VIII, for political motives and possibly as an agent of Roman policy, systematically destroyed the Greek scientific community in Alexandria (see Polybius, *History*, 24:14). A 180-year efflorescence of science, virtually extinguished not only as an historical phase, but also in historical memory: such is Russo's theme. Fragments of and clues to this efflorescence survived, but often distorted by later misinterpretation. The science of the seventeenth century, according to Russo, came about in considerable measure as a recovery of Hellenistic science.

Prongs of this thesis that may astonish are the following. Russo, while granting that articulate rationality was an achievement and legacy of the Greeks of the fifth and fourth centuries BC, and that this rationality was an essential element in Hellenistic science, argues that it evolved into exact science only in the time of Euclid. (This claim presupposes Russo's definition of "exact science," which I'll come to shortly.)

Secondly, exact science according to Russo emerged out of the interaction of Greek rationality with the technical cultures of the empires that Alexander conquered—above all the Egyptian empire. Egyptian technology at the time, an affair of lore and tradition that had accumulated over centuries, was superior to Greek technology. Alexander's aim was to Hellenize the world. In the newly conquered empires, as ruled by him and after his death by his successors, Greek intellectuals were put in administrative roles. Under the beneficient feigns of Ptolemy I Soter (ruling from 323 to 283 BC) and Ptolemy II Philadelphos (ruling from 283 to 246 BC), Greek administrators were challenged to rationalize and improve upon the technological processes they were required to supervise. Russo sees the invention of both exact and experimental science as a response to that challenge.

Thirdly, according to Russo, as the Romans, scarcely out of barbarism and totally ignorant of science, conquered the Mediterranean world, the maintenance of scientific culture became increasingly difficult and finally impossible. The transmission of scientific culture by oral instruction was thus interrupted; and when post-Hellenistic writers like Vitruvius, Plutarch, Pliny the Elder, and Seneca came to write of science and scientists, they failed to understand the terminology and methodology of the Hellenistic works that had come down to them, and remained largely ignorant of their true mathematical and experimental content. It is a mistake, in Russo's view, to paint a picture of a unified Greco-Roman culture, as Plutarch did in his *Parallel Lives*.

To argue for such a sweeping thesis, Russo has to be a textual archaeologist, a detective who ferrets out and connects together a multitude of small clues. There is much in his book that is admittedly conjectural; much that the reader may wish to investigate further. In the following, I give references to relevant pages to help the reader in locating the pieces of Russo's argument referred to.

An early question must be: What does, Russo mean by "exact science"? His definition is rather restrictive (31ff). "Exact science," as he proposes to use the term, has to do not with concrete objects of the real world but with specific theoretical entities, for instance angles and segments in geometry, or temperature and entropy in thermodynamics; these do not exist in nature, but are cultural products, originating in human activity. The theories of "exact science" have a rigorously axiomatic-deductive structure; they start from a few fundamental enunciations (postulates or axioms or principles), and deduce consequences from them by strictly logical inference. Expertise in a theory of this kind means being able to pose and solve exercises or problems formulated within the theory's ambit, and agree on the correctness of the solutions; the existence of such expertise is an identifying characteristic of "exact science." Finally, applications to the real world are made possible by "rules of correspondence" between the entities of the theory and concrete objects. Technology and exact science, in Russo's understanding, are distinct but go hand in hand.

Is there a documented use of "exact science" in classical, pre-Hellenistic times? Russo says No (39). Eudemus, a disciple of Aristotle, wrote a history of geometry (known only through references to it by Proclus in his commentary on the first book of Euclid's *Elements*), in which he assigned the discovery of certain geometrical theorems to earlier thinkers (51f). For instance, he makes Thales the discoverer of the theorem that triangles with one side and the adjacent angles equal are equal, because Thales is said to have used this proposition to calculate the distance of ships observed at sea. But, counters Russo, one could use this proposition, accepting it as true, without having formally proved it. More generally, Russo doubts there was a motive for demonstrating apparently obvious truths (like the equality of vertical angles, or the equality of the halves into which a diameter divides a circle, "theorems" that Eudemus also assigns to Thales) before it was discovered that deductions could lead to quite unsuspected truths. The deduction of theorems probably began in Plato's time, but the notion of a unified axiomatic-deductive system hadn't yet solidified (55). None of Euclid's five postulates is attested in pre-Euclidean geometry, nor are there any alternative sets of postulates documented as having been set forth before Euclid (65).

It is thus Russo's view that one author, Euclid, with a unitary aim, put together Euclid's *Elements*, which became the paradigm of a deductive system for all succeeding ages. But he suspects that the text as we have it contains interpolations. Among these may be the definitions of "point," "line," "straight line," "surface," and "plane surface," at the start of Book 1, which possibly stem from Hero of Alexandria, of the first century AD (235-44). These definitions are "realistic," implying the existence of their objects in the world, rather than nominal like the other definitions of Book 1. Euclid, Russo believes, had a "constructivist" rather than a "realistic" conception of his geometry (71-75). The first three postulates state the allowable operations: to draw a line from any point to any point, to extend the line, and to describe a circle. These operations are the clear and explicit transposition to the level of mathematical theory of the operations normally executed on papyrus, using pen and ink, straight edge and compasses. To be sure, there is an enormous difference between mathematics and engineering design; but the postulates "model" what the actual instruments do, so that it is perfectly clear whence the postulates derive and what the "rules of correspondence" are that permit the mathematics to be applied. By means of his unitary starting point and strict adherence to deduction, the mathematician becomes independent of both philosophical speculation and immediate engineering concerns (57). But the relevance of mathematical science to engineering applications, Russo believes, was something Hellenistic society was unlikely to let the mathematician forget.

According to Russo, it is an indication of Euclid's constructivism that he uses the term *semeion* (sign, mark, token) for "point"; the earlier term had been *stigma*. The latter term, Russo states, has like the Latin *punctum* the realistic sense of a prick or stab, whereas a *semeion* or sign is a cultural construction or an interpretation of one thing as meaning another. Archimedes, Apollonius of Perga, and apparently Hipparchus continued to use *semeion* for "point," but in the post-Hellenistic period there was a return to the earlier term with its realistic connotation (72-73).

Another indication of Euclid's constructivism is that all entities in the Elements are constructed on the basis of the initial postulates (73-74). His inclusion of the fifth or parallel postulate among his postulates, without any attempt to demonstrate its truth or otherwise comment on it, is consonant with a high level of sophistication. He had evidently found the postulate to be necessary for the derivation of the properties of the figures constructed. His formulation of the postulate is sparing, ascetic; it does not speak of infinity (cf. 61, 331). The ascetic choice of basis suggests an artfulness that has reached mature self-awareness, that knows it can proceed only by construction.

But what about the often alleged "platonic" inspiration of Euclid's *Elements*? Let us *distinguere,* as the schoolmen used to say. Mathematicians, in pursuing their deductions, discover truths hitherto unsuspected. One such discovery was that there were five and only five of the so-called "platonic bodies," the convex polyhedra that the Demiurge in Plato's *Timaeus* uses for the imagined construction of the universe. Do such "platonic bodies" have an "independent existence," whatever that might mean? Not independent of the constructivist starting point, if Russo is right. He is urging that the Hellenistic mathematicians were conscious, and perhaps even proud, of their independence of metaphysical speculation.

At the same time, Russo believes, they were sharply aware of the engineering relevance of their propositions. Euclid's *Elements* had widespread application to the solution of architectural and engineering problems. The standard method for solving a problem was to draw a diagram, and apply the propositions of Euclid's Book 2 (which Russo, like Heath, calls "geometrical algebra"). The method can be comparable in accuracy with the slide rule, the standard engineering calculative device before the age of the digital calculator. The modern complaint about the ancient 'aesthetic' preference for straight lines and circles is thus, according to Russo, badly misplaced (57-60). In the technological context in which Russo places it, the constructivism he sees in Euclid's *Elements* fits comfortably. Russo finds evidence of the same constructivism in Apollonius of Perga (72, 113-14).

Russo's sense of the technological context is supported by the surviving Hellenistic mathematical treatises that are explicitly concerned with sciences we would call "applied." Euclid's *Optics* is a deductive science having to do with visual perception, and permitting the quantitative determination of the apparent sizes of objects seen from a given distance and given standpoint (79-83). The fundamental "elements" to be considered here are *opseis,* straight lines collocated in a cone with apex at the eye. It is by considering how objects are placed in this cone that we understand how they appear perspectively. The term *opseis* in Euclid's treatise has usually been translated "visual rays." Beginning in the fourth century AD, we find complaints about Euclid's claiming that vision occurs by visual rays issuing from the eye, rather than by rays of light entering the eye. But this, says Russo, is misinterpretation. Euclid is not proposing a physical cause of vision, but a geometrical theory of how things appear, which requires a consideration of the visual cone, the *opseis* or lines of sight with apex at the eye. The use in paintings of geometrical rules of perspective is documented for Euclid's time, but not earlier; geometrical perspective was rediscovered during the Renaissance in the context of an interest in Hellenistic culture (81).

The treatise "On the equilibrium of plane figures" by Archimedes (ca. 287 BC - 212 BC) gives us what we call the "law of the lever," and shows how to determine the "centers of gravity" of variously shaped plane figures (91-94). Archimedes's choice of "plane figures" rather than "weights" is perhaps a way of insisting on the quantifiability of his subject matter. In this treatise he lays the basis for the determination of the "mechanical advantage" of simple machines, making possible the quantitative design of machines for the lifting of a given weight through a given distance with the available force. Aristotle had asserted that a single man could not move a ship; a century later, Archimedes showed how by an appropriate machine the single man could do it. Don't believe, says Russo, the picture that Plutarch gives us of Archimedes as an otherworldly theorist; Plutarch was writing two and a half centuries after Archimedes was killed by a Roman soldier, and knew little or nothing of Archimedes' mathematics.

Similarly, Archimedes's treatise "On Floating Bodies" laid the scientific basis for naval architecture, leading, according to Russo, to a major increase in size of ships (94-97, 125-27). Other Hellenistic sciences had similar engineering relevance: hydraulics, for instance, for the delivery of water under pressure to elevated citadels as in Pergamon (130-33); and catoptrics for the construction of parabolic mirrors, employed in the lighthouses that now came to be built in the larger ports of the Mediterranean, including the great lighthouse on Faro in the harbor of Alexandria, one of the seven wonders of the world, projecting a light that could be seen for forty-eight kilometers (127-29).

The Hellenistic Age, Russo believes, saw an unprecedented development and application of technology. He reviews the evidence in his Chapter 6 (179-200); it comes mainly from administrative records and twentieth-century excavations. A good deal of it was assembled by Rostovzev in the early decades of this century (see his *Social and Economic History if the Hellenistic World* (Oxford: Oxford University Press, 1941). The early Ptolemies energetically pursued a policy of increasing agricultural production. The areas cultivated under royal control were increased by the draining of swamplands and the irrigation of the edges of the desert. New machines, with iron parts (possibly including geared wheels, which are first documented for this time), were introduced for the raising of water and the sowing and reaping of grain. New kinds of wheat were imported; old and new kinds were hybridized. That this policy was successful is suggested by the following facts (190):

(a) In late Pharaonic (pre-Ptolemaic) times, the population of Egypt is estimated at three million.

(b) In 1836 it was estimated that the agricultural lands of Egypt, if utilized to the maximum, could support a population of eight million; by 1882, as a result of economic reforms, the population of Egypt had increased to 6,800,000.

(c) Alexandria, founded by Alexander in 331 BC, had in, 50 BC about 500,000 inhabitants; it was the largest city in the Mediterranean world. In the first century AD, the population of Egypt was estimated on the basis of fiscal records to be 7,500,000.

And Egypt during the Hellenistic period not only consumed grain but exported it on a large scale, along with many industrial products such as glass, ceramics, paper, textiles, perfumes, and pharmaceuticals (190-95).

Economic prosperity led to a large increase in wealth, not only of the Hellenistic rulers but of an expanding middle class. In the first century BC Diodorus Siculus described Alexandria as "the first city of the world, by far superior to all others for elegance, size, riches and luxury." Its broad avenues were illuminated by lamps throughout the night; its houses were supplied with running water; its theaters and baths provided entertainment and recreation. All the genres of painting that would later be developed in the 17th century—portraiture, landscapes, still-lifes—appeared here in the third century BC.

And by royal policy scientific research was richly supported. The Ptolemies created at Alexandria the Museum, the first public institute of research known to history (182). There, just as at our Institute for Advanced Study, meals were served in common, and mathematicians and scientists exchanged ideas. At the disposition of the Museum's guests was the famous Library. Ptolemy II Philadelphus not only bought books from merchants but requested copies of new books from all the states with which he had diplomatic relations. Ships docking at Alexandria were required to list all the books they carried and to allow them to be copied. Within a few decades the Library contained a half million books. A separate section of the Library was open to the public, constituting the first public library. Pergamon under the Attalid dynasty (263-133 BC) followed similar policies, instituting a library and art collection, and fostering interchange of ideas between mathematicians, scientists, and engineers. At Antioch, too, mathematical and scientific research was supported under the Seleucids.

In the sciences of hydrostatics, hydraulics, and pneumatics, experimentation was a prerequisite for formulating postulates or principles, since the latter did not emerge as obvious truths from ordinary experience. Again, in the breeding of plants and animals, experimentation was the only avenue for advance. Active experimentalism, according to Russo, emerges naturally out of a constructivist "exact science," self-consciously pursued.

Russo makes much of the experimentation of the anatomist and physiologist Herophilus, active in Alexandria at the start of the Hellenistic period (154-69). He it was who first introduced the taking of the pulse as a diagnostic procedure. For this purpose, an accurate clock was required. At just this time, his associate at the Museum, Ctesibius, had modified the ancient clepsydra or water clock by making it a constant-level device (by means of constant inflow combined with an over-flow trough), so as to insure a constant pressure on the water issuing from the orifice at the bottom (121-22). This orifice was lined with gold or gemstone, to avoid corrosion or incrustation. The water issuing from it was collected in a second receptacle and measured by a float that moved an indicator over a graduated scale. With such an instrument Herophilus determined average pulse rates for persons of different ages.

Herophilus dissected human bodies and carried out physiological experiments. A century after Aristotle had declared the function of the human brain to be that of cooling the blood, he had recognized it as the central organ of the nervous system; and by experiments (perhaps on condemned prisoners—horrible thought!) had distinguished the motor and sensory nerves issuing from the spinal cord. He had discovered the reticular structure of the retina, suggesting the discrete nature of the visual receptors; and this discovery may underlie Euclid's assumption, in his *Optics*, of a discrete set of opseis as forming the cone of lines of sight. Denying the adequacy of ordinary language—an adequacy on which, according to Russo, Aristotle had depended—Herophilus invented special names for various anatomical organs that he discovered, such as the epididymus and the Fallopian tubes; many of these organs would be rediscovered only in the 16th century. Galen (second century AD) expressed disapproval of Herophilus's raising strong objections to every proposed cause of a physiological phenomenon, then proceeding to make use of such causes; to Galen these actions seemed self-contradictory. In Russo's view, Herophilus was merely being critical and self-aware in his use of hypotheses (165-66). The school of Herophilus persisted into the first century AD, but was then extinguished amidst growing incomprehension of Hellenistic science.

Russo's account of Hellenistic astronomy (99-109, 251-84) is not the least daring of his reconstructions. Only two works of Hellenistic astronomy, both minor, have come down to us: Aristarchus's *On the Dimensions and Distances of the Sun and the Moon*, and Hipparchus's commentary on Aratus's poem *Phenomena*. Neither tells us anything about the motions of the Moon or planets. But a passage of Archimedes's *Sand-Reckoner* states that Aristarchus (fl. 280-264 BC) had produced "demonstrations of the [planetary] phenomena." Plutarch (first century AD) in his *De facie quae in orbe lunae apparet* states that Aristarchus had sought to "save the appearances" by assuming the Earth to have motions of rotation and revolution; Aristarchus's theory was evidently heliocentric. This theory appears to have made little immediate stir, and it has been generally assumed that it dropped out of sight for 1800 years, until revived by Copernicus. Not so, says Russo.

One myth to be rejected is that Aristarchus, because of his heliocentricism, was accused of impiety; the story comes from a misreading, by the seventeenth-century philologist G. Ménage, of a passage in Plutarch's *De facie* (104, note 101). That absolute rest is not detectable by observation was a commonplace among Hellenistic writers: Euclid in his *Optics* states that the appearances of motion depend solely on relative motion between observer and object observed; a similar statement by Herophilus is reported by Galen; and Lucretius in *De rerum natura* makes much of the same point. Heracleides of Pontus and others in the fourth century BC had proposed accounting for the daily apparent westward motion of the stars by assuming an eastward diurnal rotation of the Earth; such an assumption already challenges the naive belief that we can observe absolute rest, since it implies that the inhabitants of Mediterranean cities are being whirled eastward at speeds of around 1000 km/hr. Mathematicians, being constructivists, would not have made a realistic or exclusive claim for the heliocentric theory; its excellence for them would have been in the elegance with which it accounted for the appearances. Archimedes mentioned Aristarchus's heliocentric theory without raising physical objections to it.

We know that Archimedes constructed a planetarium; its design, Russo argues, must have been essentially heliocentric. Cicero saw it two centuries later, and said of it that "the invention of Archimedes is to be admired in that he thought out how a single conversion could reproduce dissimilar, inequable, and contrasting motions" (104). Cicero's stress on the unicity of the "conversion," Russo urges, is incompatible with a mechanism of Ptolemaic type; it suggests a single center about which the revolutions occur. Assume a mechanism in which all planets revolve about the Sun in the same direction; to show what the appearances would be for a terrestrial observer it would suffice to hold the Earth fixed while the 'conversion' continued. Another planetarium is credited to Posidonius of Rhodes (first century BC), and Russo believes that it, also, is likely to have been heliocentric in the sense just explained. By contrast, an apparatus of the type described in Ptolemy's *Planetary Hypotheses*, incorporating a totally separate epicyclic mechanism for each planet, would be difficult to construct and unrevealing as an object of contemplation.

Russo believes the Hellenistic scientists went beyond the mere proposal of a heliocentric scheme, to a consideration of mechanical concepts that would be compatible with it. Let me cite some key pieces of the evidence he brings to bear.

Hero of Alexandria, now dated to the first century AD, is viewed by Russo as reflecting Hellenistic ideas and discoveries. In his *Mechanica* (1.4, 20-21) he announced: "We shall demonstrate that weights placed on a frictionless plane can be moved by a force smaller than any given force," The demonstration consists in approximating the horizontal plane by means of a series of inclined planes with ever smaller inclinations (251-52). Here we would seem to be close to Galileo's experimentation with inclined planes, and to the modern idea of inertia, according to which a body set in motion continues its motion without the application of force. This discovery crucially depends on the recognition of friction as a force that under ordinary terrestrial conditions brings motion to a stop.

That gravity causes acceleration was well-known to the ancients, including Aristotle. Strato of Lampsacus (d. ca. 270 BC), successor of Theophrastus as head of the Peripatetic school in Athens, according to a report of Simplicius (commentator on Aristotle, fl. ca. AD 530), noted that this acceleration was observable in a stream of falling water, which after a certain point in its fall breaks into separate drops. In the pseudo-Aristotelian *Mechanical Problems* (not datable, but presumably late-or post-Hellenistic), immediately after the explanation of the parallelogram rule (for what we would call the vectorial combination of motions), it is observed that a point moving in circular motion is subject to two simultaneous displacements: one described as according to nature *(kata phusin),* along the tangent, and the other contrary to nature *(para phusin),* directed to the center (*Mechanical Problems,* 849a14-17).

Plutarch's *De facie* is a dialogue in which one of the interlocutors takes the side of the mathematicians while the other opposes the mathematicians' "paradoxes." According to the first of these interlocutors, "Certainly the Moon is kept from falling by its own motion and the rapidity of its rotation, just as objects in a sling are kept from falling by circular motion. In fact, motion according to nature guides each body, unless it is deviated by something else. For this reason the Moon does not follow its weight, because it is equilibrated by the effect of the rotation." Plutarch's source evidently assigned weight (heaviness toward the Earth) to the Moon, and appears to have had a dynamics based on the principle of inertia. Russo (254-55) gives evidence tending to date the theory reported by Plutarch to the time of Hipparchus (second half of the second century BC).

A passage in Seneca's *Naturales quaestiones* (7.25, 6-7) can be interpreted as expressing the same theory, but now extended to all the planets: "We have met with those who say to us: You err in judging that any star can stop in its path or go in reverse; the celestial bodies cannot stop or turn away; all advance; as they are once launched, so they proceed. . . . Should they be stopped, the bodies now conserved by their regular motion would fall the one on the other. What is then the cause why some seem to turn back? A falling in with the Sun, and the nature of their circular paths, so positioned as for a certain time to deceive the observers, impose the appearance of slowness, on them. Thus ships, although moving under full sail, yet seem to stand still." The passage, Russo admits, may allow of a non-heliocentric interpretation, but he believes the heliocentrism of Seneca's source is detectable in it. And that source may well have been Hipparchus, who may also have been the source of the mathematical "paradoxes" discussed in Plutarch's *De jacie*. Pliny in his *Naturalis historia* (2.95) says that Hipparchus had an ingenious theory for explaining planetary motions, and laments the failure of anyone to exploit the Hipparchan legacy.

Archimedes's discovery of the buoyancy principle, Russo argues, made Aristotle's theory of gravity untenable (273). The discovery showed that there was no such thing as absolute lightness, opposed to gravity. All terrestrial bodies are heavy, but bodies float in water or rise in air because they are less dense. In face of this discovery, Aristotle's theory of the elements becomes unsustainable, and a major reason for supposing celestial bodies qualitatively different from terrestrial ones is removed. In his treatise on floating bodies, Archimedes derives the sphericity of the ocean's surface from the symmetry of gravity about the Earth's center. Later, as reported by the historian Diodorus Siculus, it was suggested that the Earth was initially fluid, and that the Earth's spherical form is the result of the gravity of its parts. The apparently spherical forms of the Sun and the Moon were explained in the same way; thus Plutarch in his *De facie* reports the notion that the Sun draws to itself the parts of which it consists, just as the Earth does (274).

According to the geographer Strabo (first century BC), Eratosthenes (third century BC) objected to Archimedes' claim that the form of the oceans was exactly spherical; the tides, he believed, were due to the Moon's attraction, which thus altered the spherical; form (275). Also according to Strabo, Seleucus, a mathematician of the first century BC, studied the tides of the Arabian Sea, and found an annual cycle: the two daily tides differed maximally in size when the syzygies (full or new moon) occurred in the solstices (in midwinter or summer), and differed least when the syzygies occurred in the equinoxes (in spring and autumn). The effect suggests that the Sun's attraction as well as the Moon's is involved in producing the tides, and that the effect of these bodies is maximal when they are at the zenith or the antipodal point (see Appendix).

Pliny in his *Naturalis historia* (2:212, 213, 216) seems to be referring to Seleucus when he says that the Moon and Sun cause the tides, and mentions that the difference between the two daily tides becomes nil at the time of the equinoxes. He also remarks on the retardation of the tides with respect to the positions of the Sun and Moon: a remark that makes sense only in the context of a gravitational explanation. That Seleucus believed in the Earth's motion is shown by a passage in Plutarch's *Platonicae quaestiones* (1006C), where he mentions the idea that "the Earth was projected, not confined and stable, but revolving and rotating, as successively affirmed by Aristarchus and Seleucus, the first only assuming it by hypothesis and Seleucus rather proving it." Russo's suggestion is that Seleucus based both his theory of the tides and his proof of the heliocentric theory on the hypothesis of a gravitational interaction with the Sun (276-78).

Hipparchus (second century BC) was a younger contemporary of Seleucus, and according to Strabo accepted Seleucus's findings on the tides (278). Hipparchus is known from Ptolemy's *Syntaxis* to have discovered the precession of the equinoxes, determined the distance of the Moon as 59 Earth-radii from the Earth's surface (a good value), and developed an accurate theory for the Sun's motion, and a theory giving accurately the Moon's positions in the syzygies. From other sources we know that Hipparchus wrote on gravity, and Russo is of opinion that Hipparchus's theory was one of universal gravity, in which the Sun kept the planets in their circular courses by attracting them out of their inertial paths.

Of the several items of evidence that Russo assembles in support of the idea that such a theory had been developed in the late Hellenistic period, I shall cite only one, a passage from the *De architectura* of Vitruvius, a Roman engineer of the first century BC. The passage (in *De architectura* 9.1, 12-13; Russo, 267) is obscure; Russo assumes that, like much else in Vitruvius, it stems from a Greek source (possibly Hipparchus), one that Vitruvius found difficult to interpret. Here is the translation of the passage given in the Loeb Classical Library:

> [T]he mighty force of the Sun extending its rays in the form of a triangle draws to itself the planets as they follow, and, as it were curbing and restraining those which precede, prevents their onward movement and compels them to return to it and to be in the sign of another trigon.
>
> Perhaps it will be asked, why does the Sun cause delay by these heats, in the fifth sign away from itself rather than in the second or third? I shall therefore explain how this seems to happen. Its rays are spread out in the firmament on the lines of a diagram of a triangle with equal sides. Now each side extends neither more nor less than to the fifth sign.

This text, it will be granted, is not very illuminating. Russo believes that the meaning of the original Greek passage can be reconstructed if we realize that Vitruvius's *signum*, by which he seems to understand a zodiacal sign, can have been his translation of the Greek word *semeion*, and further that the ordinal numbers "second," "third," and "fifth," would in Greek be expressed by the second, third, and fifth letters of the Greek alphabet, *B, Γ, E*. On this understanding, we can suppose that the original Greek passage had to do with vertices of triangles labeled "point B," "point Γ," and "point E." (In another passage at 9.7, 3, Vitruvius shows that he did not understand *semeion* in its technical geometrical sense as a point: he gives "signum et littera C" apparently as a translation of *"semeion Γ,"* taking *semeion* to refer not to a point but to the concrete letter written on the page.)

The original diagram, as Russo envisages it (Figure 1), would have the Sun *(H)* at the center, and the planet initially moving *kata phusin* along the straight line *AB* but being drawn back *para phusin* toward the Sun, and so arriving at *Γ* rather than *B;* the same composition of motions would be repeated

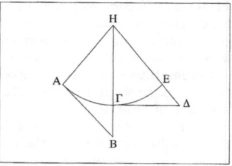

Figure 1

in the next adjacent triangle, so that the planet moving along *ΓΔ* would, arrive at E rather than *Δ*. Thus the Sun, by pulling the planet toward itself, would cause it to follow a circular path, and the intent of the diagram would be to show how the circular orbit is produced dynamically. Russo believes that we can see here the foundation of celestial mechanics, eighteen centuries before Huygens and Newton.

It is generally maintained that Ptolemy's *Syntaxis* rendered earlier works on astronomy obsolete, so that they ceased to be copied (see, e.g., G. J. Toomer, *Ptolemy's Almagest* [Princeton: Princeton University Press, 1998], 1). This view, according to Russo, gives too little weight to the interruption of an active astronomical tradition between the time of Hipparchus and that of Ptolemy. In the Syntaxis, the last observation of Hipparchus cited is dated 126 BC, and the next astronomical observation cited, taking them in chronological order, is an observation of the Moon by Agrippa dated AD 92—an interruption of 218 years. According to Russo, Ptolemy's adoption of Aristotelian 'realistic' cosmological premises, as in his insistence on the Earth's being at rest at the center of the cosmos, is evidence of a deep gulf of incomprehension separating him from his Hellenistic predecessors. Russo derives a similar conclusion from other aspects of Ptolemy's work (259-62).

(Russo has given his reconstruction of late Hellenistic astronomy in a lengthy article in English published in Vistas in *Astronomy* 38 [1994]: 207-48: "The astronomy of Hipparchus and his time: a study based on pre-Ptolemaic sources.")

Russo's book is a multiply-pronged argument with many parts, and it is impossible here to give anything approaching a thorough review or critique of the whole. It is to be hoped that an English translation will soon appear, and lead to detailed assessments of the several parts of Russo's argument by competent specialists. Here a few general conclusions will be ventured.

If Russo is right, the mere survival of some books into the post-Hellenistic period was not sufficient for the maintenance of an ongoing scientific and mathematical tradition. Pace Hegel, the choice of books for copying and preservation showed little in the way of geistreich intelligence; one has only to think of Heiberg's accidental discovery, in 1906, of a palimpsest giving Archimedes's *On method* to know that important works did not necessarily survive. Science itself was a fragile growth, dependent on oral tradition and financial support from beneficent rulers. Such bits of Hellenistic science as made their way into the works of the literati of the imperial age were transmitted only in distorted or ambiguous form.

If Russo is right, Hellenistic science discovered many experimental facts and reached many deductive conclusions that have previously been regarded as special achievements of modern science as founded in the seventeenth century by Galileo, Kepler, and Descartes. The assumption, often made within the St. John's community, that modern science presupposes certain philosophical doctrines first articulated in the seventeenth century, thus comes into question. The question deserves the radical inquiry called for in a recent lecture at the Annapolis campus (Grant Franks, "'Everything Aristotle Has Said is Wrong': The Authority of Texts and How We Got This Way," February 6, 1998).

As we have seen, Russo attributes to his Hellenistic mathematicians a self-conscious, methodological constructivism, leading them deliberately to avoid "realistic" definitions. Is the attribution of so sharply defined a methodological stance to ancient authors—whose statements about methodology, if any, have not survived—an imposition of modern views? In a recent article Russo has argued in detail for the hypothesis that Euclid's first seven definitions are an interpolation, perhaps due to Hero of Alexandria ("The Definitions of Fundamental Geometric Entities Contained in Book I of Euclid's *Elements," Archive for History of Exact Science* 52 [1998]: 195-219); and I believe he has made a cogent case. He is himself clearly a passionate methodological constructivist. He views the widespread realistic acceptation, in the schools and in the press, of such concepts as "elementary particle" and "black hole," as a transmogrification of science. I am inclined to agree: we understand these concepts only when we understand both that and how they are human constructions. But had the Hellenistic mathematicians and engineers arrived at so sophisticated a view? I believe it is a fascinating possibility, worthy of our consideration.

Appendix

Why do successive tides at the solstitial syzygies differ most in height, while those at the equinoctial syzygies differ least? Russo gives the explanation on 277, note 139. Suppose the Sun to be initially at the zenith of a point A on the Tropic of Cancer, while the Moon is at the zenith of a point B on the Tropic of Capricorn (see Figure 2). If we ignore the delay of the tides, we can say that there will be high tides at A and B. Twelve hours later, the Earth will have rotated through 180°, carrying the meridian CEB into the position formerly occupied by the meridian AE'D; we shall have high tides at C (not B on this meridian as twelve hours earlier) and at D (not A on this meridian as twelve hours earlier). Thus in the solstitial syzygies the point of high tide on the meridian CEB shifts from B to C and back again; and the point of high tide on the meridian AE'D shifts from A to D and back again. At a particular point on the Earth's surface like A, successive tides will therefore differ in height. In the equinoctial syzygies such a shift does not occur, because the Sun is aligned with the Equator, and the Moon is either accurately so or not more than 5° off the Equator.

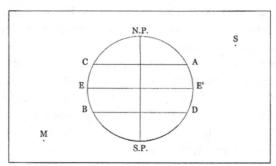

Figure 2

Monster Am I
(2011)

(Book Review of Paolo Palmieri, *A History of Galileo's In-clined Plane Experiment and its Philosophical Implications* [Lewiston, New York: Edwin Mellen Press, 2011.])

The project of this book, Palmieri tells us, emerged slowly, in close re-lation with the attempt to reenact certain of Galileo's exper-iments, in particular the inclined plane experiment. Galileo's adventure with balls rolling down inclined wooden beams was not a single event to which a date can be assigned, nor yet a set of operations described in sufficient detail to admit of mere copying. It was a sequence of explorations last-ing nearly a lifetime, involving difficulties and puzzles that Galileo struggled to resolve, with less than uniform success. A love-affair, Palmieri calls it.

Our author seeks to touch the very nerve of Galileo's endeavor. He challenges the assumption—beguiling to some Galileo scholars—that armchair philosophy can plumb the complexities that Galileo met with in the inclined plane experiment. He seeks to put himself in Galileo's actual problem-situation, with its puzzles on both the experimental and the theoretical side. Experimental work and theoretical explanation, carried on in tandem, pose questions of each other. The result, Palmieri reports, is liberating: the experimenter-theoretician-scholar probes more feelingly, with a new intensity. He becomes a participant in a rev-olutionary endeavor.

The earliest writings we have in Galileo's hand appear in Vol.I of the Edizione Nazionale under the title *Juvenilia* assigned by the editor, Antonio Favaro. He took them to be a compilation from unidentified sources (they remain unidentified today). Their late-medieval character is striking. Paragraphs frequently begin with *Advertendum quod. . .* ("It is to be noted that. . ."), a scholastic verbal tic implying that the fol-lowing sentence is accepted on authority. Is our scribe a mere copyist? Palmieri detects indications to the contrary. In places Galileo appears to be paraphrasing or summarizing; here and there he leaves blank spaces as though for later comment or elucidation.

Included in the *Juvenilia* is a discussion of "the intension and re-mission of forms," a much-treated topic in scholastic discussions from the fourteenth century to at least the early sixteenth century. It concerns increase or decrease in the intensity of a quality. The hotness or hard-ness of a body may vary from point to point or from instant to instant at a point. There is no evidence that the fouteenth-century schoolmen attempted actually to measure such variations, but they introduced lan-guage for describing them as measurable *secundum imaginationem.* One of the qualities thus dealt with was the speed of a moving body.

To the uniform variation of the intensity of a quality, the schoolmen applied a special rule, now dubbed "the Merton rule" after the Oxford college where it seems to have originated. It states that a quality vary-ing uniformly in intensity over a spatial distance or interval of time is equivalent to the unvarying or uniform quality of the mean degree stretching over the same extension, spatial or temporal. Suppose, for instance, that the hotness of a body varies uniformly from two degrees at one end of the body to eight degrees at the other, the degree being an imagined unit of intensity. This "latitude of form" was said to be equivalent to a uniform hotness of five degrees from one end of the body to the other.

Applied to the intensity of motion or speed, the Merton Rule was interpreted as saying that the distance traversed in a motion uniformly accelerated from an initial to a final speed was equal to the distance traversed in a uniform motion having the same duration and the mean speed of the accelerated motion. Compare this with the crucial first proposition of Galileo's treatise "On Naturally Accelerated Motion" in the Third Day of his *Two New Sciences* (1638)—do not both come to the same result? The Merton Rule is not mentioned in the *Juvenilia* or in any of Galileo's writings. Nevertheless, a number of historians of science, Pierre Duhem and Marshall Clagett prominently among them, have concluded that Galileo's theorem was a redaction of the Merton Rule.

The jury, Palmieri objects, is still out: no evidence has yet turned up to show that Galileo actually encountered the medieval enunciation of the Merton Rule. Palmieri suggests that such influence as the me-dieval discussion may have had on Galileo was likely indirect, through the *Geometry of Indivisibles* (1635) of his friend Bonaventura Cava-lieri. The trajectory of Galileo's thinking, Palmieri urges (following Favaro), is best determined from his writings and the experiments he sought to carry out.

At the very time Galileo was writing the *Juvenilia*, and thus be-coming acquainted with the scholastic conception of the natural world, he was also annotating Archimedes's *On sphere and cylinder*, a strict deduction of mathematical consequences from premises. The analytical thrust of Archimedean thinking, Palmieri believes, peeps through the text of the *Juvenilia*. Galileo sweats to understand the medieval four-element physics of the sublunary realm, and how all qualities are to be derived from the four 'prime qualities' or 'alterative qualities,' hot, cold, dry, and wet. Are motive qualities and speeds grounded in this fundamental Aristotelian ontology? The question is not explicitly addressed in the *Juvenilia*, but shows itself in Galileo's *De Motu,* dating to ca. 1590.

The *De Motu*, Palmieri observes, is polemical. Galileo denounces his teachers for the way they teach. When introducing the elements of physics, they bring in Aristotle's other works, quoting from *De Anima, De caelo,* or *Metaphysics,* as though their pupils already knew everything or else will accept all on faith. Galileo pledges to proceed differently, following the mathematicians in advancing solely by deductive steps derived from explicit premises.

A central question posed in the *De Motu* is: How do the weights and speeds of the same body, descending along planes differently inclined but of equal elevation, differ? By considering the forces needed to equilibrate the weight on the different planes as weights applied to a lever, Galileo shows that the forces are inversely as the lengths of the planes. (Unbeknownst to Galileo at the time of the *De Motu,* this same result had been given in the thirteenth-century *Scientia de ponderibus* attributed to Jordanus de Nemore.)

Galileo in the *De Motu* also attempts to deal with the question why a falling body accelerates. The body's heaviness is a *virtus impressa* (impressed force) that acts downward. Were it acting alone, Galileo assumes that it would produce a constant speed of fall. But to this impressed force downward, Galileo adds an accidental lightness or levity, imparted to the body when we raise it from the Earth. When we release it, its motion downward accelerates because the impressed lightness exhausts itself over time. The downward acceleration is thus explained in terms of the Aristotelian qualities of heaviness and lightness, with the important additional assumption that the accidental lightness decays with time. Galileo is here entangled in the fundamental ontology and categories of the *Juvenilia*, along with a misconception, widely accepted up to the time of the publication of Descartes's safari *Principles of Philosophy*, that every velocity has to be maintained by an impressed force. How did he free himself—as he unquestionably did—from the medieval mind-set and its stultifying questions?

Palmieri proposes that certain life-worldly learning experiences—among them, finding how to make glass goblets sing and brass plates howl—taught Galileo a lesson about the fine structure of nature. By patiently repeated experimentation, the young Galileo learned how to attend to and control the fine detail of what happens in the production of these effects. The beginnings were in the workshop of his father, Vincenzo Galilei. During the 1580s, Vincenzo, a professional lutist, engaged in musicological controversy. Opposing the Pythagorean claim that numerical ratios are the cause of musical intervals—that the ratio 2:1 is the cause of the octave, the ratio 3:2 the cause of the fifth, and so on—he claimed that these intervals are to be determined by the ear alone. One of his prime exhibits was the singing glass, a goblet containing water which, on being struck, gave forth a musical tone. The pitch depended on the amount of water. Years later Galileo Galilei in his *Two New Sciences* told of a further result, not previously reported: the goblet could be set singing if its rim were stroked with a wet finger-tip. Concomitantly, a standing wave was produced on the surface of the water. Sometimes the tone shifted up an octave, at which moment the number of waves per unit length in the standing wave doubled.

Palmieri—apparently the first among Galileo commentators to do so—has replicated this experiment. Success requires practice, and it is best to begin with a large goblet (Palmieri used a brandy snifter). One must rub the rim rhythmically, while repeatedly wetting the finger and watching for the evanescent wave pattern. The wave pattern is more readily produced in the brandy snifter than in a smaller goblet, but Palmieri found it possible to obtain Galileo's result also with the latter.

The howling brass plate is another of Galileo's experiments that Palmieri has replicated. As Galileo reports in the *Two New Sciences,* while scraping a brass plate with a chisel to remove stains, he found himself producing sounds. Sometimes they were musical tones, and in such cases the chisel left evenly spaced marks on the plate. On one occasion two tones sounded in succession, forming the interval of a fifth. In the two sets of marks formed on the plate, the numbers of marks per unit length were as 3 to 2. Getting these results was helped by a bit of practice, but was easier than obtaining the standing wave in the glass goblet.

A correct interpretation of these experiments presupposes the physics of sounding bodies, which Galileo himself lacked as have some of his recent commentators. A sounding body vibrates predominantly with certain frequencies that depend on the shape and mechanical properties of the material. These frequencies are the body's 'natural modes' of vibration. For a body of regular and relatively simple shape, the predominant frequency modes are harmonically related, e.g. as octave or fifth, etc. The vibrations are reflected from the boundaries of the body, and the reflected waves, combining with the original train of waves, form a standing wave pattern. For a given speed of propagation (which is determined by the medium), the wave-length is inversely as the frequency, and thus the wave-lengths of two standing waves have ratios inverse to the integral numerical ratios of the corresponding natural modes. The natural modes therefore account for the emergence of the Pythagorean ratios in these two experiments. By confirming Galileo's experimental results, Palmieri has put them beyond doubt. They form a beautiful early confirmation of the theory of natural modes.

Important as this conclusion is, it is a different point that Palmieri aims primarily to make. Galileo's experimental results are obtained only with patient attentiveness to the fine structure of experience. They yield an experience in which hearing, touching, and seeing are integrated—a holistic experience. Such experience can direct consciousness away from false expectations and towards new facts. This kind of learning, Palmieri proposes, assisted Galileo in liberating himself from the medieval mind-set with its pre-established categories.

What about the inclined plane experiment? Here also, besides the visual sight of a ball rolling down the plane, a complex of other sensory data is offered—sounds, vibrations that can be sensed through skin and bone as well as the ear, changes in sound as a bronze or wooden ball descends along the wooden track. Did Galileo attend in a focused way to these effects? We know only of the cases already cited, in which he focused on the details of experimental happenings with attentiveness and care. In the Third Day of the *Two New Sciences,* in the section *On Naturally Accelerated Motion,* he focuses on the kind of acceleration that nature employs for descending bodies—on this, its consequences, and not on causes. The latter question as posed by the schoolmen has been set aside:

> [W]e decided to look into [the properties of this kind of motion] so that we might be sure that the definition of

> accelerated motion which we are about to adduce agrees
> with the essence of naturally accelerated motion.

It is this that he is now seeking the essence of—naturally accelerated motion itself.

Galileo's adoption of this new focus, Palmieri believes, can have been triggered by the very intensity of the auditory and vibratory experience of the ball rolling down its inclined track. To receive this non-visual experience as fully as possible, Palmieri placed his forehead in contact with the underside of the beam serving as inclined plane, and grasped the beam with his hands around its sides so that his fingertips could sense the upper side of the beam. An assistant then released a ball to roll down the inclined plane. As it rolled, Palmieri's fingertips picked up the vibrations induced in the beam, they were also transmitted through his cranial bone, and he heard the sound through his ears as well.

The resulting experience Palmieri calls holistic auscultation. It is no mere juxtaposition of different effects, but an integrated effect. It powerfully suggests that through our senses we can delve deep into the fine-structure of physical reality. The experience is markedly stereoscopic. The experimenter, hugging the plane at a particular location, is first aware of the ball's starting to move far up behind his head, then hurtling close by, and finally fading away in the distance. The descending ball produces a sound that varies as the ball speeds up. Sound and speed grow uniformly together, and this togetherness takes center stage. The arresting character of this experience, Palmieri proposes, can have derailed the young Galileo's ambition to reduce changing speed and sound to effects of the qualities dubbed primary by Aristotle and the schoolmen.

> In the scriptorium [where the *Juvenilia* was produced], the hot-cold-dry-wet chemistry of pitch and speed can only be thought-through. But it is possible to leave the scriptorium, visit workshops, and turn life into a tastier affair. . . . We reach a new balance between knowledge and values when we learn how to reconfigure life-worldly objects while letting our senses be affected by them.

In *The Two New Sciences* Galileo stresses the *simplicity* of the means nature adopts—in the case of descending bodies, the increase of speed in proportion simply to time elapsed. Reenactments of

Galileo's inclined plane experiment, however, yield at best rough confirmations of this relation. In multiple repetitions of the experiment, Palmieri and his students used a water clock of the type Galileo describes, weighing the water released during the duration of the descent to obtain a measure of elapsed time. In a descent of the whole plane, compared with a descent of one quarter of it, the expected ratio of the times is 2:1. In five trials of a bronze ball one inch in diameter, running on the groove cut by a router into the beam (so that the ball was running as though on rails), the numbers obtained were 2.18, 2.19, 2.15, 2.09, 1.97, averaging to 2.12, hence with 6 percent error and a root-mean-square dispersion from the mean of 0.08. In five trials of a bronze ball seven sixteenths of an inch in diameter, running in the groove, the numbers obtained were 2.04, 1.90, 1.95, 1.90, 1.84, averaging to 1.93, hence with 3.5 percent error, and a root-mean-square dispersion from the mean of 0.067.

Palmieri records twelve more sets of five trials each. The errors are dramatically larger for decreased inclinations of the plane, especially in the case of smaller and thus lighter bronze balls. Five trials with a bronze ball seven sixteenths of an inch in diameter and an inclination of 1.36 degrees gave an average of 2.74, hence with 37 percent error; but five trials with a bronze ball one inch in diameter and the same inclination yielded an average of 2.17, hence with 8.5 percent error. An increase of the inclination to 3.8 degrees for these two balls reduced the errors to 18 percent and 6 percent respectively.

The deviations from expected theoretical ratios do not easily admit of a detailed explanation, nor does Palmieri attempt one. Of the factors likely to be operative we mention two. Human reflexes cannot be relied upon to open the water-clock precisely when the ball is released to start rolling, or to close it precisely when the ball hits the stopping block. And, throughout the run, friction is no doubt operative. Friction is an action at or between surfaces. Seeking to find what schoolmen were saying in Galileo's day concerning friction, Palmieri examined the *Juvenilia* and a book on natural philosophy by Galileo's contemporary, the Paduan professor Giacomo Zabarella (d.1589). He found discussions of "reaction" and "resistance," but not of friction. The presumptive role of friction in the inclined plane experiment, Palmieri believes, was a potent riddle leading Galileo to abandon scholastic explanations in favor of an atomistic ontology. Galileo knew and undoubtedly consulted Lucretius's *De rerum natura*. It gives a psychophysical explanation of pleasant and unpleasant tastes in terms of smooth and rough or hooked atoms. The shapes of atoms could similarly account for friction in the sliding or rolling of one surface over another. Friction would thus be a "fight" between particles of different shapes. The amounts of friction would no doubt differ with the extent of contact between ball and trough, with the shapes of atoms, and with the speed of the ball. Such factors may be the causes of the deviations between observed and expected time ratios above reported. But it is hard to imagine how this hypothesis could be tested quantitatively. Besides, Galileo may have shied away from openly entertaining a hypothesis deriving from Lucretius's philosophy—such a move on his part could have led to a new charge of heresy.

One of Lucretius's doctrines appears to have played a seminal role in Galileo's thinking about falling bodies. Lucretius states that, since bodies falling in the void meet with no resistance, all fall with the same speed. He attributes the observed differences in the rates of fall to the checking action of the medium, which hinders the motion of lighter bodies more than that of heavier bodies. Galileo in the *Two New Sciences* will reach an analogous conclusion, but with a crucial difference: all bodies falling in the void fall, not with the same speed, but with the same *acceleration*. We recall that earlier, at the time of the *De motu*, Galileo had thought that the rates of fall would be as the specific gravities (weights per unit volume) of the bodies. That assertion differed from the Aristotelian position, which made the speed of descent proportional to the body's *weight*. Galileo rejected the latter position on the basis of the following argument. He hypothesized that any heavy body that falls has its speed, or (if accelerated) its degrees of speed, fixed by nature, so that the speed or the acceleration cannot be increased or decreased without violence. (Thus the argument applies whether the body falls with a constant or an accelerated speed.) He then imagined two bodies equal in volume and weight, e.g., two bricks. If let fall together, their speeds or accelerations are equal, and they remain side by side. If tied together so as to form the double weight, the result does not change: they still fall with the same speed or acceleration; neither "burdening" the other. Hence speed of fall cannot be proportional to weight.

By the time he wrote the First Day of *The Two New Sciences* (probably in 1634), Galileo had concluded that all bodies that fall begin by accelerating, and he was hypothesizing that all bodies in the void, in-

334

dependent of their specific gravities, accelerate with the same acceleration. The reasoning leading to this conclusion, as given in the *Postils to Rocco* (marginal notes on a work in which Antonio Rocco attacked Galileo's arguments in *Two New Sciences*) proceeds as follows. He imagines two equal spheres, one of gold and the other of cork, that are let fall from the same height. Since both are surrounded by air, both are buoyed up by the same force, equal to the weight of the volume of air they displace (the buoyancy effect identified by Archimedes). Each body in its motion will also be slowed by the viscosity of the air, and this effect, since it derives solely from a property of the air, would likewise be the same for both. Friction, which Galileo explains as due to the sticking of particles of the medium to the asperities of the body's surface, can also be imagined to differ negligibly in the two bodies (both could be covered by the same surface material).

Finally, there is the resistance to the speed of each body, which is greater for greater speeds. Galileo does not imagine that this resistance can be eliminated practically (as it was a few years later, after Galileo's death, in experimentation with Torricelli's mercury barometer and with von Guericke's air pump). But experience, Galileo tells us, suggests that this resistance is entirely an effect of the medium. In a fall of the gold and cork spheres through 100 braccia (150 feet?) in air, the gold, he asserts, will precede the cork by two or three braccia. In a fall of 1 or 2 braccia the difference all but disappears. If in a thin medium like air, differences of speeds all but disappear, then, Galileo claims, we are entitled to hypothesize that in a vacuum the speeds would be identical. For this conclusion, we note, Galileo can claim only a hypothetical status.

Galileo's Third Day of his *Two New Sciences* is in one respect strange—even peculiar! The First and Second Days, dealing with the strength of materials, are clearly about real material bodies. The Fourth Day, likewise, deals with real projectiles, actual bodies moving through the air and resisted by it in their motion. The mathematical part of the Third Day, by contrast, is presented as about *points* descending along inclined *lines*. Real bodies are absent, as are real planes along which they could descend. Friction is nowhere mentioned. The idealization of bodies is at least as drastic as that in Euclid's *Elements*. Galileo's adoption of this extreme idealization may owe something to the seeming impossibility of eliminating the effect of the medium, and the difficult of quantifying the effects of friction.

335

Among the numerous theorems proved in the Third Day is the "expansion theorem": points falling simultaneously along variously inclined lines starting from a single point as origin are all at each instant on the surface of an expanding sphere. Galileo has the interlocutors of his dialogue engage in a considerable discussion of this theorem. It may be, Palmieri suggests, a relic of an earlier project to elaborate a Lucretian cosmogony starting with point-atoms—another dangerous project which Galileo may have relinquished to avoid further conflict with the papacy.

Early in his book Palmieri cites John Dewey's *Experience and Nature* for its "take" on Galileo's quest for a science of nature. To Dewey, Galileo's turn to active, controlled experimentation represented a radical challenge to the Graeco-Christian spectator-theory of knowledge. In Aristotle's philosophy, as co-opted by Christian philosophers like Thomas Aquinas, high value was placed on detached contemplation of the world. The human being was seen as situated at the center of the cosmos, empowered to survey and understand its parts. But in Dewey's pragmatist perspective, true knowledge can be gained only by intervening in the world and attempting to bring disturbing or confusing situations under control. Detached contemplation is "out of touch," powerless to penetrate the intricate mysteries of the world. Only by intervening, risking mistakes and failure, can we begin to learn the world's ways.

Was Galileo self-conscious about the revolutionary break he was making with the methodology of earlier natural philosophy? Palmieri in his Chapter 7 adduces three pieces of evidence suggesting that he was.

During his three-year professorship at the University of Pisa (1589-1592), Galileo had as friend, tutor, and colleague the professor of Platonic philosophy, Jacopo Mazzoni. Mazzoni was a syncretist, seeking to show the compatibility of Plato and Aristotle with each other and with Christianity. He was, indeed, the very model of a late sixteenth-century Graeco-Christian philosopher. Yet there was one opinion, apparently shared by Plato and Aristotle, that Mazzoni took issue with, the opinion that theoretical mind was categorically distinguishable in kind from practical mind. Every branch of philosophy, Mazzoni insisted, has both a *theoria* and a *praxis*. Each of these incorporates operations directed toward particulars. In *theoria* such operations are for the sake of *propping up* the truth; in *praxis* they are for the sake of *attaining* the truth, and of finding the essence of things in the order of existence. Such praxis Mazzoni saw as dangerously rushing towards particulars, plunging the seeker after truth into the perilous world of the unstable, of the disturbed situation where action can make the difference between failure and success. In this characterization Galileo, struggling to make sense of his results in the inclined plane experiment, might have recognized himself.

Another glimpse into Galileo's discomfort with the suffering of experiential learning may be gathered from his *Considerazioni al Tasso.* This consists of notes criticizing Tasso's epic poem, *Gerusalemme liberata.* The publication of the poem in 1581 had sparked a lively debate among Italians as to its merit relative to Ariosto's *Orlando furioso* (1516). Pro-Ariosto by taste, Galileo was vehemently anti-Tasso. He rejected Tasso's treatment of the human passions. Whereas in Ariosto's poem there was a metamorphosis of *res* into *verba,* leading to a satisfying outcome, Galileo judged Tasso's poetry to lack poetic inspiration, and to end up cobbling together fragmented *concetti* (imaginations) lacking continu-ity and reciprocal dependence. The result was thus like marquetry, in which colored pieces of wood are fitted together, and the border lines between pieces always remain sharp and crudely distinguishable. Tasso had failed to realize that the passage from *res* to *verba* must be dynamic, transformative.

Aficionados of Tasso found in his poetry a new conception of human feeling: feeling as a force originating from deep sources in the senses and the body, so strong at times as to overwhelm the mind. Imitating the pathos in Tasso's poetry became a project for composers of madrigals like Monteverdi. The resulting works were among those sharply criticized by Galileo's father Vincenzo. Vincenzo saw "modern music" as mixing together voices and modes, diverse words (in polyphony), different rhythms and tempos, and thus giving rise to disparate and confusing emotional reactions in the intellect of the listener. The future of music, Vincenzo urged, lay in resolving the polyphonic "confusion" of voices of the madrigal into a monodic style of singing.

Vincenzo's criticism of polyphonic music as fragmented and un-intelligible is closely parallel to the younger Galileo's criticism of Tasso's poetry as marquetry. Palmieri sees Galileo's disdain for Tasso as a disdain for "the real, oblique, polyphonic nature of experiential learning." Galileo's preference for Ariosto is "a preference for an ideal of experience in natural philosophy in which he had been inducted by Mazzoni's teachings." But "Galileo's radically new practice of philosophy . . . had brought him face to face with the reality of experiential learning." A deep rift runs through Galileo's mind, as Palmieri reads him: on the one hand, Galileo ardently strives after a science of nature—which requires dealing with the reality of experiential learning; on the other hand, he would like that science to conform to the ideal sketched by Mazzoni. Perhaps, Palmieri suggests, Galileo was salvaging something of Mazzoni's ideal in his sanitized accounts of his experiments.

Finally, Palmieri adduces a sonnet written by the dying Galileo, giving it both in the original Italian and in a translation by Dennis Looney. The latter runs as follows:

Enigma

Monster am I, stranger and more misshapen
than the harpy, the siren, or the chimera.
There is not a beast on land, in air or water
whose limbs are of such varied forms.

No part of me is the same size as any other part;
What's more: if one part is white, the other is black.
I often have a band of hunters behind me
who map out the traces of my tracks.

In the darkest gloom I take my rest,
For if I pass from the shadows to bright light,
Quickly the soul flees from me, just as

The dream flees at the break of day.
And I exhaust my discombobulated limbs
And lose my essence, along with life and name.

Palmieri interprets the sonnet as a meditation on experiential learning, caught between the polarities of individuality and universality. The metaphor of the monster captures the jagged contour of experience. The darkness is Galileo's blindness, physical as well as metaphorical. The hunters are his persecutors, real and imagined. Only after death, with the loss of individuality, will the light of truth shine forth. Thus Galileo recognizes that knowledge is not coextensive with human experience. His sonnet refracts as through a prism his life-long pursuit of truth; an active engagement with the life-world, turning up more difficulty and more unsolved conundrums than he has been able to cope with. The strife, the tension, as he still re-lives it towards the end, is tragic in its intensity. The only resolution is limitless relinquishment.

**Some eulogies of colleagues
and a few other occasional discourses**

Remarks at the Memorial Service for Ford K. Brown
(1977)

I have known Ford Brown since I came to St. John's in 1948. During one of those early years, I sat in on a freshman Greek class he taught. He had then as always a way of putting things that printed them unforgettably in memory. He was marvelously informed about many things. He it was through whom I learned about Jespersen's works on grammar, and about the studies of Yugoslav epic-singing, and the studies of the intonation of Sanskrit, and how they were being applied to the interpretation of the Homeric poems. He it was who made forever clear for me the difference between accuracy and precision, by pointing out that one could be precisely wrong. I remember, too, his insistence that when you were to write an essay the first thing to do was to make an outline of what you wanted to say; and then you should begin the essay by saying what you were going to say, and then you should say it, and finally you should say that you had said it. It is a wonderful rule, and I should like to say that I had followed it, but the truth is I have never succeeded in doing so, owing to simple weakness of mind. Still, Ford's rule has always stood out for me in the distance as the ideal to be aimed at. I could at least try to achieve, toward the end and after much destruction of foolscap, the straightforward clarity and outline that Ford had asked from us at the start.

Ford was a man who regarded the recognition of concrete fact as necessary, and the failure to recognize it as reprehensible; but he knew, too, that such recognition must be in a proper style, free of despair or of unwarranted or unbecoming enthusiasm. He much regretted the College's abandonment of Latin, and consequently of certain Latin authors, who he felt had a special appreciation of the practical and the concrete, a worldly wisdom that had nothing grubby or petty about it, but was combined with an appreciation of greatness. Those qualities were his, as well.

Some of his lectures became legend. Alumni from the 1940s speak repeatedly, with wonder, of his lecture entitled "Shakespeare," which turned out to be, with entire appropriateness, a lecture on the Cathedral of Chartres. He knew that there were certain things that couldn't be said, or should not be attempted to be said, and he knew that there were others one should go to the trouble of spelling out in detail. I remember

341

with especial clarity his definition and illustration of the imagination, given in this very room many years ago. We would know what the imagination was, he said, if we could visualize atop the piano here a blue sphere, twelve feet in diameter, and poised atop it, another sphere, green, eight feet in diameter, and atop that, still another sphere, this time red and four feet in diameter, and atop that—but I fear the image will no longer have the meaning it once did—atop that the then superintendent of buildings and grounds, Archibald McCourt, five feet high, with his impenetrable Scotch brogue, giving orders to the hired hands.

It is in the last thirteen years that my friendly acquaintance with Ford Brown has meant the most to me. In '64 and '65 we were together in Santa Fe, during the opening year of the College there; and during that year, he was the grand old man of the campus, loved by the students, sought out by them, appreciated by them as one of the secure and dependable facts of the College and indeed of life itself. And so he has remained for me. A jocular, ironic wit; gentlemanliness and generosity of spirit; an old-school way with words, combining the courtly with the clipped, the anecdotal with the trenchant; a habit of observing, without unnecessary insistence or any trace of bitterness, that there is a proper way of doing things here—"here" being just anywhere you happened to be: all these were inimitable features of Ford's style and way of being in the world and observing it all and making his comment and thereby teaching. Forgive me for mentioning that in these last years I was especially touched by the praise he directed to myself, which was given out of no knowledge on his part of my just deserts, but out of, first, the pure, apriori conception that there were things proper for a dean to be doing here, and out of, second, the purely generous assumption that I was perhaps doing some of those things rather than other ones. I was grateful to him for his not wanting to know the grubby, embarrassing details, and grateful, too, for his gentlemanliness and generosity and unfailing wit.

Ford had a great spirit. It was clearly somewhat difficult for him at the last. Using his walker he made it to the coffee shop almost every day this past summer, till the shop closed on August 11. He did for himself what he expected others to do for themselves, to exercise a bit of self-discipline and to have a bit of courage. And so we should. We shall miss his familiar figure. And we shall not forget.

Remarks at the Memorial Service for Jacob Klein
(1979)

During a number of years—exceedingly memorable years for many of us—the first Friday-night lecture of the College year began with a statement that is likely to seem unremarkable enough when repeated now, but which I have never heard made by anyone else but the lecturer of those years, the then Dean of the College. Nor can I imagine its being made, with the same meaning and effect, by anyone else. I cannot quote exactly, but with some huffing and clearing of the throat, the lecturer would say something like this: I have to begin by saying that, before the immense, the immeasurable difficulty of my task, I am filled with trepidation. The task he was speaking of was, of course, to speak of the task of education, the task of this College.

He proceeded to do that with a certain soberness, slowly, deliberatively, choosing words with care. The words were not fancy or technical; they were simple and arresting. We were asked to think—and first of all to think on the fact that our understanding of ourselves and, hence, of everything else was, of necessity, bound to intellectual traditions. And these traditions, while preserving the traces of the original insights and experiences from which they had sprung, necessarily did so only in a veiled and distorted way. That was an unavoidable consequence of our dependence on language, and of the evanescence to which every understanding is inevitably subject. Hence this task: to penetrate through the layers of forgetfulness and distortion, to recover the foundations of our views and attitudes, and to assess, as far as possible, their truth.

An enormously difficult task; extravagantly ambitious, incompletable in a lifetime. Yet not to undertake it is to remain prisoner, chained, with vision confined to the shadow-play on the cave's wall. Hence an inescapable tension. On the one hand, there is the finitude of our human condition, of which we must not lose sight. And on the other there is the extravagant goal which is implicit in, yet transcendental to, every attempt to say the truth. This is the peculiar two-foldness in the Delphic Oracle's enigmatic command: know thyself.

Jasha Klein, who often remarked with a special appreciation that someone or something was in some way remarkable, was himself the most remarkable person that many of us are ever likely to know. His gifts of imagination and intelligence and judgment were truly extraordinary. But more remarkable still was their union with qualities of heart-spiritedness, warmth, a spontaneous and irrepressible energy, abundant enjoyment and delight in many things. And with such qualities of mind and heart, during many years, he devoted himself unstintingly to the teaching and learning at this college, believing, as he said, that the annoying and time-consuming efforts and tribulations that necessity imposes upon the teacher are—I am quoting—"beyond measure compensated by the insights he gains in the struggle with young and vigorous minds and in the witnessing of intellectual growth."

Teacher, guardian, guide, friend: it is impossible to think that he is gone, impossible not to imagine that he is looking on with a certain quizzical smile that some of us know. As for us, we shall pay proper tribute only if we recollect the meaning of the task of which he spoke so unforgettably, its difficulty, the constant danger of misconstruing it or reducing it to something other and less, its fearsome immensity.

Remarks at the Memorial Service forWilliam O'Grady
(1986)

"Not ignorant of evil," Dido says when she first meets Aeneas, "I know at least one thing—to help the afflicted." It was characteristic of Bill O'Grady to insist that what he knew or was clear about was something simple, some one thing amidst many others unknown, put forth usually with an apology for its inadequacy to the full complexity of what there is, yet put forth with a certain insistence. And then the one thing, with further articulation, would break into a wondrous multitude of things, like light through a prism; for his fellow-feeling, and his sense of the suffering of others, was combined with a remarkable acuity of discernment and intellectual force. The dominant quality was the concern with the wisdom of the heart; a wisdom that he sought with moral intensity as much in everyday relations as in the study of Homer or Sophocles or Nietzsche.

Bill O'Grady has been such a unique presence in this College; that presence has had so much to do with the soul of the College; and he has been taken from us so suddenly, and so sadly. For months I have known that, for me, this particular week in January would be packed with busyness; I did not know that in the midst of other doings I would be stumbling to find the right words to characterize so unique a presence as Bill O'Grady has been; nor have I any confidence that with more time I could be even moderately successful. In fact, the situation is one that Kierkegaard would characterize as a repetition: it is exactly the situation that Bill O'Grady has repeatedly found me out in; that is to say, he has repeatedly found me busy, involved in trying to decipher intellectual puzzles whose solution would not necessarily touch the heart. I am grateful to him for never blaming me for my acquisitive curiosity. His presence was inevitably a demand, a kindly but insistent demand, that the human heart be taken into account. I have wanted to respect that demand, to live up to it. Others, I imagine, have felt something similar.

I'm grateful to Bill for a number of insights into a number of books. I am grateful to him for a letter he wrote me, when he was on the Instruction Committee and I was dean, and he had discerned how exhausted I had allowed myself to become, and how dispirited; the only thing to be done with me, I knew, was to put myself in wraps until the animal spirits regained their resilience. Such mere mechanisms were not in his thoughts. He wrote me a letter, a lovely letter of comfort. So few of us manage to do things like that. More than once, when I was dean, Bill helped me by telling me what the real situation was that I was confronted with, how it was that the words that X had spoken meant the opposite of what they seemed to say, or what it was that Y would have said if Y had had the confidence to speak to me, or what some student was undergoing that I knew nothing about. Such revelations are precious to any mere administrator.

I know that there are uncounted students who owe debts of gratitude to Bill O'Grady, back through the years. I remember how during his first and only sabbatical year he was to be seen at the oddest hours, in the coffee shop, paired in earnest conversation with one senior after another, over senior essays, reminding me, no doubt inappropriately, of Lear's line near the end of the play about two birds in a cage.

No, it is not within my reach to say here the right words, about what Bill O'Grady has meant to so many of us, and will always mean. I have in my hands a letter to Bill from his father, written when Bill was about to finish his first year of college at Notre Dame. Joie has given me permission to read from it.

"Hope the exams have been going as well as possible, but don't worry about 'em. As the man said, your main objective is to become a sophomore, and I believe you have that just about accomplished. . . .

"There is one thing I would like to say in this last epistle. The year, I am sure, has been a rewarding and stimulating one for you. It has possibly opened up many new byways for you to explore . . . you have learned that, in addition to the selfish and the self-seekers and the just plain no-damn-goods in this world (and there seem to be too many), there are also intelligent, brilliant, selfless people, who love their fellow men (and there are more such people in the world than we realize).

"But I am sure, too, the year has also brought to you disappointment in various measures, a certain kind of loneliness, a difficult adjustment to college life, to people of all kinds, a diminishing, perhaps, in some areas of your ideals as feet of clay become more apparent, and, in short, all the trials and tribulations that go to make up this none-too-perfect world. All of this while you were facing, up to this time, at least, the most difficult adjustment of your life.

"And despondent though you must have felt many times, you never showed it in your letters, in your visits home, and in your own blithe spirit.

"In other words, you were the epitome of plain old-fashioned guts, the kind that never shows up on an athletic field, on a battlefield, or in other facets of life.

"It is the kind that is unseen, unknown (for the most part) and always unheralded.

"You displayed rare courage, true courage. You displayed maturity in its purest sense . . . the kind we all strive for but seldom accomplish.

"I am so very, very proud of you.

Love,

Dad"

The last line, applied to Bill's all-too-short life as a whole, speaks, I am sure, for many that are here. We are so very proud of him.

Remarks at the Memorial Service for Winfree Smith
(1991)

The words that came of themselves, when I heard of Winfree's death, were "Dear Winfree." I seem to have heard others address him thus, though I was embarrassed to do so myself. Such an unusual person he was: modest, unpresuming kind, gentle, ever courteous. of unimpeachable integrity, and with a sterling sense of justice. A Virginia gentleman, yes, but sui generis, his own person. Many of us have drawn on his moral capital, which was a treasure.

There was a reserve about him, and I can remember only once having heard him utter a negative remark about a person of our acquaintance, and then it was because of being called upon to make a judgment in the Instruction Committee. And in this case, I especially remember his restrained vehemence, as he said of the tutor under scrutiny. "But he lied!" Winfree hated dishonesty.

Yet, believing as he did that there is none of us who is not subject to original sin, he was deeply compassionate. Moral superiority was no claim of his, not even the inverted kind that in effect claims superiority in claiming non-superiority. At our memorial service for Simon Kaplan he remarked that Simon Kaplan came as close as anyone he had ever known to being a living refutation of the doctrine of original sin. The doctrine is not an easy one to take, but there seemed to be no strain in Winfree's acceptance of it. Our self-centeredness goes deep, and is not to be extirpated by constructions of thought or feeling. Winfree's wisdom, I believe, came from that recognition. Sometimes the best response is silence. He knew silence, the value of it, the fact that sometimes it was better than anything else.

Not the least memorable of his qualities, on the other hand, was a love of language, an enjoyment and relishing of classic formulations. The case of Scripture was, of course, special: when reading the lesson, he read slowly, almost monotonically; every word counted. He had studied and valued works of higher criticism like Von Rad's commentary on Genesis; but higher criticism never robbed the scriptural texts of the weightiness and the sacredness of each word.

A fabulous memory: he remembered verbatim huge quantities of the philosophy and mathematics he had studied. The exercise of this memory brought with it a certain enjoyment, and in some instances, amusement. He was the senior tutor in my first seminar at St. John's in 1948-49, and I remember his citing Aristotle's remark that only the male of the species is beautiful as evidence for Aristotle's sense of humor. Whether it was Winfree or Aristotle who was pulling our collective leg, I don't know.

Once, in one of Winfree's classes in junior mathematics which I audited in 1949-50, Winfree used the word "truth," and a student whom I suspect of not having done his homework, asked Pilate's question, "What is truth?" Without hesitation Winfree intoned, "Truth is the adequation of thought to thing." The formula is not one I get much mileage out of, but it was probably the right riposte.

I must admit that, as a California boy with a couple of years of overlay of New York experience, I found the Virginia contingent at St. John's, not to mention the European emigré contingent, rather overwhelming. At some party in my first year at St. John's, Winfree corrected either me or somebody else who had referred to the Civil War. Strongly, sonorously, he insisted: The War Between the States. I, whose paternal grandfather marched as a drummer boy with Sherman to the sea, was struck dumb. Last December, at my wife's invitation, Winfree came to our house for several nights running to watch the second showing of the Civil War series on TV. Winfree had missed the first showing for lack of a TV, and we had missed it because of being out of the country. And we shared the experience of that series together, mostly in silence, taking in the terrible human cost of that war. Winfree began reading Shelby Foote's history of it, making great progress, but I doubt he had quite finished it when he died. I went back to Walt Whitman's Specimen Days, which I had always wanted to return to after a first read some fifty years earlier. I felt there was a bridging of a gap.

It would not have occurred without my wife; it was through her that Winfree and I were kept in touch. Winfree's sister, Margaret Smith Lauck, had been my wife's friend in Fredericksburg, and when the Laucks came to Annapolis at Becky's urging, it was through Becky that Margaret became secretary to the dean, Jacob Klein. When I succeeded Mr. Klein as dean in 1958, I inherited Margaret as secretary. And through her I came to see Winfree not just as one of the pillars of St. John's, but as a member of a family with remarkable qualities. Margaret, too, had that sensitive intelligence combined with great modesty and tact.

Winfree's loyalty to St. John's was unquestionable, but it was also eminently sane. He once sermonized against the formula "my country right or wrong." Similarly, I felt, he opposed narrow or cliquish conceptions of what the College stood for. He once remarked to me that Scott Buchanan had been too detached—from the College that is; this suggested that one could also be too attached. College Avenue was tangent to Church Circle, but circles and their tangents were not to be confused.

Once, when I was speaking to seniors at their senior dinner, I quoted Walt Whitman: I have nothing, Whitman chanted, against institutions. The line is no doubt barbed, and Whitman was meaning that institutions could be oppressive, but that he was not going to let himself be oppressed by them. Winfree thanked me afterwards for my remarks. There seemed to be nothing of Whitman's self-dramatizing "barbaric yawp" in Winfree, but he too knew that institutions though necessary are inevitably imperfect. That would go for any college or any parish church.

There was a sober balance in Winfree of reason and faith. He stood on his own feet insofar as humans may, and treated all others as peers. So he would speak of members of the St. Anne's parish, or of students, staff, and fellow faculty members at the College, with a near-perfect sense of their equality and uniqueness. Such recognition is rare. For those of us who knew him, his influence and his presence will not disappear. He is missed. The love he evoked will remain.

Commencement Address
(1964)

Members of the Graduating Class:

Permit me, first, to wish you well and to express the hope that you will carry through, with courage and with all the intelligence you can muster, the new tasks which you are about to undertake. For many of you, I expect, there will be times when you find yourselves in restrictive circumstances where the chances for intelligent thought and action seem terribly limited. I hope that you will manage to navigate these narrows, and to discover, in the longer run, a satisfying exercise of your powers.

In considering what I should say to you, I felt that I would have to be engaging, however tediously, in moral discourse, and speaking of qualities that I feel ought to be preserved in the world. I thought I should accept the situation without evasion. What I shall be trying to point toward, or make some gesture in the direction of, is an intellectual conscience. If I can point at all, it is because I find myself a member of a community of teachers and learners who live now or who have lived and written or taught before. I cannot pass over entirely in silence the topic of Obligations to Society. The two kinds of obligation, intellectual and social, are by no means disconnected.

What is frequently said, that the age we live in is apocalyptic, seems to be true. It is a world of tumbling change. One hundred years ago, in November of 1863, C. S. Peirce stated in an address:

Let us bring here the sublimest intellect that ever shone before, and what would Dante say? Let him trace the rise of constitutional government, see a down-trodden people [he means the English] steadily bend a haughty dynasty to obedience, give it laws and bring it to trial and execution, and finally reduce it to a convenient cipher; let him see the most enthralled people under the sun [the French] blow their rulers into a thousand pieces and establish such a terror that [and now he is quoting from Revelation 6.15] "all the kinds of the earth, and the great men and the rich men and the chief captains and the mighty men and every bondman and every freeman hide themselves in the dens and in the rocks of the mountains"; let him see the human mind try its religion in a blazing fire, expose the falsity of its history, the impossibility of its miracles, the humanity of its revelations, until [again, quoting Revelation 6.14] the very "heavens depart as a scroll when it is rolled together"; and then let him see the restless boundary of man's power extending over the outward world, see him dashing through time, conversing through immense distances, doing violence to the lightening, and living in such a fire of activity as less salamandrine generations could not have endured; and he who viewed Hell without dismay would fall to the earth quailing before the terrific might of intellect which God has scattered broadcast over this whole age. [From an oration delivered at the Cambridge High School Association, November 12, 1893.]

What, one wonders, could Peirce add if he were to return today and read Melvin Calvin, a molecular biologist writing under grant from the Atomic Energy Commission, and claiming that the next stage in human evolution is the exploration of outer space; that this stage *must* be undertaken, because evolution depends upon the development by each organism of *every* potentiality; that it may suit man in the future to change the orbit of the moon; that there may be other organisms doing similar things at some millions of other regions in the universe; that our view of the place of life and of man is hereby completely inverted, from being that of a trivial to being that of a major cosmic influence; and that, finally, we come to this view "entirely upon the basis of experimental and observational science and scientific probability."

Peirce's next paragraph reads:

This century's doings taken apart are merely jugglery—clever feats—but this age is that in which [quoting once more from

Revelation] "the sun becomes black as sack-cloth of hair, and the moon becomes as blood, and the stars of heaven fall unto the earth even as a fig-tree casteth her untimely figs, when she is shaken of a mighty wind."

My reaction to all this is less brainy than visceral. I tend to think of man as a frail creature, a reed, though a thinking reed, as Pascal says. Phrases like "the conquest of space" or "the conquest of nature" seem inappropriate. And I suspect that there are problems more urgent than that of taking ourselves and our machines off the surface of the earth. I believe that you and I can accredit ourselves as having a bit of the available sanity, and in modest ways may be able to do something toward keeping the earth habitable: by opposing the waste of natural resources, the wanton destruction of trees, and the contamination of the atmosphere; by resenting violence and injury to persons, including our own, and insisting on the enforcement of civil and criminal law; by recognizing the complexity of the situations with which public policy is called upon to deal, and putting our weight in the scales against fanaticism and barbarism; by supporting high standards in elementary and secondary education; and in other ways which will occur to you or be forced on your and my attention now and later. Some of the details are arguable, but the end is not. The preservation of human life and orderly society need not be the highest conceivable or most heavenly aim, but it is one which most of us expect each other to affirm, and if we are to be logically consistent, we must legislate the means to it as obligations for ourselves. If a picture of the universe is presupposed here, it is not a model to scale, but it is drawn in perspective, with human beings in the foreground—and the stars are as small as dimes.

Now I come to my main topic, but with difficulty and trepidation. There are so many voices, more every day. They speak in such various tones, of invitation, of warning, of sober sensibleness, of threat and blame, of high aspiration and poetic ecstasy, of stern realism, denying the dreams of youth. All will agree: there is an obligation to truth. All will agree: there must be no grinding of axes or polishing of apples, where truth is concerned. Speaking of intellectual honesty or integrity or probity, all will imply that truths are to be recognized and followed up, independently of whether they conduce to emotional comfort or practical success. Or perhaps I should say: *perhaps* all will agree. There is the radical nihilist, whom you have met; I do not know what he will say. Let him accompany us here like a shadow if he likes, and if he wishes to speak, he can gain from me the admission that, whenever I speak or think, I do so from within a framework or horizon that is not fully explicit. That is the reason I must speak of an intellectual conscience, and not of truth.

But fine sentiments, Gide says, make bad literature—and presumably bad speeches; he fails to mention that there may be bad literature and bad speeches without fine sentiments—but that would spoil the epigram. But I must try to heed the warning and turn to what is simple and before us. There are things to learn. There are subject matters, developed over a period of time by a series of investigators, and accredited as trustworthy, as sound knowledge. In high schools, if I recall correctly, they used to be called *solids,* and you could take more than five. I am concerned only with the solid solids. Frequently the first initiation was, and is, somewhat confusing and disappointing. You have to learn something badly before you can learn it well. The course is not just a series of magnificent or charming discoveries. Nor is it merely a process of running through classificatory schemes, or deductive developments. It is a process of becoming familiar with and making things lucid for oneself. If the subject has any depth at all, if it is really a solid and not a surface, one has to learn repeatedly that, where one had thought one was clear, one was muddled. Some of one's presuppositions get pared away: as, say, a student of physics may realize with a little shock that energy cannot always be localized in space, as he had somehow previously assumed it could; or as one can suddenly find that a sonnet he had always read in one way can be taken in another. This kind of re-examination and reflection can go on for years, yet there is a stage which we can call connoisseurship and mastery. I speak of the scholar who knows his way about, as does a gardener in his garden.

This image, of course, suggests the gardener's love of his shrubs and flowers, and it suggests the garden wall with possibly a gate. There may be problems about the gate and the wall. But let me say a word for the faithful scholar. A scholar must dwell patiently with particulars. It is something to know just about everything that can be known about partial differential equations, or to have deciphered the Babylonian astronomical tablets, or to have followed the behavior of bees or spiders in every detail, or to have traced, through endless obscure sources, the patterns of ritual and play of archaic peoples, or to have studied Philo Judaeus or Kant till the loci of thoughts and questions are all familiar. Such scholarship has its private satisfactions, independent of scholarly fame; and it has its conscience. There is a virtue in just being right, about matters that anyone could get straight if he knew the language, had the patience, and could take the time.

Please understand: I am not advising you to become university scholars, unless you want to. I am saying: whoever you are, try to know some of the things you know. You cannot know all that you know; there is not time. But to pierce the veil of a formula, of a catch-phrase, of knowledge-at-second-hand, is to sharpen the intellectual conscience. I propose no recipes, no rules for the direction of the mind. Here knowing is doing and you will detect the muddle, the swindle, the inadvertence when you detect it. Always you will be asserting or presupposing tacitly more than you know.

And what if you do enter the university? Amidst a conglomeration of hothouses with closed doors, and a variety of exhibits that may mislead, you are expected to choose. It can be perplexing. But do not be too harsh on specialization. It is for the sake of progress, particularly of science, that research he been so divided up and parceled out. It is so that the scholar can contribute to progress before he is nearly dead, before he has mastered any great depth or breadth of subject matter. And specialization *does* make for progress, as a thousand of the more technical journals testify. And it provides, in the more recondite subjects, a scale of reliability of information. But there are embarrassing side-effects. With progress, we become shamefully ignorant. And if the providence of God can no longer be seen in the anatomy of a louse, it does not follow that vanity and pretension diminish. And it may be, as Tolstoy argued, that the idea of progress is itself baneful. The ancients, he argued, could die old and full of life because life underwent no change, because they believed they had experienced everything life could provide; but modern man, confronted with a feverish advance in ideas, knowledge, and questions, can only be tired of life, so that for him death has lost its sense. The scholar as scholar, as I have described him, cannot answer this reasoning. Perhaps I can say, with Melville, gulp down your tears and hie aloft to the royal-mast with your hearts. Perhaps I can add: take with you your wit, your imagination, your poetry. I turn from the scholar to the poet.

What astonishing claims have we here! Listen to St. John Perse:

> By means of analogical and symbolic thinking, by means of the
> far-reaching light of the mediating image and its play of corre-

spondences, by way of a thousand chains of reactions and unusual associations, by virtue also of a language through which
is transmitted the supreme rhythm of Being, the poet clothes
himself in a transcendental reality to which the scientist cannot
aspire. . . . By his absolute adhesion to what exists, he keeps us
in touch with the permanence and unity of Being. And his message is one of optimism. To him, one law of harmony governs
the whole world of things. Nothing can occur there which by
its nature is incommensurable with man. The worst catastrophes
of history are but seasonal rhythms in a vaster cycle of repetitions and renewals. The Furies who cross the stage, torches
high, do but throw light upon one moment in the immense plot
as it unfolds itself through time. [From his speech on accepting
the Nobel Prize for Literature, December 10, 1960.]

This is not new. The poet is older than the scholar. He antedates
science. He was the ancient *vates,* the inspired one, the thaumaturge,
the prophet, the seer, the sacred poet. For archaic man, knowing was
magical power. All particular knowledge was wonder-working wisdom, directly related to the cosmic order. To safeguard the orderly procession of cosmic events, it was necessary to know the secret names,
and the origin of the world. This knowledge, too, had to be safeguarded
and attested to, and so there were initiations and competitions in esoteric lore. So were ordained the sacred riddle contests. Listen to the
questions:

I ask you about the uttermost ends of the earth; I ask you,
Where is the navel of the earth? I ask you about the seed of the
stallion; I ask you, Where is the highest place of speech? [From
the *Rig-Veda.*]

Whither the moons and the half-moons, and the year to which
they are joined? Whither the seasons—tell me the source of
them! Whither in their desire hasten the two maidens of diverse
form, day and night? Whither in their desire hasten the waters?
Tell me the source of them! [From the *Atharvaveda.*]

Frequently the ritual riddles were capital riddles; you solved them
or forfeited your head. It was accounted the highest wisdom to put a
riddle no one could answer. King Yanaka holds a theological riddle-
solving contest among the Brahmins attending his sacrificial feast, with
a prize of a thousand cows. The wise Yajnavalkya, certain of victory,

has the cows driven away for himself beforehand, and proceeds to defeat his opponents. One of them, unable to answer, literally loses his head, which separates from his trunk and falls into his lap. Finally, when all are silenced, Vajnavalkya cries out triumphantly: "Reverend Brahmins, if any of you wishes to ask any questions let him do so, or all of you if you like; or let me ask a question of any of you, or all of you if you like."

In Greek tradition, the seer Chalcas dies of grief or kills himself out of chagrin, when bested in a riddle contest against Mopsos. His followers attach themselves to the winner.

The Venerable Nagasena, conversing with King Menander in the presence of 500, Ionians and Greeks and 80,000 Buddhist monks, throws out each of his dilemmas with a triumphant: "Find your way of that one, your majesty!"

We are not far from the sophist here, who booked such fabulous successes in fifth century Greece. But the word still has sacred, liturgical power, and he who knows it, can live his life. For the cosmos, like the riddle contest of the seer and poet, is agonistic. Strife is the father of all things; love and hate, attraction and discord, rule the universal process. And the Furies must be appeased. All things, of course and finally, must perish in that same principle from which they arise. For, says Anaximander, they have to render expiation to one another and atone for the wrong they did according to the ordinance of time. But the sacred poet is he who, with play of words, vouchsafes the divine decision.

Now all this would be nonsense, were it not for the fact that it is not; the reason it is not may be that we are all poets—bad poets. I propose that, with our words, each of us makes a world which he inhabits. It is hard to say what a word is, but in the first instance it is a gesture. Like all gestures, it presupposes an intention, and a muscular activity. Unlike other gestures, its goal is to symbolize, according to a pre-established convention, an idea or universal.

The word has two parts: a center and a penumbra. The center is composed of two things: the intended meaning on the one hand, and the mere sound on the other. The central meaning has lifeless fixity; it is the same for all men who speak the same language at the same epoch; or if it changes from one context to another, it does so by jumps, from one identity to another; this kind of fixed identity allows of immutable combinations, syntactical structure. The mere sound is inanimate; this becomes evident to me if, repeating the word to myself several times, it happens that I suddenly observe it in its starkness; it then appears as bizarre, arbitrary, dried out, dead. The penumbra, on the other hand, can be called living; it is the totality of memories associated with the word; it is different for different men. like all living things, it is disintegrated and re-integrated incessantly. The memory is *not* a storehouse of images and associations; rather, when a memory stirs, I have an image which resembles an earlier one but is nevertheless new. A word, or a sentence, or a book, evokes in me a world, threatening and discordant, or consonant and affirming, but at the center of which I live, believing myself distinct, with suffering or joy.

It is with this penumbral world that the good poet works: playing with dissonances, that the world may be enlarged; showing us what is *there* in the human heart; perhaps affirming some ultimate and mysterious harmony.

"It is enough," says St. John Perse, "for the poet to be the guilty conscience of his time." Let it be so.

I have been trying to deepen the notion of a conscience in intellectual matters, beyond mere submission to certifiable knowledge, and scholarly rejection of the false, the muddled, the misleading. And it may be an unwarranted delay to have thus far not mentioned logic, the morality of thinking. No logic, no knowledge.

Whitehead says: "we may conceive humanity as engaged in an internecine conflict between youth and age. Youth is not defined by years but by the impulse to make something. The aged are those who, before all things, desire not to make a mistake. Logic is the olive branch from the old to the young, the wand which in the hands of youth has the magic property of producing knowledge." [From a lecture to the British Association for the Advancement of Science, September 6, 1916.]

I turn from the poet to the mathematician, the maker and master of deductive systems.

Here all is admirably clear. This may be, a mathematician will perhaps suggest, because he has created the entire structure himself. It is a curious fact that, while poets discuss their role in terms of knowing and truth, the mathematicians, in evaluating what they do, use words like "elegance" and "beauty."

But I am more concerned at the moment with the fact that the mathematical method, by its deliberation and purity, can reveal every jump in an assertion. The orderly demonstrations carry me forward step by step, each step turning, as on a fulcrum, round the law of identity. We can ask: How is this formality constituted? How is it situated in the totality of the life of consciousness?

Mathematical objects are not given as such in the field of perception, nor do they belong to a purely intelligible world with which we can put ourselves in an immediate relation. They are given in certain *evidences* by means of appropriate *constructions;* for instance, geometrical diagrams or algebraic symbols to be manipulated. These constructions depend in one way or another on the data of perception. The evidences obtained through the constructions are enchained in itineraries that one can somehow traverse. The domain of mathematical objects is not before us like a spectacle of which all the parts are equally accessible. The domain is unveiled progressively; each new development opens new perspectives. Study of the properties of a given object or structure may lead to the discovery of other objects and structures. When the mathematician has gained sufficient knowledge of these new objects or structures, he may be able to define them in an autonomous way, without appeal to the earlier theories in which they had their origin. The conceptions are not fixed once for all. Mathematical experience has an internal life.

The various elements of mathematical experience are indissolubly related to one another. Analysis of any one of them returns step by step to all the processes which have led to its appearance, and finally, to the whole development of mathematical thought from its beginnings.

But note this: to perceive its object as it is, mathematical thought must succeed in detaching it from the living experience in which it first appeared, and represent it in a form which borrows from that experience only quite elemental data, easily controllable and verifiable. It must shell out, from a confused penumbra of associations, a core of fully determinate meaning. It must formalize. The project of formalization is inherent in the mathematical enterprise from the beginning, and becomes ever more apparent as more general structures are discerned.

What would be the final aim of this project, if it were to be carried to a logical conclusion? To put it all too briefly: Would it not be to elaborate a system which was fully autonomous, which would recapitulate all previous acquisitions, and contain in itself the key to all further development? Such a system would constitute the *total* mathematical object. Mathematical being would here be simply and immediately present, as identical with its representation. The representation, elaborated in experience, would no longer be in continuity with that experience, but would go beyond it and substitute for it. We are here before the limit-idea of a perfectly closed system, silent and full, existing by its own proper intelligibility, uniting paradoxically the characteristics of the thing and those of thought, constituting the objectified model and paradigm of all discourse.

Now I think you have heard that such a system cannot be achieved; that precisely by means of formalization, it has been found possible to *prove* that such a total system is unachievable. There is no substitute for studying in detail the various theorems of limitation that are here involved, and one might claim that, beyond stating the precise content of these theorems, there is nothing more to be said about their significance. But I think that, stepping resolutely back from the enterprise of formalization, and trying to reflect on its role in the life of knowledge, one can see the limitation theorems as mirroring the *temporal* character of thought. I believe this view can be followed out, for instance, in a discussion of the relation between the potential infinite and the actual infinite. The former is the notion of an unending series of constructive steps, each of which can in turn be performed effectively; the latter is the result of a *pseudo*-totalization, not necessarily harmful to thought, and even perhaps necessary, yet differing essentially from an effectively realized performance.

Both poetry and mathematics construct, and project, as into an exterior space, the formulae of their achievements. Yet they do so on the most widely separated mountains. Both have their forms of authenticity and rigor. There is the poetry of mathematics, and the mathematics of poetry. For both, there is an incessant dialogue with the world. Both attempt, though in widely different ways, a totalization, a completion. What are their roots?

I am coming to my final point, which is not a single one perhaps, except—perhaps—as a gesture towards a more conscientious conscience. The mind's work seems in a way to be a simple thing; and indeed the mind just *is* its work. What it does is a process of attending to and grasping affairs presented, tracing out the connectedness of a manifold of elements. The governing idea of serious work in always wholeness, completeness, adequacy of grasp, correctness of the demonstrative links.

But this movement is engulfed in time. Whatever is grasped at any instant escapes from the immediate coincidence of self with self, which defines the present, into the evanescence of the past. There is no moment which would be in itself the recapitulation of all the others, a present which would absorb into itself the past as well as the future and which would expand itself to the dimensions of all experience.

What is grasped at any instant involves elements that are accepted passively, taken as pre-given. Words are encountered as familiar; we understand them in the sense of being vaguely, passively conscious of their meaning. Where do these meanings come from? There is a certain opacity in the direction of the past.

On the other hand, what is achieved at any instant remains open to modification or reinterpretation. Consciousness, occurring only in the mode of consciousness, something at the thin edge of the present, is open to the future, to the non-actuality of what is not yet. There is an opacity in the direction of the future.

Temporality, I would propose, is here, in the realm of intellect, the ground of moderation.

And yet, though thought be time's fool, consider this: the past, both as personal life and as a longer history, is not merely behind and extraneous to us but is contained and implied in our present. Amidst the general confusion as to principles which seems so characteristic of our time, it is yet possible to question our past as it presents itself. It presents itself primarily as the present, as the world in which we live, the world which is there for *us*. It is the whole world, and it is a world of meaning, of what has had, what has, or what can have sense; an outside, some unknowable thing-in-itself, is precisely *non-sense*. The world has had to come to be *for* us, has had to constitute itself in original experience *for* consciousness; and therefore we can raise the question of its origins, its roots, its hidden history. To raise this question, and to pursue it, seeking an attunement to the beginnings, is intellectual courage.

Intellectual moderation and intellectual courage, twinned together, are the virtues of the aspirant to truth.

Let this be my gesture—on this occasion I would have liked it to be different and better—and let me conclude with Augustine's command:

> *Noli foras ire; in teipsum redi; in interiore homine habitat veritas.*
> (Do not go out; go back into yourself; it is in the *inner* man that truth dwells. [*De vera religione,* 39, 72.])

Commencement Address
(1977)

Members of the Graduating Class:

You are glad, I hope, to be crossing the finish-line. If I am not mistaken, to get there has taken a bit of pluck and patience and good will. For those qualities in you, I praise you, and also thank you. They are important qualities anywhere; they can be more important in intellectual work than is sometimes realized; but they are especially important for the community of learning that St. John's tries to be.

We at St. John's are a tiny band. We believe, and at our best act on the belief, that we have been entrusted with something special. Recently a candidate for admission, after visiting the College, wrote of the students he had met and talked with here: they carry a sacred flame. To an older tutor the flame is not always visible, but he must nevertheless acknowledge that he would not be so alive or so moved to exert himself in his own efforts to learn and understand, did he not confront these students day by day, with the light of intelligence and quizzicality in their faces, and the desire and the demand to know in their hearts. Together we have aimed at the high goals of the program. In old books we have sought the wisdom whose price is above rubies. For you and for us, self-education has been a matter of passionate personal concern. That concern, and the quixotic quests it has led us into, join us together in a certain complicity today. We would be hard put to explain it to our visitors. In that respect as in others, Commencement resembles a family affair.

But the bell tolls, and it is as if the great world were calling us to account. Your request and the President's invitation that I should speak here is a great honor, but it has also caused me some trepidation. Of the great world, I have little to say that is not cheap wisdom, readily available elsewhere. I must rather speak primarily of what you have been primarily doing here: learning and thinking. And all that I have to say is simple in the extreme.

For four years you have given a good deal of your time to reading and thinking, and to speaking and writing about what you have read and thought. A certain amount of information you have no doubt acquired; but much of that you have forgotten or will forget—a fact over which you should not fret, for the shadow of things forgotten will protect you from many illusions. If you go now to one of the professional schools, you are likely to have to shift intellectual gears, and to start taking in information at a considerably more rapid pace, in order, I assume, to be protected from yet more illusions. But here at the College you were primarily asked to do something different: to think. To ponder on certain books and paragraphs and sentences; to meditate on certain questions that were posed to you, or that you posed to yourselves; to speak and to write on these books and paragraphs, sentences and questions. What was this engagement of yours, and what did it do for you?

To describe thinking is well-nigh impossible. John von Neumann, the mathematician, compared thinking to riding in an airplane, and now and then going up front to do a bit of steering. We can do this without being able to know what keeps the airplane aloft or moves it forward.

Thinking involves a temporal process. Whatever is accomplished in thinking, whatever is grasped or understood, is grasped or understood on the basis of successive steps. At a certain moment I believe myself justified in saying: "I see; I understand." What led to that? The thing that is understood is a complex of elements, with their properties and relations; if it were only a solitary thing, without any internal complexity or any relation to anything else, we would speak not of understanding but perhaps of trance. Understanding is therefore always of something that is somehow many. And so in coming to understand a situation, I have presumably been tracing out the relations between the elements of the situation, presumably one by one, and then I say: "I understand now." But at this moment in which I say that I understand, it does not seem possible that all these relations are present to me at once, in their full significance, and it becomes a problem as to *how* they are present. It is evident that the acquisition of any understanding involves *necessarily* a kind of evanescence; different aspects of the situation to be understood have to fall successively into the background, into the past. And when I try to understand my understanding, to grasp reflectively what has gone on in the process of understanding, it seems I must either reactivate the original process, step by step, or else I am liable to fall into superficiality or false generalization.

The problem of evanescence goes beyond any single process leading to any single act of understanding. *All* intellectual work is based on previous acquisitions which have become as though embedded and submerged in one's thinking. Previous acquisitions, in order to be transmittible from one person to another, or even to remain accessible to one person, have in general to be framed in words, written or spoken. And written or spoken words exercise a seductive power. Increasing familiarity with certain words and patterns of words makes possible a certain kind of passive and superficial understanding, which carries us forward to another stage without our necessarily having grasped the full meaning. Even thinking that has seemed satisfying and adequate always involves an interlacing of what is grasped centrally and with a degree of clarity and distinctness, and what is accepted passively as pre-given. Learning never starts from a zero-situation, complaining members of the teaching profession to the contrary notwithstanding.

Now this understanding of understanding has consequences for one who would understand. Whenever I understand anything, I am dependent in my understanding on the manifold of relations and elements that are passively there, in the background, waiting in the wings and ready to appear on stage, should the further progress of my thought require it. Were my understanding not thus surrounded and supported, it would consist of entirely isolated thoughts, which would be incapable of being improved and enriched by further reflection. At the same time, this dependence on a passively maintained background is a danger for thought. What is now passively accepted may be the residue of thought undertaken earlier, thought that has been intense and that has taken account of some of the complexity of things. On the other hand, it may be the residue, and much of it *has* to be the residue, of what has been passively and unthinkingly accepted with the learning of language and with the words of others, that I have somehow absorbed and come to use, from childhood onwards. My dependence on this passively accepted background means that only with effort can I escape the ill or misleading effects of cliché and jargon, of conventional vocabulary or of the deeper assumptions of the culture in which I was raised. To begin to take responsibility for my thought, what is required are circumstances in which my tacit assumptions can be challenged, as rigorously

362

as possible, and then, on my part, not anger or rejection but a special attentiveness and effort to be thoughtful.

A further consequence of this understanding of understanding is that whatever I accomplish in thinking remains open to modification or qualification by a series of future thinkers, including me. If I claim to understand or grasp anything with any kind of completeness, there yet remains the open possibility of grasping further relations, determinations, connections, so that what has thus far been grasped appears as a special case of something else.

Now suppose that, with this understanding of understanding taken for premise, a program of studies is to be founded, its aim being not simply or primarily the acquisition of information, but rather chiefly the improvement of the mind. The founders will surely know that it is not possible for us to start from a zero-situation. They will know that it is necessary to begin in the middle of things. At the same time, they will warn the participants, namely us, against taking anything for granted; they will ask us to take the vow of poverty of knowledge. This means, not that we should become sceptics, settled in our ignorance, but that we should begin to exercise an alert *skepsis,* a looking that seeks to know *that* and *what* it does not know.

What should be studied, given that there are no absolute starting-points, that we have to begin in the middle of things? Surely we must avoid current fashions and jargon, surely we must allow ourselves some distance, surely we must find the means that will enable us to probe most deeply and to see farthest, that will make us most thoughtful. Is it not evident, then, that we should choose books and paragraphs, sentences and questions, that first-rate minds have rated as first-rate, and that have significantly shaped language and thought? Is it not clear that we should turn to them-questioningly, indeed, and yet with tentative trust, entertaining the possibility that the sense they make could be deeper than we have yet dreamed possible?

So we embark on a course of reading and thinking, speaking and writing. With our companions and our books, we enter into a sequence of interactions. A play of ideas begins.

Always, for each of us, there is the central core of what is attended to: it stands out in a clear pattern against an all-enveloping vagueness. But for each of us, the core is fringed with faint patterns that can be brought to distinctness. Queer things go on in the fringe, sometimes related, and sometimes not, to what is occurring at the center. The activities in the fringe include the elements of the play of myth, the goat-play that becomes tragedy, the apparently silly associations of words, the childish *hubris* that appears in every rage. The obscure shapes in the fringe undergo strange evolutions, and there are flashes of significance which emerge from the fringe or from beyond it, the relatedness of ideas emerging from images or out of the blue.

There is a play of opposites. The ideas generally come in pairs, substance and accident, the fixed and the flux, form and function, the actual and the possible, subject and object. If we try to catch these opposites in act, they change shape and multiply, reappearing at different levels.

In the course of our conversation and thought, configurations of ideas form and dissolve. Each configuration is a whole of a kind, but also partial and relative, because there is always more in the fringe and beyond. But one configuration can lead to another, and there are relay effects, configurations which can serve as stepping stones, if we can somehow capture them in words or images. For we need the word and the image to fix our thought, if we are to advance. We cannot be sure that we know what we think till we see what we say.

Do we in fact advance? Changes in us and in our thinking do, I believe, occur. We are no longer so quick to accept or reject. We become more soberly and critically thoughtful. It is with a changed and no longer eristic spirit that we now engage in the battle of words. We have learned to listen more patiently, to wait, to consider what the progress of the conversation demands. We have become more adept at noting and bringing forth what is in the fringe, or as someone has put it, at "thinking aside." Increasingly, our thinking and our speaking can be governed and guided by the idea of totality, that is, of the world—of a whole not encompassed but to be encompassed. Increasingly, thoughtfulness and wonder can be at the center.

Now and again, in the play of thought, configurations arise which are accompanied by a sense of resolution or completion. This is discovery. It can be preceded by arduous labor, but it cannot be regarded as the simple result of that labor. And the completion of a whole, when carried into full awareness, is accompanied by a sign or mark which may be called beauty or radiance.

Such is my attempt at a description of the process in which we have been engaged.

Now liberal education that is worthy of the name is a quiet affair; it does not operate under klieg lights, or with blaring P.A. systems, or in-circuit television, or anything fancy. It is not easily characterizable. It will never answer its critics once and for all. That is because the issues are perennial and go deep.

Callicles in the *Gorgias* says that liberal education is a kind of play that is good for children, but shameful for grown-ups; so much so, he adds, that when he sees a man continuing the study into later life, and not leaving off, he would like to beat him. Now I disagree with Callicles. Yet, contained in his position is a proposition that I believe to be correct. This is the proposition that education occurs under the aspect of play. It is an old thought; in Plato's *Laws* it is pointed out that the words *paideia,* education, and *paidia,* play, are almost the same, Both are derived from the word for *child.* What, Callicles is demanding, do *paideia* and *paidia* have to do with adult life? And he is implying that the answer should be—nothing at all. Adults through the ages have tended to view play as *mere* play, neither serious nor useful. When we become grown-up, we put away childish things.

Yet this opposition between play and the grown-up leads to a play of opposites such as I referred to earlier; it is dialectical, and is capable of transformation in manifold ways. The notion of play is not adequately defined by the opposition to the serious, the useful. the adult. We can be warned of this if we notice how the business and politics of the world are pervaded by terms that are derived from play. Business men take risks and play the market; sophisticated descriptions of economic behavior are couched in terms of the mathematical theory of games. Politics is the game of politics, and includes the staging of scenarios, beneficent or nefarious; logrolling, and public contests for votes and approval. War itself has been conducted within recent memory on one side as a game in the allocation of men and materiel, against an enemy that would not play that game, but played another game of wit and will. Playfulness was even attempted, though awkwardly, in the protests against the war. The playful finds its way into art and music and every cultural form. In families and professional life, adults play roles, sometimes grimly and sometimes with flair. Also, they play at play, or work at play, sometimes in ways that may be too emotionally demanding for health and wholeness. So play and the playful are by no means absent from the grown-up world, but appear in manifold forms, sometimes frightening, sometimes pleasing,

We can see further into the paradox if we examine the play of children. How serious and intent it can be! Child's play is utterly serious because it is an exploration, an advance into a field of forces and possibilities where chance must be coped with, where the unknown and the unexpected have play, where one's powers and guesses are tested. In Plato's *Laws,* the paradigm of child's play is leaping. Leaping is not done in empty, undifferentiated space. It is only possible in a sufficient gravitational field; it could be very dangerous on one of the small satellites of Mars; I launch myself, I hope, only after having learned that there is that *against* which my leap is launched, and which suffices to bring me back. The mastery of leaping is part of the mastery of the erect bipedal posture, a major fact in human development, a gift of evolution that has to be appropriated anew by each individual human being. With it, one acquires an orientation with respect to what is before and after, above and below, in specifically human ways. The spatially up and down, right and left, forwards and backwards, become metaphors for what is valued and sought, sinister or righteous, rejected or left behind. Mastery of the bipedal mode of locomotion, to the point of free and unrestrained advance, becomes a metaphor for similar advances into quite different realms, where the forces in play are of different kinds.

Wherever play and the playful appear, whether within designated playground or elsewhere, we may expect that there is a field of forces and possibilities within which the play takes place; that chance *is* involved, and risk, and the unknown, and tension, and possibly joy. If we are dealing with the unknown—and when are we not?—then the playful or something like it is likely to come into play. Walter Lippmann, in his book *Public Opinion,* while criticizing the notion that reality is a fixed thing to which we must adjust ourselves, points out why the unknown is inevitably present for us.

> The real environment [he says] is altogether too big, too complex, and too fleeting for direct acquaintance. And although we have to act in that environment, we have to reconstruct it on a simpler model before we can manage with it. . . . [So we must recognize] the triangular relationship between the scene of action, the human picture of that scene, and the human response to that picture working itself out upon the scene of action. . . .

The range of fiction extends all the way from complete hallu-
cinations to the scientists' perfectly self-conscious use of a
schematic model. . . . The very fact that men theorize at all is
proof that their pseudo-environments, their interior representa-
tions of the world, are a determining element in thought, feeling, and action.

And Einstein, writing about such theorizing, says:

All our thinking is of the nature of a free play with concepts;
the justification for this play lies in the measure of survey over
the experience of the senses which we are able to achieve with
its aid.

With these quotations, I am seeking to support the notion that
thinking and learning involve necessarily the playful, the imaginative
arrangement and rearrangement in thought of relations and possibilities, with a view to taking account of the whole of them.

We should hope, then, and strive to make our hope come true, that
at the center of our lives there would be an openness to possibility, a
playful and inquiring thoughtfulness. Liberal education should be a training in arts and habits that continue through life: the habit of attention;
the art of "thinking aside"; the readiness to assume at *a* moment's notice
a new intellectual posture; the art of entering quickly into another person's thoughts; the habit of submitting to censure and refutation; the art
of indicating assent and dissent in graduated terms; the habit of precise
accuracy where this is required; the habit of working out what is possible
in a given time; the art of carrying conflict to high levels of rational articulation; above all, the virtues of intellectual courage and intellectual
sobriety. These things we call practice and continue to practice, and so
learn to be playful with the playful, as life requires of us.

So learning, we may also learn to be serious with the serious. The
playful and the serious, Plato tells us, are twin sisters. We can disengage
ourselves from the magic circle of play only by recognizing the playful
as playful, and by turning toward the ultimate—two movements that
join finally into one. One recognition of the ultimate and the serious
that has remained memorable for me occurs in a story told of the aged
Immanuel Kant. A few days before his death, two strangers arrived for
a visit. They wanted him to stay seated, for he was feeble. He insisted
on rising to greet them, despite the effort, the embarrassment, and the
time it required. Recovering his breath and composure, Kant explained:

"The sense of humanity has not yet left me." So, with the ritual act of greeting, Kant expressed his recognition of that which must ever be treated as an end, never as a means. With some such recognition, I think we could learn to be truly playful with the playful, and serious with the serious. That is what I wish for you.

Go forward, then, and do not look back; and, to echo Socrates at the end of the *Republic,* may you fare well both now and in the journey of a thousand years.

A Toast to the Republic
Delivered at the President's
Dinner for Graduating Seniors
(1986)

The custom of giving this toast at the President's dinner is a rather old one, having been initiated in the early years of the New Program by Scott Buchanan, and continued without interruption down to this time. My task is to try to say what it means.

St. John's is a rather unusual place, and one of its peculiarities is its peculiar mixture of traditionalism and questioning—questioning that can become radical. When I first came, I was shocked by many of the questions, especially those of the *ti esti?* variety. In the second seminar I attended here, the opening question was, What is a hero? During the course of the seminar, the leader who first posed the question, repeated it, in measured tones, in truly Aeschylean style, no less than three times. It was not a question I knew what to do with; it was not one I had hoped to hear. I really had no use for heroes. In the several uncomfortable silences, the only thing I could think of was that line in Bertolt Brecht's play, *Galileo:* it is a sad time, Galileo says, when a country needs heroes. The Brecht line has long ago lost its charm for me. The staggering question that Winfree Smith asked remains. And in general, the most important questions for me have become not those which admit of systematic reply—though I am very fond of working out anything algebraic and easy—but those which arrest the mind in the presence of the awesome, the beautiful, or the utterly mysterious. And so I am grateful to colleagues who ask unanswerable questions, and insist on our stretching our thought toward the height and depth of them. Underlying the toast I must give tonight is such a question.

368

What do we mean by the republic? It has to be *the* republic, not this one or that one. Robert Bart first explained that to me. The Latin term *res publica* means that which is public, the public thing. Early on, it was used to refer to the contents of the public treasury, the common wealth in the most literal sense. Later on, its meaning was extended so as to include the whole body of institutions and traditions and knowledges by which the lives of the citizens were shaped, and on which they depended in order to have a realm of freedom and action.

The corresponding Greek term was *politeia.* Cicero used *res publica* to translate the term *politeia,* the title of a dialogue by Plato. *Politeia* meant originally citizen life, the life of the free citizen. The life of the citizen within the *polis* was always sharply contrasted with everything private—with everything having to do with family or kinship or household, everything having to do with the necessities of survival, with food-getting and childbearing. *Politeia* was a realm of speech and freedom and action. It was a higher realm. Only in this realm was great speech or great action possible.

Such are the origins of the term *republic.* In the last quarter of the twentieth century, all of this can seem rather far away from us. The *polis,* the political community based on face-to-face relations, is no longer. Citizen virtue is harder to summon up. We are members not of a *polis* but of society. Society, the sociologists tell us, is a web of relations. The relations are so intricate and far-reaching as to be well-nigh untraceable. We find it hard to see where our responsibility begins and ends. The distinction between the public and the private becomes badly blurred. And to say the worst that can be said, in society no one really acts, and instead of speech and action and freedom there is behavior, which is studied by the behavioral sciences. Society is ruled by an invisible hand, that is to say, by nobody. And as Rousseau well knew, we members of society, offended terribly by society's invasion of our privacy and its attack on our dignity, nevertheless bear within ourselves a guilty conscience.

I exaggerate, perhaps, in painting the dark picture.

There are sparks of light in the darkness, and they are to be cherished and protected. But I think it not so bad a thing that the dark prospect should be before us, and turn us back into ourselves. We have to assess what we are about. There is a justifiable *contemptus mundi,* or turning from the world, and a justifiable *amor mundi,* or turning to the world, and we must learn to balance and use them aright. There is

369

an inner court of the soul, in which ends and means have to be judged and sorted out and arranged in hierarchy. We cannot do everything. We must find the center from which we go out, and to which we return. These inner tasks are imposed on us with especial urgency, because we live when we do, and not in a simpler time. Carrying them out, we may come to see that we are not islands unto ourselves. In being what we are and becoming what we become, we are dependent on a heritage of tradition and knowledge; we are dependent upon friends who can ask unanswerable questions, and now and then show faith in our possibilities. And by a calculus that may be simpler or subtler, we may determine to pay back something for what we owe.

> For Who [says Donne] is sure he hath a Soule, unlesse
> It see, and judge, and follow worthinesse,
> And by Deedes praise it? Hee who doth not this,
> May lodge an Inmate soule, but 'tis not his.

This toast includes Plato's *Republic* or *Politeia,* a work which as much as any other addresses itself to the question of the relation of knowledge and citizen virtue. Citizen virtue was always problematic, because it would not explain its altruism to itself. Plato's *Republic* gives citizen virtue a new grounding, not in knowledge but in the quest for knowledge. And I think this knowledge that is sought, if the heights and depths of it are recognized to be as high as they are and as deep as they are, will lead to a new kind of civic virtue, not doctrinaire, but more gentle and more humane.

This toast includes the Republic of Letters, by which is meant the community of writers living and dead, who have sought the truth, responding to one another, laboring to complete what others have begun, maintaining their faith in the possibility of human knowing through the use of language. The toast, in effect, pledges allegiance to the maintenance and fostering of that republic.

The toast includes the Republic of the United States of America. The American dream, people were saying a few years back, was becoming a nightmare. We should pledge our alertness to try to keep that from ever happening. Much is wrong. But there are freedoms of thought and action which we enjoy, not merely because of favorable economics, but because of political traditions, and they are not enjoyed widely in the world. To the maintenance of those freedoms, we should also pledge our lives and our honor.

Finally, this toast includes the Republic that is St. John's College. This college is an attempt to provide circumstances under which the great and unanswerable questions can be asked, in which we can be stretched to the height and depth of them, in which we can discover and rediscover the center from which we go out, and to which we return.

Ladies and gentlemen, I give you the Republic.

Commencement Address
(1988)

Members of the graduating class:

It is an honor to be asked to speak to you on your Commencement Day. A recent book bears the title *How to Survive Education*. I haven't read it, and do not know how it's done; but you've done it, and we congratulate you. I myself, who came here forty years ago next August, have not quite made it through: I have still the junior and senior language tutorials to do.

The St. John's program does a lot of stretching and spreading thin. If a sheet of plastic is stretched to its tensile limit, it is likely that a hole will open in it here or there. In trying to cover as much as we do, we are quixotic. Quixoticism has been our keynote from the beginning.

Fifty-one years ago Scott Buchanan and Stringfellow Barr brought forth on this campus a New Program, dedicated to the proposition that liberal education is possible. The enemy was the elective system, introduced by Charles Eliot in his inaugural address as president of Harvard in 1869, and since then adopted in colleges and universities generally. Before Eliot, the curricula were all-required, but—and this was Eliot's objection—they had failed to incorporate the modern natural sciences. Year by year the natural sciences were transforming the world and our picture of it. Eliot, by licensing the natural sciences as elective possibilities, aimed to broaden, deepen, and invigorate education. One consequence, a good one, was that the universities became nurturers of natural science. Another was that liberal education lost its sense of identity and direction. Adding to the muddle were the new social sciences, so-called. College education became a grab-bag of choices ranging from the sublime to the ridiculous.

371

Barr and Buchanan set out to do the impossible. We were all, students and tutors, to become autodidacts, tackling the reading of a veritable Everest of books. The books were to be chosen for their classic quality, for being first-rate. The great works of mathematics and natural science would be included, starting with Euclid and continuing through Newton, Maxwell, and so on. We would become literate in the classics of the scientific tradition, as well as in the classics of literature and philosophy. In my one leisurely conversation with Buchanan, in 1965, he wanted to know how we were getting on with Maxwell and Josiah Willard Gibbs. Gibbs, if you haven't heard his name before, was a nineteenth-century American, one of the great physicists. I had to confess that we hadn't managed to cram Gibbs's works on thermodynamics and statistical mechanics into the program.

This program that Barr and Buchanan conceived was accused of being Epimethean, medieval, backward-looking. In fact, it was as Promethean as all get-out. Like John Dewey, Buchanan came from Vermont, and like him mixed Yankee pragmatism with a visionary dream of what this country could be. He expected the college's new program to become the seedbed of an American Renaissance.

As for the natural sciences, he had the idea of a hands-on approach. He saw the laboratory as the distinctively modern institution for the acquisition of knowledge; and so he dubbed one whole part of the program "the laboratory." At different times he considered organizing it round medicine, or the airplane engine. A good carpenter himself, he wanted the students to get handy with hammer and saw, plane and drill. And in the laboratory, both intellect and manual technique were to be brought to bear.

All this was to be done in a marvelously original, eclectic way. For organizing principles, Buchanan turned to medieval tradition. He had the library collection reorganized under the rubrics of the seven liberal arts, so that physics, for instance, came under music. What a librarian's headache! Buchanan was aiming to tweak all our complacencies.

Of those heady times, which were before I arrived, I have gathered intimations from here and there. One of the earlier tutors told me how, in the autumn of his first year, his senior math students—all male, of course—took him to a local pub, sat him down with a beer, and commanded him to stop considering himself responsible for their education. They were responsible; if they needed help, they would ask for it.

In the middle of the academic year in which I arrived, 1948-49, Jacob Klein became dean of the College; he was to remain so till 1958.

It was a time of consolidation. A new president, Richard Weigle, proceeded to put the College on a tenable financial basis. Coeducation was introduced, a step, it was hoped, in the direction of civility. The faculty, weary of more chaos than it was comfortable with, became more insistent that students acquire something more like ordinary competence. As the dean put it, the glories of fifth-century Athens must now be succeeded by the more pedestrian achievements of an Alexandrian age. The obbligato of social revolutionary fervor dropped to a whisper.

From my first decade here what has remained most clearly in memory are the dean's opening lectures. Huffing, clearing his throat, Jacob Klein would begin by speaking of the trepidation he felt, in attempting to formulate our task. All knowledge, he would be telling us, was our province, and to keep the wholeness of it in mind was enormously difficult. For it was ever our tendency to make ourselves comfortable by limiting the view.

He would speak of the babble and unexamined jargon of everyday speech. Are we not prisoners to it? In Plato's simile of the cave, we must somehow be freed of our chains and turn round, looking away from the shadows to what causes them. Surely there is something right about this image of education, whatever the mystery or confusion concerning the rest of the story, the Sun and the upper regions and the fourth part of the divided line. But what was mainly being impressed on the hearers of those lectures was a simple idea and a sobering one: the thought that to take account of the wholeness of things is difficult and demanding.

Of these lectures the one that I remember best was Jacob Klein's final lecture as dean, delivered in September, 1957. The text has not survived, but the title was "The Delphic Oracle and the Liberal Arts." It dealt with the ambiguity of the injunction "Know Thyself."

One meaning is this: Know that you aren't god. Know that you are a finite, mortal being, dependent on your fellow human beings, prone to error, prone to *hubris,* the error of overstepping your boundaries. The lesson to draw is modesty, sobriety, circumspection, a sense for our equality with our fellows. To a student this could mean: doing the homework, learning the paradigms. To a scholar it could mean: getting the footnotes right.

The other meaning rested on the recognition that everything is connected with everything. From the farthest reaches of the cosmos to the depths of the human psyche, nothing is simply isolatable, so as to be

373

fully understandable by itself alone. Hence, to know myself, I must know everything. The quest for self-knowledge is thus inherently incompletable. But the oracle, under this interpretation, enjoins it.

In what I have been recounting, there is an aspect I do not want you to miss. We at this college are a bunch of crazy autodidacts, holding madly onto two horns of a dilemma. If you did not quite know what you were getting into when you first came here, surely the truth has dawned on you by now.

In the last thirty years the program has not changed in essentials. Some things we do better. But I myself, conniving with others, have helped add to the madness, in seeing to it that we get to Maxwell's equations, relativity, and quantum theory. Mature physicists admit to having been discombobulated in their first encounter with these theories. To get them to seem familiar, the only way is to trace and retrace the routes that have led to such odd consequences: undulations where there is nothing to undulate, events strictly correlated yet separated in space, with no message passing in between, and so on. Imagination has been instrumental in leading to each such result, but the result transcends and contradicts the imagination. As J.B.S. Haldane put it, the world is not only queerer than we have imagined, it is queerer than we can imagine. My discomfort over what we fail to do would be greater if you had not met, at least briefly, with this encouragement and rebuff to our analogizing.

What I have said is no excuse for remediable faults in the program. As to what could be improved, I have ideas, and so do some of my colleagues, but our ideas are not all the same. But the main point I have been making is that the program here is an unfinishable affair.

The remaining words I have for you are by way of homily. To your stack of books, you must now add what Descartes called the great book of the world. The image is not in every way apt, but it is preferable to thinking of the world as an unalterable harsh mechanism, to which you are required to adjust your misfitting shapes. A book can be read, and that, as you know, is an active, formative process. Try observing the world; there is much that is thus to be learnt. But the stance of the altogether detached observer that Descartes projects of himself is probably neither possible nor really desirable. A better simile for you is that of organic adaptation. You must adapt to the world; and, in ways that, to begin with, may not seem as important as they are, the world will have to adapt to you.

374

For four years you have been discussing works of literature and philosophy, writing essays, analyzing plays and poems and arguments. These exercises have developed in you a number of skills that should be prized: the habit of listening carefully, of being attentive to a question and seeking out its sharp edge; the habit of readiness to enter into another person's thoughts, and to assume a new intellectual posture in response to new facts or ideas. Here and there, by bits and pieces, these habits should prove transferable from one context to another.

For God's sake, don't say, as one graduate put it on a resumé a few years ago, that at St. John's you learned to think. These hyperboles will harm you. There is no point in merely astonishing strangers with our strangeness.

Mind that some of our patterns here are all-too-easily turned into caricature. We inculcate the respect for great books, the taking seriously of the texts we read, as possibly revealing truth. A good habit. Works about our authors we eschew, telling you that these should not be your authorities, that you should think for yourselves. A good pedagogical ploy. But it does not mean that a biography of Cervantes, say, might not reveal something important about the book begun in the prison of Seville; or that the composition of Shakespeare's audience might not have something important to do with the composition of his plays and the wondrous mix of the high and the low that he achieves. Our authors were creatures of flesh and blood, and their works, in many cases, were prompted by the paradox of real-life situations. They were not always merely chatting with one another.

We tutors, who are paid, I guess, to defend all these books, are not to be imitated in all our sophistries. Not everything in the books we defend is defensible. Kant gets from the first to the second Critique by leap rather than logic. Newton fails in Book I to prove satisfactorily a crucial proposition on which Book III depends. Not every failure of logic in the Platonic dialogues is necessarily to be explained away by reference to the *mythos* of the dialogue.

Graduates, I gather, tend to become nostalgic for St. John's-style conversations. Well, you can have them again, and better, in new circumstances, when you have completed more homework. You will need friends, of course, and to cultivate friends is difficult in this age of endless mobility and of work-days that stretch on into the night. Don't fail to cultivate friends. There are conversations waiting out there to take place. It may not be very easy to find where and when they can occur. Tentatively, you can begin to take a bit of initiative.

You must find your own footing, your point of vantage and vision, freed at last from both the comfort and the annoyance of pedagogical authority. After having been stretched in so many directions, you must begin to assess, and reassess, where your own redefined center of gravity may be, and where your powers can take you, and what you can discover.

I wish that we at the College had managed to give more attention to heuristic. The word comes from a Greek verb. Archimedes used its first-person singular perfect as he leapt from the bath, on making his great discovery. To discover is indeed a perfect thing, in the sense that it brings elements together, makes a new whole. How do you go about making discoveries?

There are no very particular rules. In general you need to be asking a question, focussing on it, wrestling with it. You have to have the bravura to suppose that the parts of the answer can come together for you. You have to have faith in your powers. The evidence for there being something we could call unconscious thinking seems to me very strong. A manifold of processes must go into the recognition of a face or any other gestalt, but it seems to come instantaneously. So with the "Aha!" experience of discovery.

In discovery there is a kind of interweaving of the old and the new, the Epimethean and the Promethean, the traditional and the innovative. In learning anything, we learn it less well before we learn it better. Our ignorance frequently consists in knowing what is not so or not quite so. Discovery is the finding of the new or the different within the matrix of what is already present, whether in latent memory or conscious thought. There is no point in merely clinging to what is past. What you learned and forgot can come back after years to haunt you, or to fill in the gap in an uncompleted gestalt. As Buchanan put it, we learn to swim in winter, and to ice-skate in summer. You must have faith in possibility and in what is hidden within yourselves. The quest must be to find the question and to persist in the questioning.

In a book about Chinese brush-and-ink painting, I found some precepts that are eminently applicable to discovering. Here they are:

Follow tradition in basic design.

For powerful brushwork, there must be ch'i or spirit. The brush should be handled with spontaneity.

Be original, even to the point of eccentricity, but without disregarding the li of things (li means principles or laws or essences).

Learn from the masters but avoid their faults.

Now I am supposing—is this a mere academic's dream?—that such rules are not applicable only to mathematical discovery or originality in painting. The questions to which heuristic is applicable need not be high-falutin or esoteric. With a bit of good will and heuristic, with patience and pluck, you might be able to transform the daily routines in schoolroom or office, or the community for some miles around. Oh, if you must, save the world, or write another great book! But I have thought it wiser to wish you a simpler destiny, neither tragic nor comic, but similar to the one that Odysseus chose at the end of the Republic.

I think I have about finished. Oh yes! do read and re-read some of Montaigne, when you get the chance.

Such is my homily for you today; it is from the heart. May you fare well.

Eulogies for Curtis Wilson

These eulogies were delivered at a memorial service held at St. John's College in Annapolis on 30 September 2012.

Remarks by Nancy Buchenauer

It is a great honor to speak about Curtis Wilson. I first met him in my early years in Santa Fe when I was an utterly ignorant tutor and he was an almost mythical figure. Those of us new to the College passed his lectures and manuals around to one another, sure we would find crucial insights and interesting ways to help both ourselves and our students. To us in Santa Fe Curtis Wilson was one of the great men of Annapolis, who had been part of its formation in the old days when it was rising and finding its way. He was a person universally spoken of with respect and affection, as the best colleague one could imagine.

With great pleasure I came to know the man himself after I moved here in 1997. The generous hospitality I met in the home of Becky and Curtis was paradigmatic of the active, living intellectual community that one heard of from the college's early days. They brought together wonderful groups of people from in and outside our college with whom I would not otherwise have become acquainted. In Curtis I encountered a man who seemed never to age, but to be truly one of the immortals, always ready to share whatever question in astronomy he was currently engaged with, the conferences or study groups he was attending, the books and articles he was reading or himself about to publish. He spoke with a kindly, sparkling-eyed wit, as would a man so full of the delight in learning and discovering that he could not help but pass it on to those around him.

Curtis stood for something, a kind of moral perceptiveness and intellectual integrity. He showed himself genuinely unselfish and completely honest. Invariably his words were deeply considered and deeply human in the highest sense. In conversations with me once about the deanship he expressed the urgency to continue explicating and fostering the liberal arts, which he saw to be the fundamental task of the Dean as it is of the College. His own lectures enacted what this meant, by presenting an irresistible question in which the students might find themselves already immersed. In "The Archimedean Point and the Liberal Arts," he wrote:

> [Man] can become aware, as by a sidelong glance, of his own linguistic activity, and raise it to the level of conscious artfulness, liberal artistry.

And further:

> Man exists at the horizon between appearance and idea: his
> being is intermediate, metaxy, as Plato would say. . . . And the
> task of education, starting in the middle of things, is to use
> the appearances, the images, the names and the sentences, to
> produce a development toward hierarchy and wholeness
> which uses all the terms. . . . So questioning and responding,
> both to himself and others, man becomes a responsible being,
> a moral being.

Curtis thought much about science, what it is and what it is not, and unfolded his understanding with delightful clarity and precision. He exposed the stereotyped project of modern science to present "the completed description of the world and of ourselves as an assemblage of spatially and temporally located, deterministically interacting parts—a machine," calling this "a bad metaphysical dream, a world of bare fact from which problems and persons, learning and knowing and valuing are absent." ["On Knowing How and Knowing What."] In his own thought and writing, persons, learning and knowing and valuing, were never absent.

Curtis was one of the fine scholars in our community, bringing us a touch of the academic glory of the outside world, and showing that it could find a place in our midst. Owen Gingerich in the notice soon to appear in the *Journal for the History of Astronomy*, calls Curtis "the most highly regarded historian of astronomy of this generation," and adds:

> It was no surprise that, when the L.E. Doggett Prize for contri-
> butions to the history of astronomy was established by the His-
> torical Astronomy Division of the American Astronomical
> Society, Curtis Wilson was unanimously selected as its first re-
> cipient in 1998.

Still, for all his excellence as a thinker and teacher, Curtis will always be dearest to me as a kind and generous friend, one who could smilingly correct an error or just ask the right question to further another's work. I am grateful to have known this wonderful man.

Remarks by Joseph Cohen

We are here to celebrate the life and to honor the memory of Curtis Wilson. I believe it is no exaggeration to say that in the larger world of the St John's College community, Curtis Wilson was universally admired and loved by everyone whose life he touched as tutor, or dean, friend or colleague, including those who knew him mainly as a renowned scholar of the history of astronomy.

It is perhaps a truism that one's memories and thoughts arise from one's own experience and perspective. So I will naturally be speaking about my own relationship to Curtis and this College. Indeed, until his death, Curtis was the last survivor among those of my colleagues who had also been my tutor. Today I shall recall others of this community as well.

Looking back to the time when I first came to this college in 1952, newly graduated from high school, I have always felt that being here was being home, first as a student for four years and then as a tutor for nearly 50. Throughout this long and happy period, Curtis was for me a significant and beneficent figure.

During my undergraduate years Curtis was my tutor in junior mathematics, in senior laboratory and, with Jacob Klein, in senior seminar.

My own career as a tutor here officially began when Curtis was Dean. I was then teaching in the Basic Program of Liberal Education for Adults at the University of Chicago, and St. John's was expanding its faculty for the expected opening of the Santa Fe campus. After an interview with the instruction committee, I received from Curtis a letter of appointment effective July 1962. He then sent another letter informing me that a major instructional proposal had been approved establishing preceptorials for juniors and seniors, thereby eliminating a number of seminars in those years. (Later I was told that this proposal became a matter of confidence. If it failed to gain faculty approval the Dean and Instruction Committee would have had to resign.) This letter also said I would be teaching a Junior Seminar, and asked me to prepare a preceptorial topic. I was delighted to join Michael Ossorgin and Tom Slakey in that seminar and I'm very pleased that Tom is here today.

During those years when Curtis and I were both active faculty members there were also many opportunities to visit socially with Curtis and Becky. Included were those annual occasions when a women's book club of equally long duration invited spouses to join the discussion followed by a well prepared meal. Becky and my wife Sandy are still members of that reading group, many of whose loyal members are here today. Another example of an opportunity to engage with Curtis was initiated by Chaninah Maschler, who also asked Ben Milner and me to join in the task of explaining Daniel Dennett's highly touted book "Consciousness Explained."

When Becky invited me to speak at this event I immediately and gratefully accepted. My remarks about Curtis will include the contents of a pair of letters from him which I happened upon recently and thought appropriate to share. They reveal a piece of who he was, which, although not surprising, may not be generally known.

The letters were written in January and February of 1998, and pertained to a Friday night lecture I was scheduled to give on February 20 on the subject of Spinoza's *Theological-Political Treatise*. In the January letter Curtis wrote:

"Dear Joe, Can you put up with a bit of advice about lecturing from the doddering, the superannuated, and the hearing-impaired? For the sake of us old folks, you must speak directly into the mike, keeping it only a few inches from your lips. There! I've said it. For some of us, it makes a lot of difference. I really do look forward to your lecture. Spinoza has always seemed interesting to me—perhaps because of my not-very-well-assessed romantic tendencies. I'll be there—with ears cocked!"

When I gave the lecture, I dedicated it to Michael Littleton who had died two weeks earlier. The words of the dedication were as follows:

"Michael's virtues and talents were indeed many and wonderful. In light of my lecture this evening, it seems fitting to say, that of them all, he may have most excelled in those qualities identified by Spinoza as "love, joy, peace, temperance, and faith towards all." His memory is a blessing to his family and to all who knew him.

After the lecture, Curtis sent me the following note:

"Dear Joe, thank you for your solid, carefully constructed lecture, and thank you for defending Spinoza. It is the first defense of Spinoza I ever remember having heard in a Friday night lecture at St John's.

"I suppose that there is something to be said for [Leo] Strauss's attack on Spinoza, if one believes that the truth of miracles must be defended at all costs. Winfree Smith shocked me out of my wits when I first came to St John's by insisting on defending not only Biblical miracles but the magic in the *Odyssey* and other old stories. I have remained an unreconstructed liberal democrat, and it warms the cockles of my heart to have Spinoza defended. It was lovely of you to dedicate your lecture to the memory of Michael Littleton, another defender of peace and the liberal democratic tradition."

I should add that Leo Strauss was one of my teachers when I was a graduate student at the University of Chicago and Winfree was one of my teachers at St. John's. He was significant for me in many ways: he was the seminar leader in my Sophomore and Junior years and my math tutor in Senior year. Winfree, who was an ordained Episcopal clergyman regularly offered evening classes on the New Testament, some of which I attended; and during my last three undergraduate years we had many conversations about philosophical and theological books. When I joined the faculty our friendship deepened, and I continued to benefit from his mentoring and his friendship.

In my remarks at Winfree's memorial service in 1991, I spoke of him as "a particular object of my wonder" because it seemed to me that he was both completely devoted to the cultivation and teaching of the liberal arts, the arts of reason, and also completely devoted to fulfilling the requirements of his Christian faith. Winfree seemed perfectly comfortable with this duality because he had decided on his priorities. In one of his sermons at St. Anne's Church he stated his belief that "the authority of the word of God is infinitely superior to the authority of human reason." But Winfree also believed that the Creator God had given to his creatures the gift of reason as well as the gift of faith. It was never clear to me why the gift of reason is deemed by some to be less divine than the gift of faith.

My last conversation with Curtis occurred just before he and Becky left for the trip to Michigan where he died. It touched somewhat on this same theme. We discussed a newly published book by Lawrence Krauss with the metaphysically seductive title, "A Universe from Nothing: Why there is Something rather than Nothing." The author is an eminent writer on modern science and cosmology who thinks he has the answer to Einstein's question whether God had any choice in the creation of the universe. Curtis did not disagree with Krauss' account of the experimental and theoretical discoveries at the heart of modern physics, but he strongly objected to what might be called his aggressive atheism. Curtis reacted to Krauss' concluding assertion that "even a seemingly omnipotent God would have no freedom in the creation of our universe . . . [which] further suggests that God is unnecessary—or at best redundant." Curtis thought this account did not properly include or embrace man's spiritual dimension. For Krauss, however, the scientific enterprise is itself a profound expression of the human spirit in its desire to know. I thought that Curtis would agree with this statement, but that afternoon our conversation had to stop. I fully expected it to continue when he returned.

To help me understand what Curtis may have meant, I turned instead to Becky. She told me that Curtis had for some years been engaged in various meditative practices based on Buddhist teachings. He was seeking and finding a non-dogmatic, non-orthodox way of attaining a peaceful mode of being. Perhaps he was referring to this mode as enhancing a life of the spirit.

Curtis had also spoken at Winfree's memorial service, and I quote the final words of his remarks: "For those of us who knew him, his influence and his presence will not disappear. He is missed. The love he evoked will remain." To conclude, I will apply these same words to Curtis himself. For those of us who knew him, his influence and his presence will not disappear. Curtis is missed. He is loved. And the love he evoked will remain.

Remarks by Tom May

It was my privilege to know Curtis Wilson as a colleague, an advisor, and a friend. I met Curtis and Becky in my first week here at the College, in an after-lecture gathering at Mrs. Klein's. This led to a dinner invitation soon afterwards, a typically gracious gesture that the Wilsons extended to new faculty, a practice they continued right up to the present.

My second year here found me teaching both sophomore seminar and freshman laboratory for the first time, and a change mid-year in the teaching slate made Curtis both my co-leader for seminar, and archon of the Freshman Laboratory. In both of these contexts, Curtis was a gentle, thoughtful, authoritative presence, with a marvelous sense of humor and deep sensitivity for what whatever we were reading. Both of us were morning people who found the lateness of seminar an additional challenge; I recall Curtis's saying to me one evening as we drove home together, "I fail to see why we should all have to be held hostage to Scot Buchanan's metabolism in perpetuity." In seminar, I recall particularly his love of Dante—he led an independent study group for any-one interested in reading Dante in Italian. I myself read the *Commedia* as a whole for the first time with him that year in that memorable sem-inar of "lively" students—that was Curtis's word. On the night of our second *Purgatorio* seminar, I came into our classroom to find Curtis feverishly completing a detailed schematic diagram of the great pag-eant in the Earthly Paradise that concludes the *cantica*, as intent on its accuracy and completeness as if he were demonstrating a proposition in Apollonius or Newton. Once class began, he opened our discussion by asking why Dante loses Virgil and is returned to Beatrice in such an elaborate setting. Both here and in our archon meetings for lab, he was attentive both to the details of the readings and to the difficulties that tutors and students might encounter in reading treatises by Archimedes, Pascal and Lavoisier for the first time, and he also wanted us to appreciate the bench work, how things were supposed to work in the practica, even if it meant the ruin of a beautiful plaid shirt one af-ternoon as, absorbed in a demonstration, he brought his sleeve too close to a Bunsen burner. "So you must learn, too, how that shower works," he said, pointing to the fixture on the other side of the room.

I subsequently learned that when Curtis came to St. John's in 1948, Jacob Klein remarked to another colleague that now the College had a tutor who could really help us make sense of Newton's *Principia*, which he did for four decades—Newton and so much else besides! His coming here was initially a concealment of sorts from the larger aca-demic world, though; Curtis had the reputation of being the outstanding graduate student and scholar when he was pursuing his graduate studies at Columbia, who had disappeared to what a fellow scholar who knew him there termed "a very interesting school in Maryland," which this Columbia professor recommended to another young man named Howard Zeiderman, who was then casting about as to where to go and what he should do. Curtis as tutor here would celebrate the conclusion of the Newton sequence in Junior mathematics with a Nabisco fig new-ton party. Tutor Sam Kutler suggested to me in a conversation just this week that this initial interest and work on Newton initiated Curtis's abiding interest in celestial mechanics, leading him to his study of Galileo, Kepler, Horrocks, and on up to his last book on the Hill-Brown theory of lunar motion—"He followed the moon and its difficulties right through to the end," he said. Shortly after the College received word of Curtis's death, Robin Dunn shared a story about this last book.

It seems that one day, chatting with Curtis in the Bookstore, Robin pestered him to talk about his new tome. He replied, "Oh, you don't need to know about that; no one's going to want to buy it!" When firmly pressed, he disclosed the publisher's name and one or two other details, then said he'd hate the bookstore to carry it, because that would just be throwing a ton of our money away –no one would ever buy it, and the College needed the money for better things. Robin insisted that as Bookstore manager he thought it his duty to support authors within our community, because even if no other store in the world sold the book, people should at least be able to count on *us* to have it on hand. Curtis seemed actually to blush at this. Certainly, he gave one of his modest chuckle-cum-twinkles. "Ah well, be it on your own head!" said he, on the way up the steps outside the store. Robin's anecdote reminded me of a conversation I had with Curtis in the summer of 2007, after he returned from an Euler conference in Europe, where he had delivered a paper. I asked him if the conference had gone well. "Pretty well," he replied hesitantly, appearing to hedge a bit. I pursued further: "Were there any interesting papers?" After a brief pause, with a smile and a playful glint in his eye, he said, "Yes; one." Through Curtis I met Michael Hoskin of Cambridge University, with whom he worked over a number of years both editing and contributing to the magisterial *General History of Astronomy* published by the Cambridge University Press. He asked me if we at St. John's had any notion of what a scholar and treasure we had in Curtis. I answered that I firmly believed that we did, and I asked him if he thought he knew the many other aspects of Curtis that we here are the beneficiaries of.

I was, of course, thinking of his wide teaching throughout the program, his published lectures, his many study groups, his love of music, and, most especially, the truly formative legacy of his two deanships. At the end of his first deanship, in 1962, Curtis drafted a proposal that culminated in the establishment of preceptorials as an integral elective part of the St. John's Program. In it he wrote:

> Too infrequently does the program succeed in inducing a continuing process of independent investigation and thoughtful reflection leading outward from the student's natural and initial standpoint. The frustrations involved in confronting one after another, great or important works which are never adequately understood, and the unavoidable distress involved in finding oneself again and again on uncertain ground—these effects ap-

pear insufficiently balanced by a positive sense of achievement and of independent, ongoing inquiry." (Preceptorial Proposal, March, 1962.)

How aptly this describes the exhilaration and exhaustion that tutors and students here still feel at certain moments during the over-rich and relentless annual procession of greatness we encounter in our seminars and tutorials. How difficult and sometimes daunting it is to strike a balance between breadth and depth. I can't help but think that Curtis wrote this impassioned and succinct characterization of our common experience because this was the enduring struggle for balance he regularly felt in his own work as a tutor and scholar. Years later when a colleague asked him how he had happened to focus on the history of science and astronomy in his own work, Curtis said, "I just got tired of banging my head against Being." This year we should also commemorate the fiftieth anniversary of the institution of preceptorials, even as we celebrate the seventy-fifth anniversary of the Program.

Yet this was not all. At the beginning of his second term as Dean, in 1973, Curtis noticed a surplus of $1500 in a ledger budget line and conceived the idea of converting Mellon 202 into an art gallery, in a way that was reversible if need be. He asked Burton Blistein, who was just beginning his long term as Artist-in-Residence, to see to this, and Burt worked with Buildings and Grounds to make plywood exhibition walls, coming in under budget. This in turn allowed him the means to work with a graphic designer on announcements of the first show, which launched the St. John's College Art Gallery. Mellon 202 reverted to laboratory classroom status in Fall 1989 with the opening of the Elizabeth Myers Mitchell Art Gallery, which is now a fully accredited member of the American Association of Museums and has 10,000 visitors a year to its five exhibitions. Curtis also led two study groups on the place of the visual arts in the program, something he saw as an unanswered question posed by the original Program catalogue as proposed by Barr and Buchanan. The first was an unofficial study group, the second an official one with the support of the Dean and Instruction Committee. It remains a recurring question.

Finally, at the end of his second deanship, in Fall 1976, as chair of the Joint Instruction Committee, the question of the place of the sophomore music tutorial in the program was again addressed. The essence of the curriculum for the tutorial had been established in 1950, but the tutorial had languished as an additional demanding class in sophomore year for several decades. With the recommendation of the Joint Instruction Committee, the Sophomore Laboratory class was discontinued, its components divided up in a way that required the reconfiguration of the whole laboratory program, a challenge which continues up to the present. More recently, as Freshman Chorus took on a more vital role as part of the Music Program, and concerts by the class at the end of each semester became an institution, Curtis regularly attended these performances, and I would invariably find an appreciative note in campus mail the day following a performance. Many a colleague has told me finding such a note from Curtis after delivering a Friday night lecture or leading an extracurricular seminar. Such kindness, such support, such encouragement are what make us the community we wish to be, along with our books and our classes and our lectures and our conversations.

Curtis's involvement in special study groups here continued right up to this past summer, when he was part of the summer faculty study group reading John Maynard Keynes. Curtis and emeritus tutor Harvey Poe, a close friend, had come to Dean Kraus to urge that Keynes be put back onto the Senior Seminar reading list, and his own interest in Keynes grew out of his own concern for the growing inequalities of wealth in America. He said to David Townsend, the organizer of this summer's study group, over lunch one afternoon, "David, I think that at some point the metrics of the extremities effectively become ideas which can then oppose each other either in politics or in revolution."

Epigraph

Aristotle in his *Parts of Animals* observes:

> The joy of a fleeting and powerful glimpse of those whom we love is greater than that of an accurate view of other things, no matter how numerous or how great they are. (De partibus animalium, 1.5.)

Curtis, when anyone would pass him on campus and then ask how he was, would smile and say, "About the same." May we say the same of the rich legacy he left us here.

Dear Becky, John and Topper, all of us share the sense of profound loss you feel and extend our heartfelt condolences to you and your family, even as we celebrate the memory of this wonderful, friend, colleague and scholar. "Gladly would he learn and gladly teach."

Remarks by Paolo Palmieri

Ladies and Gentlemen,

It is a great honor for me to remember Curtis Wilson as a friend, as academic colleague and mentor, and as a scholar. I will speak to his achievements as historian astronomy.

But first of all I wish to revive a fond memory in my mind. One year ago I came to St. John's to offer a lecture on Galileo. At six o'clock, Curtis and his wife Becky knocked on the door of the St. John's guest room. I have a radiant picture in my mind of Curtis's joviality at seeing me. I also have auditory memories of his saying "Hello!" It is these recollections that I want to set the tone of my brief address.

Curtis was more than a historian and a philosopher of science. Curtis was a master of humane scholarship. He was a rare figure of humanist and scientist. His rigorous methodology was never divorced from poetic imagery. He cultivated the history of astronomy, combining the rigor of intellectual analysis with the most sophisticated elegance of exposition. In one paper, we read: "In describing the Moon's chief vagaries as due to a semi-annual wobble of the major axis of its elliptical orbit. . . ." "The chief vagaries of the Moon"! What an image! It brings to mind the English title of one of Galileo's masterpieces, the *Dialogue on the two chief world systems.* But Curtis plays on the power of metaphor to awaken deep meanings. Order is suggested by the phrase "chief world systems." But what ideas does the metaphor "chief vagaries" evoke? In my mind, it evokes the elegance of the lunar motions, the subtlety of nature in which order and harmony lie hidden behind the infinitely challenging display of eccentricity. And now I really wonder—since that metaphor does raise the question of the mystery of nature, a question that goes straight to the heart of the significance of Curtis's pursuits as a historian of astronomy. But let me pause. I will come back to this fundamental question.

First I wish to convey to you my sense of why Curtis's historical methodology is rigorous. In a brilliant essay entitled "Newton on the Moon's Variation and Apsidal Motion," Curtis resists the all-too-common termptation to translate the geometrical language employed by

Newton into some form of calculus symbolism. Newton failed to solve the problem of the motions of the Moon's apses. Why so? asks Curtis. Because the great Newton visualized the apsidal motion of the Moon as as that of a rotating ellipse. This is a tremendous insight. Newton's thought processes do not proceed from formulas to their physico-geometrical meaning. It is meaning in the form of the visual representation of phenomena that guides his mathematical procedures. Thus, by resisting the temptation to translate Newton's geometical language into empty formulas—in other words, by being historically rigorous—Curtis has given us a luminous insight into the workings of Newton's mind.

Of course I will not give you a pedantic bibliographic essay on Curtis's contributions to the history of astronomy and to scholarship. Let me only mention briefly the scope of his contributions so as to convey to you a sense of the impressive range of interests that Curtis the humanist-scientist cultivated. Examples will suffice. We find in his catalogue William Heytesbury, Pietro Pomponazzi, medieval logic, and the rise of mathematical physics; Kepler's laws, Newton and universal gravitation, Horrocks's lunar theory, and a two-part essay on perturbations and solar tables from Lacaille to Delambre. We find a fine translation from French of a most important book by François de Gandt, *Force and Geometry in Newton's Principia,* and finally we find his "magnum opus," a formidable monograph on the Hill-Brown theory of the moon's motion, on which more in a moment. Curtis was at home among the Greeks, among the scholastics, as well as among the moderns who challenged them and thus changed our picture of the universe.

The magnum opus defies description, however. You have to pluck up the courage to delve into its mathematical labyrinth, the Hill-Brown theory of the moon's motion, and, I must confess, I have not yet found a way out. But why should you find a way out? Perhaps in his magnum opus Curtis issued a last challenge for all of us. It is the mystery of mind and life that he wishes us to explore, and there is no redemption until you have found a way to reconcile yourself with the mystery. You have to learn how to savor and enjoy the mystery, for solving it will perhaps not be given to us.

"Magnum opus" was the expression recurring time and time again in our correspondence over the last few years. I would timidly ask him: *Dear Professor, might you spare some time to read the draft of a paper?* The answer was always and promptly "Yes, with great pleasure!" But I knew that the magnum opus would for some days suffer a setback.

Thus I learned another great lesson from this master of history and the humanities: that if you give of your time and of yourself to others generously. rewards will come in the most exquisite form—I mean in the form of wisdom that helps others to find their pathways to wisdom.

It is time for me to draw the conclusions and, as promised at the beginning, to offer you a recollection on the significance of Curtis's pursuits as a historian of astronomy. Thus I find myself pondering the question: Why the history of astronomy? Why the history of the universe? Why did Curtis find this question impossible to evade? It is a philosophical question, a question about the whole, about the infinite—a question about civilization, about the mystery that connects the human and the divine. I have no better words to phrase my tentative answer to this ultimate question than those once used by Father Fulgenzio Micanzio in a letter to his dear friend Galileo. Micanzio spoke thus: *The immense and the infinite gently force me to think of the greatness of the creator, and even though that immense and that infinite are nil in comparison to the greatness of the creator, I cannot express the pleasure that I derive from such fantasizing about chimeras. One thing is certain. I have been helped by such chimeras to raise myself to meditate about the greatness of the creator much more so than by what I have read in the writings of theologians.*

Curtis's scholarly chimeras invite us all to raise ourselves to meditate on the greatest of mysteries.

Thank you.

Remarks by Louis Petrich

I met Curtis Wilson only six and a half years ago, when he was fourscore and upwards, yet unspoiled by the crosses of life, still hard at work on the recalcitrant moon. So I must speak of late things, close enough to feel almost present. We met when I called him up one frisky spring day to ask if he would have the patience to talk with me about Kepler's six-year "war" on that most wandering of planets, Mars, a war whose devilish intricacies Curtis had lucidly and penetratingly discussed in a number of essays he wrote during the 1960's and '70's for scholars and laymen both. I knew enough of Curtis's scholarly reputation to have felt some diffidence when I contemplated that call, yet he was a tutor and I was a tutor, so what should it matter that his work on Kepler was forty years behind him and mine had just begun? In "general honest thought and common good to all," he met me and talked with me about the great astronomers, always with modesty and soft concern to maintain accuracy of speech and respect towards the thousands of years of achievement in this heavenly field.

After our initial meetings, I found it convenient, for he was more given to smiles than speeches, to send him questions through email after seminar around 11pm. Usually by 8am the next day I'd have a response. I once asked him his opinion of a popular book on the history of astronomy that I was reading and liking all too much, which categorized Copernicus, Kepler, and more so all the astronomers who had gone before, as "sleepwalkers." From this most gentle and measured of men came the following wakeup reply:

> The characterization of Kepler as a "sleepwalker" is unfair, especially when coming from a writer who has not understood the technical aspects of Kepler's work at all. I grant that, if I take a hard look at the human condition, I would be forced to acknowledge that "sleepwalker" probably characterizes us all. But the arrogance of [this author] I cannot abide. . . . I don't think much of put-down jobs on Copernicus either [referring here to the Canon's supposed timidity]. . . . Let him throw stones, I say, who is without sin. . . . [Besides,] it was Kepler himself who made fun of himself: "O me miserable man! When all that time the areas in the ellipse would fit the phenomena exactly!"

To spend all our time learning to see the exact fit of one thing to another—as they defy us there all the while—is what we humans have to do, as it is what the best of us have done. Curtis cheerfully pursued this object to the end of his life with a discipline and a rigor that I can only imagine. By telling Kepler's story so faithfully and clearly, he helped me to participate in the exhilarating process of discovery displayed in one of the most honest and impossible of the great books. He never turned his nose from the details or technical matters that truly make all the difference.

He represents for me as a tutor a kind of ideal: one in whom the extrovert scholar of honored name in the world is co-present in the same learner and teacher with the serious amateur of kindly local habitation. I do hope, for the sake of the College, that I am exaggerating should I repeat on this occasion, apropos of this ideal, the first words to offer themselves when I heard of his passing: "He was a man, take him for all in all, I shall not look upon his like again."

I got to know Curtis, the serious amateur, during the monthly group discussions of plays that he and Becky organized and welcomed me to attend. Once after seminar I sent him an inquiry about a play this group of ours was reading: What do you think of the several conspicuous questions raised one after another in *King Lear* about the causes of things in nature and the human heart, such as, to begin with, a fool's question: "What is the reason that the seven planets are not more than seven? What would the typical geocentrist of Shakespeare's day answer to that?" Curtis shot back an 8am answer: "I am a little worried about your use of the word 'typical.' Such a word is hardly appropriate," he went to explain, for a group of astronomers that included ones who believed that the human search for order and ultimate causes is fruitful and others who dismissed that search as idle, beyond human reason. Then he added that whatever there is to be said about causes, even if it is to be said that there is "no cause, no cause" (quoting Cordelia), nonetheless, by the end of all that questioning in *King Lear,* there is the experience of "an awakened heart and a wonder at the fact that it so wakes."

I am put in mind now of a refrain from the Chorus in another play we read together, Aeschylus' *Agamemnon*: "Cry sorrow, sorrow, but let the good prevail." I wonder, however, if Curtis would be a little worried about the use of that little word *let* on this occasion. Would he have wanted me to say instead, more appropriately, "Cry sorrow, sorrow, but *help* the good to prevail." No, he would insist on respecting what is written there. Still, it must seem to us that someone who was twice Dean of the College (not such a fun job, I hear) was indeed a person who helped what is good to prevail. It is fitting, then, for one who got to know him too little and too late, to express a general and heartfelt gratitude for his double labors at keeping this little College thriving.

One final reminiscence and I am done. Among Curtis's favorite plays was Chekhov's *The Cherry Orchard*. Apart from the fact that it is a very moving and great play, I can guess what it might have meant to him, or rather, to us as we think about him. The play is about the loss of a beautiful orchard estate long held by an aristocratic Russian family who have fallen into debt from various failures of character and circumstance and can no longer afford to keep what had always kept them together in place and in time. Chekhov writes lyrically, passionately about the salvific powers of work and family. Curtis knew how to work to keep close what is most worthy of love and to preserve that which gives continuity and meaning to human life. He never lost the cherry orchard that had been given him. May that be said of us when the time comes.

Remarks by Thomas Slakey

It is a great honor for me to be asked to speak about Curtis Wilson, but it is also a very particular challenge. Often at memorial like this, Curtis was the one who was asked to speak, and he did it superbly. He set a very high bar.

Curtis came to St John's in 1948 and became Dean in Annapolis in September, 1958. The story I heard later was that Jascha Klein had chosen Curtis to succeed him as Dean, because Curtis was the most intellectually distinguished person on the faculty. I can well believe that story. Curtis's achievements in the history of astronomy demonstrate it. But Curtis was never over-powering, never assertive. He was unusually soft spoken, quiet, and modest. For me, his outstanding quality was a kind of gentleness combined with good humor. Curtis was as aware as anybody of the folly of mankind. There was often a touch of amusement or irony in his eyes. But his response to folly was gentle.

Curtis once told me that he did not mind doing routine tasks. He certainly had a capacity for hard work, carefully and accurately done. I believe that he was responsible for much of the improvement in the laboratory program during those early years. This work was hardly routine, but it was very laborious, particularly the planning of experimental work and the writing of laboratory manuals.

Curtis was also responsible for one of the few principal modifications in the original design of the St. John's Program, namely breaking into the Seminar in the Junior and Senior years for the introduction of "preceptorials." Such changes in this Program are never easy. Curtis accomplished change with patient, careful persuasion.

As I said, Curtis became Dean in 1958. I came in 1959, perhaps Curtis's first appointment to the faculty. In any case, from the day we arrived in Annapolis, Curtis and Becky were good friends to Marion and me. We all went out to Santa Fe together in 1964. Curtis and Becky went on to California in 1966, where Curtis joined the History of Science Department at the University of California at San Diego. Curtis and Becky came back to Annapolis in 1973, when Curtis took on a second term as Dean, a mark of extraordinary devotion to the welfare of the College.

Curtis retired from the faculty in 1988, but continued to be involved in discussions of Shakespeare and other authors until the very end. One mark of the breadth of his mind was his special interest in the plays of the contemporary black playwright, August Wilson.

Curtis was one of the finest human beings I have ever known. I salute his memory now. I salute Becky and John and Topper and their families. We will all miss him sorely.

Remarks by John Wilson

Thank you for honoring my father with this service. Not for his own sake but for the sake of those he loved, he would have been grateful for the kindness and respect shown today. He would have known how much this means to his wife of fifty-eight years and to his family. And he would have wanted you to know how much he admired and cherished his wife, family and you: his friends, colleagues and community. He would not have wanted to be at the center of attention. Instead, he would want to focus on you, and to permit any merit to pass on to you and into the future.

Following the wonderful remarks shared by the preceding speakers, and anticipating my brother's, I don't think there is too much for me to say. A week after my father died, my mother announced, "The future lies forward." Such resolve would have surely made my father proud. He lived his life facing forward. He knew the importance of intention, and he leaned into the future—intent, purposed, but open to surprise. Pragmatic intentionality was a core trait. He liked projects. You have heard about many of his professional projects already. Here are a few examples from later in life.

Ten years ago, at the age of 81, he climbed onto the roof of my home in upstate New York to help replace shingles. He enjoyed it. Five years ago, he climbed a mountainous path in Southwestern Virginia. He persisted despite the occasional need to stop and rest. He wanted to get to the end, which culminated in a beautiful set of cascades. He succeeded. Among his final intellectual projects were advanced calculus and Keynesian economics. He learned them. The day before he died, he was pushing himself still during a walking tour on Mackinac

Island. He continued to learn things and follow trails right to the end. Those of you who knew him well cannot imagine him otherwise.

Although he pushed himself, he was also interested in the space between his projects, and in the space between his own thoughts. He was able to rest in this space. He practiced at it, or in it. He knew that unintended, surprising and sometimes wonderful realizations can arise by paying attention in this way. In one of his later conversations with me, he referred to this orientation as awareness of groundlessness. He practiced Tai Chi and meditation. He delighted in spontaneity that arose outside of things intended. He was ready, if you can be ready, for surprise.

His sense of humor was rich, a great sign of his ability to enjoy the surprises between thoughts. I remember him reading Huck Finn to my brother and me early in our lives. Laughter and tears of mirth poured from my father's face. He infected us with giddiness despite the fact that we, or at least I, didn't quite get the joke. Wit and playfulness became something relished deeply by my brother and me, and by our children. My father joined in with gusto. Picture this elderly scholar down on his hands and knees, bearing his then four-year-old grandson on his back, savoring the role of steed to the conquering knight waving a wooden sword in triumph.

As you know, my father had . . . *conspicuous eyebrows*. They often rose with interest, surprise or delight. These were among his most characteristic affect states. Those lifted eyebrows also may have been among his last expressions. Early in the morning of his final day, he was briefly awake and lucid in the hospital, though unable to speak. The ICU nurse informed him that he had suffered a heart attack and had occlusion in three vessels. Up went those eyebrows. Disease and death are always unwelcome, but he did not turn away. He opened up. It was striking enough for the nurse to notice.

One of my most vivid memories of this expression occurred during childhood, when my family went to a tourmaline mine in southern California. The tour there involved turning off electricity to demonstrate the complete absence of light in the depths of the mine, and the tour guide was concerned for everyone's safety.

He wanted to select someone from the group to bring up the rear. The guide spotted my father and announced, "You there. You're the one. You look like you have an honest face." Up went my father's eyebrows, down went his jaw. It was surprise, or incredulity. He later debated whether honesty could be detected on the face, but at the time viewed himself as no more honest, no more worthy, than anyone else in the group.

In fact, honesty can be detected on the face and my father had a thoroughly honest one. The tour guide was right. No one got lost in the dark that day. As I knew him, my father was never in the dark. He had the integrity to know what he knew. He presumed little when he did not know. He knew the dangers of over-confidence and self-deception. He was not enamored of his own opinions. He listened well, not only to the world around him and to you and me, but also to those spaces between his own thoughts. This kind of integrity led to a profound modesty. These were unmistakable traits, traits that gave rise to what so many people have remarked upon since his death: his gentleness, and his gentlemanliness.

I have one more note to share. After my father's death, following my mother's instructions, I took some of his clothing over to a donation center. I was met with inquiries about the occasion. I explained that my father had just passed on at the age of 91. The reply was unexpected and instructive. I was told, "Well, at 91 he was entitled." Had he been alive to hear it, I think my father would have chuckled, perhaps with a twinkle in his eye. He would also have been deeply saddened, knowing that this final entitlement was taking him away from those he loved. But he lived well, knowing the impermanence of all things, and he died well. In this sense, I suppose it could be said that he "deserved" the end of his life. He imposed himself on no one. Instead, he opened up space for others to enter and join him in listening. The merit and memories he has left behind for our future are the stronger for it.

Remarks by Christopher Wilson

Welcome everyone! Just one more speaker today. After listening to these esteemed speakers, were I in first grade, I might conclude that *my* dad really *was* better than *your* dad! However, I'm no longer in first grade and should refrain from such audacious remarks. We're here to pay tribute to dad's remarkable life and contributions to this community gathered here today. So, here goes.

Somewhere around 1980, my dad was the keynote speaker for the graduating class at St. John's. Since this was quite an honor, my mother demanded my somewhat obstinate teenage presence, and so I came, I think, without too much resistance. His advice to graduating seniors was simple: "be brief, be bright, and be gone!" He repeated that several times in his talk, so I knew it was important. I will endeavor to heed that bit of wisdom today, though it is somewhat of a challenge on this somber occasion.

Today, you've heard many descriptions of my dad: kind, gentle, patient, full of integrity, great capacity for intellectual humility, a love of humor. All of these are true. He was an amazing man, a dedicated scholar; a fine person, and a good father. I believe he learned something new every single day of his life. He was a person of great breadth.

He amazed me in many ways: I will try to string some examples together for you.

He was a Boy Scout in the 1930s. He achieved the rank of Eagle. Today, only four percent of all Boy Scouts make it to the rank of Eagle. I suspect it was in the Boy Scouts that dad learned how to cook pancakes. As a young boy, I routinely pestered dad to make pancakes on Saturday mornings. It was a great and rare occasion when dad made pancakes for dinner; I suspect this was less of a treat for my parents and more of a reprieve for mom, who may have been tired from the day's activities.

In the late 1940s or early 1950s, dad spent a year studying in Italy in pursuit of his PhD. This always amazed me because of the timeframe: World War II had just ended. Italy was then a country torn by war—surely there had to be some resentment against the Americans. It also amazed me because everyone in Italy speaks Italian; how on earth did Dad communicate?

In Santa Fe, in the mid-sixties, he built a harpsichord. He bought all the materials, the wood, strings, and keys. Cut everything to fit. He melted lead for the weights of the keys and let me watch. To a little boy, turning a metal to liquid was pretty neat. Watching all of this from start to finish was a marvel. The harpsichord sits in mom and dad's dining room today. It's beautiful.

In California, in the late sixties, while at the beach one summer day, I asked where dad was. I was directed to gaze way out beyond the breaking waves; and saw dad's white hair. He was swimming slowly but methodically parallel to the shore. I was astonished I How bold and brave he was to go where the water was so deep! At the time I thought there was no way I was ever going to go that far out!

Dad spent three to four months in Russia on sabbatical during the height of the Cold War, perhaps this was 1970 or so. We worried about him over there, in that country that was our enemy. I knew this was a place that took people away in the middle of the night, shipped them to a prison camp, and never let them out. My mind-set was that Russians were villains! Children of the 1960s need to think no further than the TV cartoon show Rocky and Bullwinkle. Their arch enemies were Boris and Natasha, scheming bad guys dressed in black with funny accents. Thankfully, nothing bad happened to dad. Upon his return, he talked of the kids begging for chewing gum. He talked of men—presumably government agents of some sort—openly following him. How adventurous and how scary!

In the late 70s, dad helped me with a lot of high school homework, in particular, Geometry and Physics. "Dad, I don't understand this problem." I always wanted the quick answer from him, but rarely got It. No matter what he was doing, he always set aside his work and patiently explained and ensured that I understood the theorem, proof, or law. Needless to say, I got A's in these subjects.

In his later life, after I had moved out of the house, he still swam, jogged, went rowing in racing shells, practiced Tai Chi and meditation. He continued to travel the world. A few years ago in Europe, he was determined to climb a mountain in the Black Forest, and did so with the help of younger colleagues. I think he was 88 or 89 years old. That is just astounding!

In recent years on Saturday mornings, Dad would get together with what Mom calls a student—Paul Stevens, who is, I believe, a high-school mathematics teacher—and they would do calculus problems—for fun! I believe this is not something a normal person does. I've told this story to many of my friends and co-workers when describing my father. It always raises an eyebrow. Perhaps this was akin to looking forward to a Sunday crossword puzzle, just on a much higher level.

Some of my fondest memories are of times my dad would read to me as a young boy, most often, just before bed. These were books of adventure, and I just ate them up: Tom Sawyer, Huck Finn, Swiss Family Robinson, and Robinson Crusoe. The best book by far, a great book in my mind at that time, though perhaps not in the eyes of this institution, was Daniel Boone. Mom would say that dad read it to me 100 times. I think she exaggerates and he only read it some eighty or ninety times. It portrayed a life of the great-out-of-doors, a life I longed for and pursued as a teenager and adult.

Dad told me once that *Don Quixote* was his favorite book, a classic comedy. A few weeks ago, while spending time helping to clean out his desk, looking at the many volumes that cover the walls of his office, almost all the books with subjects of a scientific or historic nature, I was heartened to see a copy of Huckleberry Finn. He and I both love comedy and could relate to each other well about funny scenes from books or movies. I will miss those discussions.

You may not know that outside of classical music, Dad enjoyed the marching music of John Phillips Sousa; he told me, "I grew up with Sousa." He also liked the syncopation of Scott Joplin. Indeed, in his 80s, he practiced playing Joplin's ragtime on the piano.

In conclusion, Dad was a man of great breadth. I will miss him. However, I will not and cannot think of him as being gone. Instead, I like to think of dad floating through the ethers of the heavens, having genial and intellectual conversations with others. Imagine that, if you will: Perhaps he stopped to chat with Saint Peter and asked why the Bible did not explain how life propagated from just Cain and Abel—what happened there? Maybe, he's run into Sir Isaac Newton and asked whether there really was a day—under an apple tree—where there was a thud, and then things clicked. Or can you see my father having a reasoned, scholarly discussion with Ra, the Egyptian Sun God, patiently explaining that Ra's really not the center of the Universe, and then proving it mathematically with a blue ball-point pen and white piece of scratch paper? Can you see him doing that?

We are all much better people for having known him and witnessed his example of a life well lived. On behalf of the Wilson family, thank you to St. John's College for your cooperation and help in providing everything here. Dad had his office in this building, so we especially appreciate your willingness to let us have the memorial in this building. Thank you to the speakers for providing your time, effort, and wonderful remarks, and thank you all for coming.

CPSIA information can be obtained
at www.ICGtesting.com
Printed in the USA
FSHW022153081118
53511FS

9 780692 832691